3rd International Conference on Futuristic Trends in Materials and Manufacturing (ICFTMM 2020)

IOP Conference Series: Materials Science and Engineering
Volume 1149

Greater Noida, India
7 - 8 April 2020

Editors:

Mohd Atif Wahid
Ranjit Varma
Noor Zaman Khan
Kuldeep Panwar

ISBN: 978-1-7138-3419-9
ISSN: 1757-8981

Printed from e-media with permission by:

Curran Associates, Inc.
57 Morehouse Lane
Red Hook, NY 12571

proceedings.com

Some format issues inherent in the e-media version may also appear in this print version.

Institute of Physics
Dirac House, Temple Back
Bristol BS1 6BE UK

Phone: 44 1 17 929 7481
Fax: 44 1 17 920 0979

techtracking@iop.org

Additional copies of this publication are available from:

Curran Associates, Inc.
57 Morehouse Lane
Red Hook, NY 12571 USA
Phone: 845-758-0400
Fax: 845-758-2633
Email: curran@proceedings.com
Web: www.proceedings.com

TABLE OF CONTENTS

Proceeding of 3rd International Conference on Futuristic Trends in Materials &Manufacturing 2020

Greater Noida

(ICFTMM 2020)

7-8 April 2021

Editors:

Dr. Mohd Atif Wahid, Associate Professor and Head, Delhi Technical Campus, Greater Noida

Prof Ranjit Varma, Director, Delhi Technical Campus, Greater Noida

Dr. Noor Zaman Khan, Assistant Professor, NIT Srinagar, Jammu and Kashmir

Dr Kuldeep Panwar, PhD Thermal Engineering, G. B. Pant University of Agriculture & Technology, Pantnagar

Preface

The content of the edited book is based on papers presented at the 3rd International Conference on Futuristic Trends in Materials and Manufacturing (ICFTMM-2020) held in Delhi Technical Campus, Greater Noida, Uttar Pradesh 201306 India on 7-8 April 2021 (Online mode). The scope of the conference includes smart materials, advanced manufacturing, design modeling and mechanics, thermal and fluid engineering.

Manufacturing industry plays vital role in extracting the wealth from the Earth. Manufacturing industry has been revolutionised due to the development of advanced materials and advanced materials processing techniques. Various challenges are being faced by scientists and researchers with the development of technology and new materials. Due to the development of new materials with high strength, hardness, toughness etc. it is become difficult to produce complicated geometries in such materials with the usual methods. Higher production rate, complexity in design and economic requirement demand the use of advanced manufacturing processes.

This book covers various topics related to development, processing and characterization of advanced materials, application of advanced manufacturing processes, process parameter optimization as well as modelling of manufacturing processes. It also includes researches pertaining to other important allied mechanical engineering areas i.e. design modelling, stress analysis, thermal and fluid engineering, etc. Academicians, researchers, under graduate and post graduate students associated with materials and manufacturing, design and fluid engineering will be benefitted from the material presented in this book.

IOP Conf. Series: Materials Science and Engineering 1149 (2021) 011001 doi:10.1088/1757-899X/1149/1/011001

Committee Members

Conference General Chair
Prof Konovalov Sergey, Head of the Department, Department of Metals Technology and Aviation Materials, Samara National Research University (Russia)

Conference Chairman
Prof (Dr) Ranjit Varma, Director, Delhi Technical Campus, Greater Noida

Conveners
Dr Mohd Atif Wahid, Head, Mechanical & Automation Engineering Department, Delhi Technical Campus , Greater Noida
Dr Pranay Tanwar, Dean Academics, Delhi Technical Campus, Greater Noida

Organizing Secretary
Mr Etkaf Hasan Khan, Assistant Professor, Mechanical & Automation Engineering Department, Delhi Technical Campus, Greater Noida

Organizing Committee
Dr Sanober Khan, Delhi Technical Campus, Greater Noida
Dr Kimmi Verma, Delhi Technical Campus, Greater Noida
Dr Pushpa Singh, Delhi Technical Campus,Greater Noida
Mr Radhakant, Delhi Technical Campus, Greater Noida
Mr Ankit Gambhir, Delhi Technical Campus, Greater Noida
Mr Rajiv Ranjan, Delhi Technical Campus, Greater Noida

Technical Advisory Committee

Prof Konovalov Sergey, National Research University, Russia
Dr Chaudhery Mustansar Hussain, Professor, New Jersey Institute of Technology, New Ark
Prof Mustafa Acarer, Metallurgical and Materials Engineering, Selcuk University, Konya
Dr Ernest Mnkandla, University of South Africa, South Africa
Prof Asanga Ratnaweera, Professor, University of Peradeniya, Sri Lanka
Prof Wan Mansor Wan Muhamad, University Kuala Lumpur, Malaysia.
Prof Zahid A Khan, Jamia Millia Islamia, New Delhi
Prof Sabu Thomas, VC, Mahatma Gandhi University Kottayam, Kerala.
Prof Manikant Paswan, NIT Jallandhar
Dr Irfan Ul Haq Siddique, Assistant Professor, King Saud University, Saudi Arabia
Dr Aditya Maheshwari, Project Manager, Westerbachstrasse, Kronberg im Taunus, Germany
Dr M. A. Quraishi, Professor, King Fahd University of Petroleum and Minerals (KFUPM), Kingdom of Saudi Arabia
Dr Mohsin Talib Mohammed, Associate Prof, Faculty of Engg, University of Kufa, Iraq
Dr Suha K Shihab, College of Engineering, University of Diyala, Iraq
Dr Shahid Hussain, Ministry of Foreign Affairs, Oman
Dr Subhra Datta, Associate Professor, IIT, Delhi

Dr S Anand Kumar, Assistant Prof, IIT Jammu
Dr M Felix Orlondo, Assistant Prof, IIT Roorkee
Dr DurgaToshniwal, IIT Roorkee, India
Dr Saurav Kumar Pandey, Assistant Professor, IIT Patna
Mr Vikas Kalra, Director-CETPA Infotech Pvt. Ltd.
Mr Dinesh Agarwal, Scientist, Indian Space Research Organisation (ISRO), Bengaluru
Dr Amitanshu Pattanaik, Scientist E, Defence Research and Development Office (DRDO),New Delhi
Dr T Ram Prabhu, Defence Research and Development Office (DRDO), New Delhi
Prof Arshad Noor Siddique, Professor, Jamia Millia Islamia, New Delhi
Prof Abdul Malik, Professor, Aligarh Muslim University, Aligarh
Prof D. Buddhi, Executive Dean, Research and Innovation Wing, LPU, Punjab
Prof Rita Wattal, Delhi Technological University, New Delhi
Prof Qasim Murtaza, Delhi Technological University, New Delhi
Prof Bhim Singh, Prof and Head, Sharda University, Greater Noida
Prof Neeraj Gupta, Professor and Director, SRMU, Lucknow
Prof Iqbal Ahmed Khan, Professor and Head (ME), Lingaya's Vidyapeeth ,Faridabad
Dr M. Mohsin Khan, Assistant Prof, NIT Srinagar, J and K
Dr M.S. Charoo, Associate Professor, NIT Srinagar
Dr Noor Zaman Khan, Assistant Prof, NIT Srinagar
Dr A F Sherwani, Associate Prof, Jamia Millia Islamia, New Delhi
Dr M Asjad, Associate Prof, Jamia Millia Islamia, New Delhi
Dr M Javed, Assistant Prof, Jamia Millia Islamia, New Delhi
Dr Swati Gangwar, Assistant Prof, MMUT, Allahabad
Dr Mohd Atif Wahid, HOD and Associate Prof, DTC
Dr Pallava Gupta, Assistant Prof., Amity University, Noida
Dr. Jitendra Kumar Katiyar, SRM Institute of Science and Technology, Potheri
Dr Eram Neha, Assistant Professor, Galgotia Engg College, Greater Noida
Dr. Neeraj Bisht, Assistant Prof, GBPUAT, Pantnagar
Dr. Mir Irfan Ul Haq, Assistant Prof, SMVDU, Katra
Dr. Ankush Raina, Assistant Prof, SMVDU, Katra, J&K
Dr. J K Katiyar, Assistant Prof, SRM Institute of Science and Technology, Chennai
Dr Rajiv Suman, Assistant Prof, GBPUAT, Pantnagar
Dr Sonal Khurana, Assistant Prof, Mahraja Agarsen Institute of Technology, New Delhi
Dr Owais Kidwai, Assistant Professor, GB Pant Engineering College, Delhi
Dr Zaheen Khan, Assistant Professor, Lingayas Vidyapeeth, Faridabad
Mr KM Agarwal, Assistant Professor, Amity University, Noida

Review Committee Members

Prof. G. N. Tiwari, IIT Delhi
Prof RK Mittal, BITS, Dubai
Prof Sabu Thomas, Mahatma Gandhi University, Kerala
Prof G. A. Harmain, NIT Srinagar, J and K
Prof Sachin Maheshwari, NSUT, Delhi
Prof M.F. Wani, NIT Srinagar, J and K
Prof Manikant Paswan, NIT Jallandhar
Prof R C Singh, DTU, Delhi
Prof Islam Nawaz, Jamia Millia Islamia, New Delhi
Prof Anand Bewoor, MKSSS Cummins College of Engineering for Woman, Pune
Prof Bharat Raj Bundel, Lingyas Vidyapeeth, Faridabad

Dr Savita Sondhi, BINUS School, Indonesia
Dr Kashif Irshad, KFUPM, Saudi Arabia
Dr Suha K Shihab, Senior Lecturer, University of Diyala, Iraq
Dr Raof Ahmad Khan, Associate Professor, GCET, Safapora, J&K
Dr A. X. Amal Rebin, Associate Professor, HITS, Chennai
Dr Ashok Dewagan, Assistant Professor, NIT Delhi
Dr Piyush Gupta, Associate Professor, AKTU, Lucknow
Dr Ashutosh Mishra, Assistant Professor, MNNIT Allahabad
Dr Pallva Gupta, Assistant Professor, Amity University, Noida
Dr Vijay Chaudhary, Assistant Professor, Amity University, Noida
Dr Mohd Javaid, Assistant Professor, Jamia Millia Islamia, New Delhi
Dr Amir Qureshi, Assistant Professor, Jamia Millia Islamia, New Delhi
Dr Zaid Ansari, Assistant Professor, Jamia Millia Islamia, New Delhi
Dr Pankul Goel, Associate Professor, IMS Ghaziabad
Dr Eram Neha, Assistant Professor, Galgotia Engg College, Greater Noida
Dr Yogesh Srivastava, Assistant Professor, Galgotia Engg College, Greater Noida
Dr Mona Aggarwal, Assistant Professor, NCU, Gurgaon
Dr Kuldeep Panwar, GBPUAT, Pantnagar
Dr Bisma Mannan, Assistant Prof, Sharda University, New Delhi
Dr Namrata Gangil, Assistant Prof, Ajay Kumar Garg Engineering College, Ghaziabad
Dr Shadab Alam, Assistant Prof, Jamia Hamdard, New Delhi
Dr Owais Kidwai, Assistant Prof, GB Pant Engineering College, Delhi
Dr Zaheen Khan, Assistant Prof, Lingayas Vidyapeeth, Faridabad
Dr Naseem Ahmad, Assistant Prof, IILM, Greater Noida
Dr Divya Mishra, Associate Professor, Delhi Technical Campus, Greater Noida
Dr Pushpa Singh, Associate Professor, Delhi Technical Campus, Greater Noida
Dr Tina Chaudhary, Assistant Professor, IGDTU, New Delhi
Dr Khalid Nafees, Assistant Professor, IPAC, Gzb
Dr Md Meraj, Jamia Millia Islamia, New Delhi
Mr KM Agarwal, Assistant Professor, Amity University, Noida
Ms Shanu Sharma, Assistant Professor, Amity University, Noida
Mr Ozair Arshad, Assistant Professor, Galgotia Engg College, Greater Noida
Mr EH Khan, Assistant Professor, DTC, Greater Noida
Mr Abhipraya, Tata Steel, Bhubneshwar
Mr Satyaveer Singh, Assistant Professor, IIMT College of Engg, Greater Noida
Mr Rajiv Ranjan, Assistant Professor, Delhi Technical Campus, Greater Noida

IOP Conf. Series: Materials Science and Engineering 1149 (2021) 011002 doi:10.1088/1757-899X/1149/1/011002

Peer review declaration

- **Type of peer review:** Single-blind

Single-anonymous: authors' identities are known to the reviewers, reviewers' identities are hidden from authors

Double-anonymous: author and reviewer identities are hidden to each other

Triple-blind: author and reviewer identities are hidden to each other, and from the Editor(s)

Open: author and reviewer identities are known to each other

- **Describe criteria used by Reviewers when accepting/declining papers. Was there the opportunity to resubmit articles after revisions?**

  A reviewer form was shared to all the reviewers which contains different criteria's i.e. originality, technicality, scientific merit, sufficient discussion on work context, presentation, organization etc.
  Yes, there was the opportunity to resubmit articles after revisions

- **Conference submission management system:** The authors were informed to submit the papers in IOP template by email at icftmm2020@delhitechnicalcampus.ac.in

- **Number of submissions received:** 52

- **Number of submissions sent for review:** 48

- **Number of submissions accepted:** 32

- **Acceptance Rate (Number of Submissions Accepted / Number of Submissions Received X 100):** 61.53%

- **Average number of reviews per paper:** 03

- **Total number of reviewers involved:** 39

- **Any additional info on review process (ie plagiarism check system):**

Research papers received for the conference "ICFTMM-2020" were reviewed through a single-blind review process. Peer review of papers was undertaken through processes administered by the organizers. Authors were instructed to strictly adhere to the guidelines of the conference and publisher and prepare a high-quality research article. Papers were reviewed independently by a minimum of two members of the conference review committee. After plagiarism check for maximum limit of 16 % by Turnitin software, papers were then sent to the reviewer by making sure that the topic of paper best matches with reviewer expertise. Assigned reviewers have submitted their anonymous reviews. After submission of all reviews, a reviewing panel made the final accept/reject decisions based on the reviews. The reviewing panel was made up of members from technical advisory committee and review committee, Conference General Chair, Chairman, Convenor and Editors. Authors were notified the final decisions via email.

Key Highlights

- Peer review of papers has been done through processes administered by the organizers and proceedings editors.
- Papers were checked for plagiarism before sending for review. Papers with less than 16% plagiarism were sent for review.
- Minimum two independent reviewers have reviewed and approved a paper.
- A maximum of 3 papers have been assigned to a single reviewer.
- Unbiased consideration was given to all manuscripts offered for publication in the proceedings regardless of race, gender, religious belief, ethnic origin, citizenship or political philosophy of the authors.
- Referees were asked to rate the paper based on the originality, technicality, scientific merit, sufficient discussion on work context, presentation, organization etc.

- **Contact person for queries:**
Dr Mohd Atif Wahid
hodmae@delhitechnicalcampus.ac.in
Covener, ICFTMM 2020

Effect of Ti/SiC Reinforcement on AA5083 Surface Composites Prepared by Friction Stir Processing

Md. Ziyaur Rahman[1], Arshad Noor Siddiquee[1], and Zahid A. Khan[1]

[1] Department of Mechanical Engineering, Faculty of Engineering and Technology, Jamia Millia Islamia, New Delhi 110025, India

er.ziyaurrahman@gmail.com

Abstract. Al alloys because of their desirable characteristic properties have made them predominant materials in lightweight design structures such as the aviation, automotive, and shipbuilding industry. However, their applications are limited due to their lower strength at higher operating temperature and difficulty in their welding. One of the alternatives to enhance the surface properties of aluminium alloys is coating metallic substrates. In the present experimental study, the effect on surface properties by introducing titanium and silicon carbide reinforcement into AA5083 alloy using friction stir processing (FSP) was investigated. Additionally, bead geometry such as the width of SZ, depth of SZ, and area of SZ in all the specimens were also analyzed. FSP was performed at different combinations of rotational speeds, traverse speeds, and shoulder diameter sizes using Taguchi L4 orthogonal array. Experimental results indicated that the highest value of hardness was observed at tool rotational speed of 355 rpm, traverse speed of 63 mm/min, and shoulder diameter size of 17 mm. Moreover, at such combination of process parameters, the lowest value of width to depth ratio was observed.

1. Introduction

Al alloys find potential applications in aerospace, automotive, and shipbuilding industry owing to their desirable characteristic properties such as lightweight, high specific strength, better corrosion resistance, and appreciable toughness. However, their applications are limited due to their poor surface properties such as lower wear resistance and lower hardness. On the other hand, aluminum alloys lose their strength/functionality at higher operating temperatures due to their low melting point temperature (656 °C) [1-2]. To encounter such problems, one of the alternatives to enhance their properties is to introduce hard components such as SiC, W and Ti to the surface of base alloy [3]. A significant study related to introduction of hard particles in the base material (BM) surface was performed by researchers using a fusion route. However, this technique was associated with various limitations such as the formation of undesirable phases and separation of micro phases during processing [4]. On the other hand, techniques such as conventional casting and powder metallurgy are effective in the fabrication of metal matrix composites (MMC) than fusion routes but are subject to certain limitations such as lack of uniform distribution of reinforcements [5]. To address, the limitations of such MMC fabrications process, friction stir processing (FSP) is one of cost-effective alternatives for fabricating MMC. FSP is an alternative to friction stir welding (FSW) and is widely used to tailor the surface properties of BMs. In FSP, the reinforcements are introduced into the well-packed groove of predetermined size. Thermo-mechanical and stirring action of FSP tool plasticizes the BM and leads to uniform distribution and embedment of reinforcement into the matrix. The plasticized MMC is subsequently forged at trailing side of tool due to axial force applied by FSP tool.

Significant literature is available attributed to the enhancement of surface properties of AA5083 using FSP [6-7]. In contrast, little research work has been performed on the influence of process parameters on the surface quality during FSP of AA5083 [8-9]. However, no study is available in open literature exploring the influence of process parameters in friction stir processed (FSPed) Ti+SiC reinforcement on surface properties. The central aim of the present experimental investigation is to obtain optimum process parameters for obtaining the best surface properties such as hardness. Moreover, the bead geometry of the stir zone (SZ) was measured at various process parameters. Taguchi L4 orthogonal array was designed for characterizing the surface properties of friction stir processed AA5083 using three predictors, tool rotational speed (rpm), traverse speed (v), and shoulder diameter size (ϕ).

2. Experimental Set-Up

In this experimental investigation, the BM was aluminium alloy 5083 (AA5083) with chemical composition listed in Table 1. Plates of dimensions 200mmx70mmx7mm were cut along the rolling direction. Grooves of dimensions 2.5mmx2mm were made on the surface of AA5083 plates along the longitudinal direction using a computer numerical controlled (CNC) machine. Reinforcement, Ti/SiC in the ratio of 10:3 was introduced into the groove (refer Figure 1(a)). Before FSP the reinforcement in the groove was well-packed with the aid of a pin less flat shouldered cylindrical tool. After compact packing of precisely placed reinforcement, the well-designed profiled tool of suitable dimensions having cylindrical pin (with right-hand threads) (refer Figure 1(b)) was introduced axially into the previously well-packed reinforcement to fabricate the MMC (refer Figure 1(a)). The surface MMC was fabricated using a retrofitted vertical milling machine. After FSP, the specimens were cut in the transverse direction for metallographic examination. The specimens were subsequently polished with emery papers up to 2000 grit sizes and were finally polished with diamond paste using a velvet cloth. The specimens were subsequently etched with a solution of 34 ml of distilled water, 16 ml HNO_3, 12 ml HCl, and 1 ml HF, and 4.8 grams of CrO_3 for 65 sec for revealing the microstructural characteristics and distribution of reinforcement in the processed zones. The microstructural images of the processed region were revealed with the aid of an optical microscope (OM). The Vickers microhardness readings were measured in a direction perpendicular to the weld cross-section using a microhardness hardness tester.

Table 1. Elemental composition of AA5083 BM.

Elements	Mg	Mn	Fe	Si	Cr	Cu	Ti	Zn	Al
Wt.%	4.365	0.5	0.16	0.08	0.07	0.031	0.013	0.002	Bal.

In this experimental study, Taguchi L4 orthogonal array was used. The experimental plan of the present study is listed in Table 2. Two levels each of ϕ (17 and 20mm), rpm (355 and 450rpm), and v (63 and 80mm) were employed in this investigation. In this study, the optimal combination of process parameters was investigated for which the highest values of hardness were obtained. Moreover, the bead geometry of SZ of the processed specimen for which higher value of hardness obtained was investigated in detail.

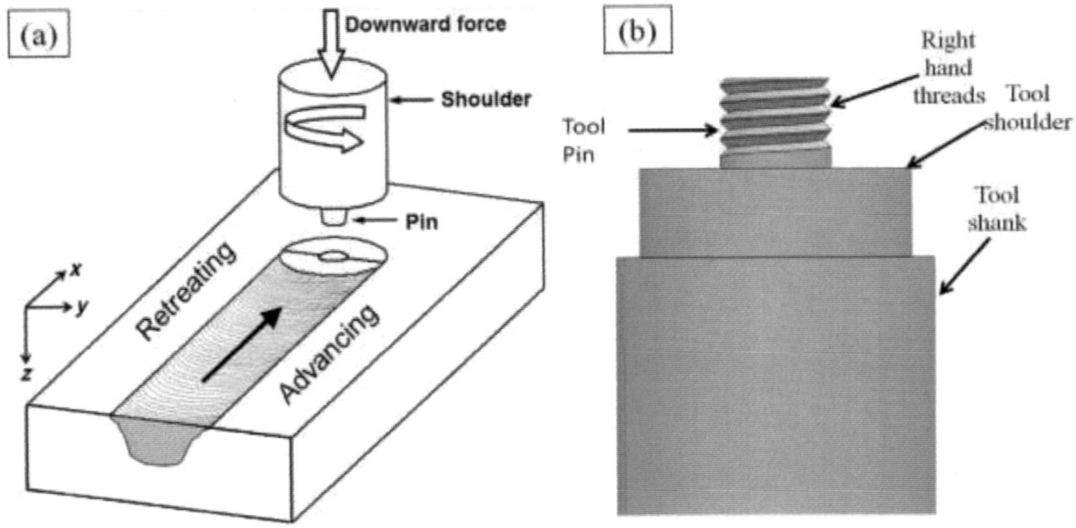

Figure 1 shows (a) the principle of FSP and (b) the FSP tool with the nomenclature

Table 2. Experimental plan

Experiment No.	ϕ (mm)	rpm	v (mm/min)
S1	17	355	63
S2	17	450	80
S3	20	355	80
S4	20	450	63

3. Results and discussion

3.1. Macrostructure and Microstructure

Figure 2 shows the macrostructure of various weld regions in S1, S2, S3, and S4 specimens produced during FSP. Microhardness indents (vertical line) can be observed on the cross sectional surface of FSPed region. In all the specimens the agglomeration of Ti+SiC powder can be observed by visual inspection. This agglomeration indicates the improper distribution of reinforced particles with the base alloy matrix. Figure 3 represents the optical microstructure of FSPed specimens fabricated at tool rpm of 355, v of 63mm/min, and ϕ of 17mm. Four characteristic zones were recognized in the FSPed region: stir zone (SZ), thermomechanically affected zone (TMAZ), heat affected zone (HAZ), and BM. In SZ, the grains were fine and equiaxed owing to dynamic recrystallization due to intense plastic deformation in this region. On either side of SZ is TMAZ having elongated and large grains owing to partial plastic deformation in this region. Adjacent to TMAZ is HAZ where grains are more or less similar to the base material and are only subjected to a thermal cycle with no plastic deformation. In general, the increase in grain size occurs as one moves from SZ to TMAZ then HAZ due to progressive removal of plastic deformation from SZ to HAZ. In SZ the highest values of hardness are observed due to the presence of fine and equiaxed grains formed during FSP. In Figure 3E, the agglomeration of reinforcement particles can be observed in SZ. This condition is attributed to the improper mixing and distribution of reinforcement particles in the base matrix during FSP.

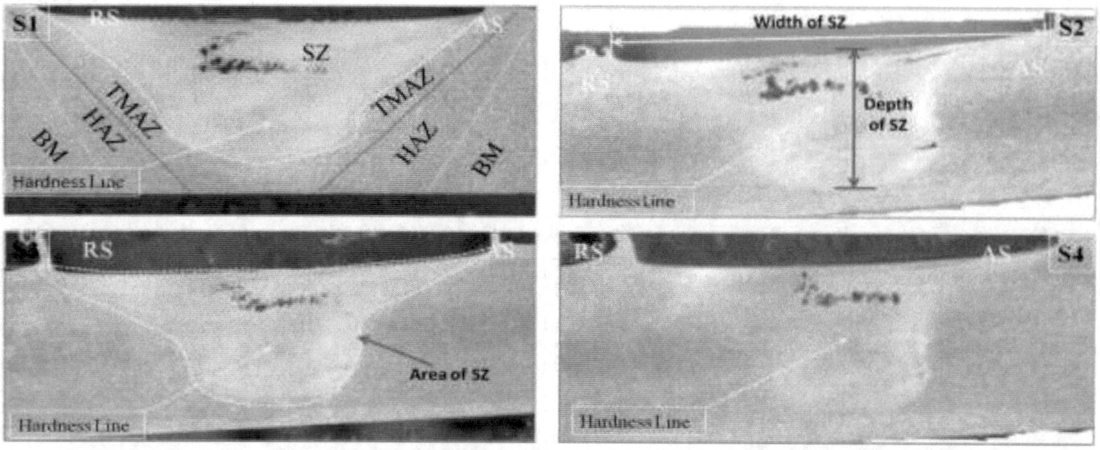

Figure. 2 shows the macrostructure and friction stir processed region in FSP specimens.

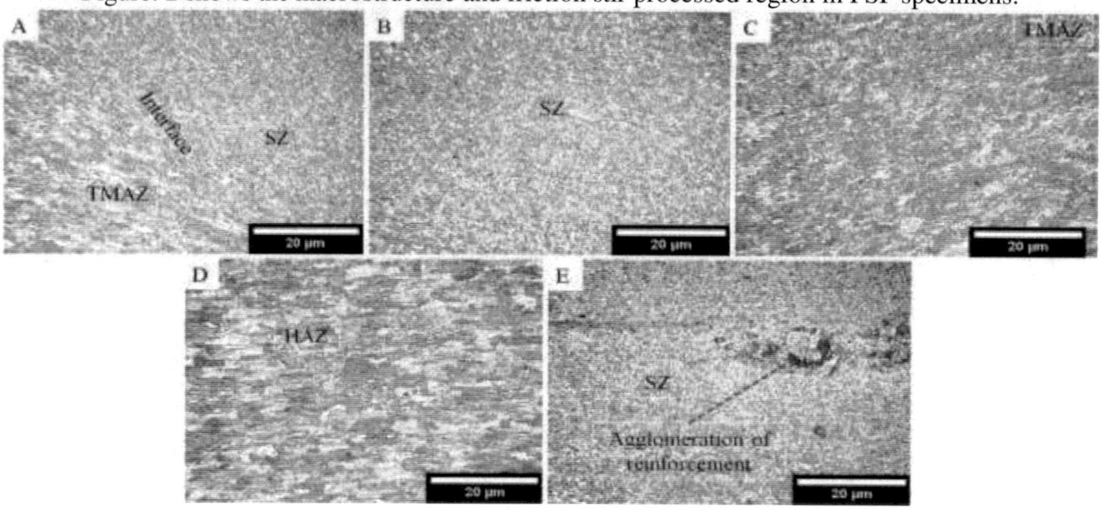

Figure 3 shows the microstructure of various zones formed during FSP in S1 specimen

3.2. Micro hardness

The average hardness value of as received base alloy (AA5083) was 80 HV. Figure 4 represents the microhardness profiles of SZ measured perpendicular to processed cross-section. The microhardness was measured at a continuous spacing of 0.4 mm under the load of 0.5 kgf acting for 15 sec using a microhardness tester. In general, the FSP does not result in softening of the FSPed specimens. The hardness profiles were more or less uniform in the SZ. There is an overall increase in hardness in SZ in comparison to base matrix which is attributed to dynamically recrystallized grains formed during FSP in this region. Besides, the hardness value is inversely proportional to grain size according to Hall-Petch Law. From the microhardness profiles, significant variation in hardness can be observed as one moves towards the bottom surface. On the other hand, the higher values of hardness can also be contributed to the uniform distribution of intermetallic during FSP.

The maximum hardness value of 124.5HV (refer Figure 4) was observed in specimen S1 with a tool rpm of 355, v of 63 mm/min, and ϕ of 17 mm. This highest value of hardness was observed at a depth of 2mm from shoulder surface. The reason for higher values of hardness is due to the distribution of reinforcement (Ti+SiC) particles which also restrict the grain growth during processing. Furthermore, the formation of plasticized zone beneath the tool shoulder results in the fabrication of MMC resulting in higher values of hardness. Besides, the existence of large force near the tool shoulder causes the severe dynamic recrystallization and formation of fine equiaxed grain leading to an overall increase in hardness near to tool shoulder surface.

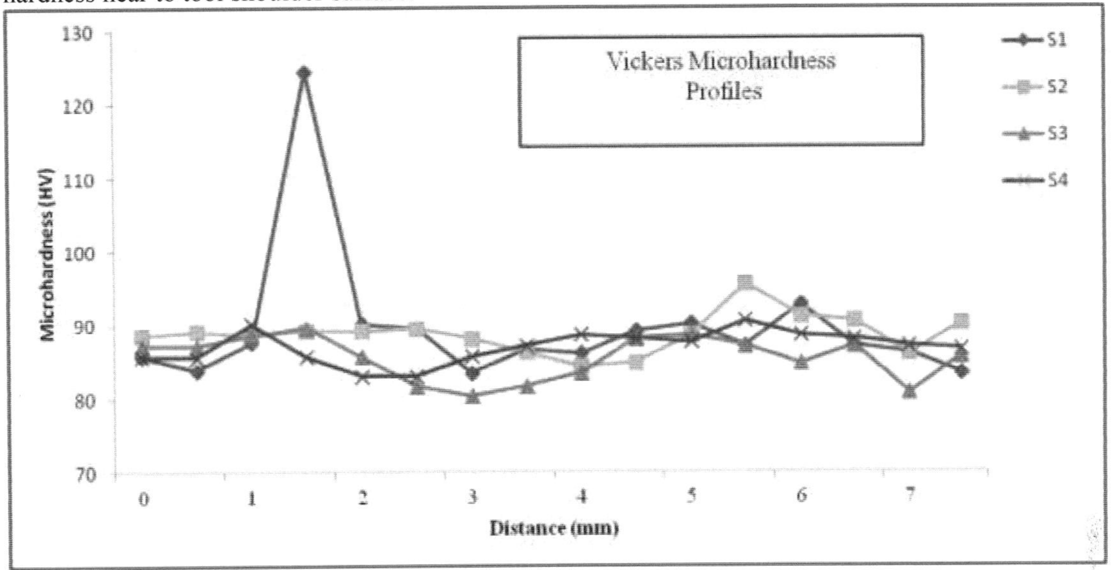

Figure 4 Represents the microhardness profiles obtained from S1, S2, S3, and S4 FSP specimens.

3.3. Bead Geometry Analysis

Figure 2 shows the macro images of the FSPed region. The bead geometry of SZ i.e., width, depth, and the area are shown in Figure 2. In all the specimens the dimensions of SZ i.e., the width of SZ, depth of SZ, and area of SZ was calculated from macroscopic images of the FSP region using ImageJ software (refer to Table 3). For specimen S1, the highest values of microhardness were obtained at tool rpm of 355, v of 63 rpm, and ϕ of 17mm. The corresponding value of the width, depth, and area of SZ was calculated as 17.23 mm, 5.743 mm, and 52.629 mm2 respectively (refer to Table 3). Moreover, the maximum value of hardness was obtained at 2mm from the top surface of SZ in specimen S1. This indicates that optimal results are obtained at lower tool rpm (355), lower v (63mm/min) and small ϕ (17mm) results in proper and uniform distribution of reinforcement within the metal matrix. Moreover, it can be observed that specimen S1 has the lowest width/depth ratio. The lowest width to depth ratio resulting in high hardness values owing to distribution of reinforcement in the specific area which ultimately leads to high density of reinforcement particles and hence higher hardness value.

Table 3. Presents the numerical values of width, depth, and area of SZ measured from micrographs using ImageJ software.

Sample No.	Width of SZ (mm)	Depth of SZ(mm)	Area of SZ (mm^2)	Width/Depth ratio
S1	17.230	5.743	52.629	3.0
S2	17.855	5.863	55.988	3.045
S3	20.164	5.707	59.272	3.533
S4	20.586	5.603	59.922	3.67

4. Conclusions

From the above experimental study, the following main conclusions can be drawn:

1. Macrostructure of FSPed specimens indicates the improper distribution of reinforced particles.
2. Microstructure of processed regions indicated the significant variation in grain size in various regions. The highest value of hardness was obtained in SZ owing to dynamic recrystallization of grains and more or less uniform distribution of reinforcement particles.

3. The highest value of hardness was obtained for specimen S1 at tool rpm of 355, v of 63 mm/min, and ϕ of 17 mm.
4. The highest value of hardness was obtained at a depth of 2 mm from the upper surface of processed region. Moreover, the maximum hardness value observed in S1 specimens bears the lowest width to depth ratio.

5. References

[1] Funatami K 2000 Emerging Technology in Surface Modification of Light Metals, *Surface Coating Technology 133,* 264–272.
[2] Statistical data of Automotive Materials Committee, *Japanese Society of Automobile Industry*, May 1997.
[3] North American Passenger Car and Light Truck Aluminum Content Report 1999, *Ducker Research Co*., Southfield Muy
[4] Vreeling J A, Ocelik J, Pei Y T, Van Agterveld D T, Hosson T M 2000 Laser Melt Injection of Ceramic Particles in Metals, *Acta Mater*. 48, 4225– 4232.
[5] Maurya R, Kumar B, Ariharan S, Ramkumar J, Balan K 2016 Effect of carbonaceous reinforcements on the mechanical and tribological properties of friction stir processed Al6061 alloy, Materials Today: Proceedings. Vol. 98: 155-166.
[6] Rana R S, Purohit R, Sharma A K, and Rana S 2014 Optimization of wear performance of AA 5083/10 wt% SiCp composites using Taguchi method, *Procedia Mater. Sci*., 6, 503-511
[7] Vignesh R V, Padmanaban R 2018 Intergranular corrosion susceptibility of friction stir processed aluminium alloy 5083, *Mater. Today* 5 (8, Part 3), 16443-16452
[8] Padmanaban R, Balusamy V, and Kishore V R 2012 Effect of Axial Pressure and Tool Rotation Speed on Temperature Distribution during Dissimilar Friction Stir Welding, *Advanced Materials Research*, 418-420, 1934-1938
[9] Padmanaban R, Balusamy V, and Nouranga K N 2015 Effect of process parameters on the tensile strength of friction stir welding dissimilar aluminum joints, *Journal of Engineering Science and Technology*, vol 10, 790-801

Vibration Analysis of Hybrid Composite Laminated Beam with Crack in ANSYS APDL

Sunita Danu[1], Nitish Kumar Saini[2], Deepak Prasad Nautiyal[3] and Anadi Misra[4]

1 Assistant Professor, Department of Civil Engineering, Dev Bhoomi Institute of Technology, Dehradun, Uttarakhand (India)
2 Assistant Professor, Department of Mechanical Engineering, Dev Bhoomi Institute of Technology, Dehradun, Uttarakhand (India)
3 Research Scholar, Department of Civil Engineering, Dev Bhoomi Institute of Technology, Dehradun, Uttarakhand (India)
4 Professor, Department of Mechanical Engineering, G.B.P.U.A.T., Pantnagar, Uttarakhand (India)
*sunita.danu2112@gmail.com

Abstract. In this paper, whole work is done for hybrid laminated composite beam which is made of three composite materials boron-epoxy, aramid-epoxy, and s-glass-epoxy. A numerical study, using finite element method is carried out to analyse the free transverse vibration behaviour of composite beam. ANSYS APDL is used as a finite element solver to simulate the free transverse vibrations. A variety of parametric studies is performed to check the effects of various parameters associated with crack geometry and laminate property on the three least natural frequencies. Parameters for the investigation are included as fibre volume fraction, ply angle for (0°/ Φ°/0°/Φ°/0°/90°/(90-Φ)°/0°/(90-Φ)°/0°) stacking sequence, location of crack geometry relative to the restricted end, depth of crack (a) and support conditions (C-C & C-F configuration) for hybrid laminated composite beam, trend of the least three transverse natural frequencies is also discussed in detail. It is noted that maximum value of natural frequency occur at Φ equals to 90° for both (C-C) and (C-F) and for relative crack depth 0.5 to 0.6, there is sharp decrement in natural frequency for (C-F). Other factors of laminate and crack geometry are also discussed in present analysis, which shows a great scope of this hybrid composite with (0°/ Φ°/0°/Φ°/0°/90°/(90-Φ)°/0°/(90-Φ)°/0°) stacking sequence in the field of aviation and prosthetics.

Keywords: Hybrid laminated composite, Transverse vibration, Ply angle, Fibre volume fraction, Stacking sequence.

1. Introduction

Composites have two or more than two phases and both the phases should not be soluble in each other and both have different chemical attributes which shows independence on Hume-Rothery rule. In the present scenario, conventional materials like metals and alloys are replaced by smart materials like composites, functionally graded materials (F.G.M.) and fibre metal laminates (F.M.L.) etc. The use of composite material became important due to its high fatigue applications as well as for its low weight to strength ratio which is required for the static and dynamic problems. The past history of various carried studies represents that the main cause of failure of any composite structure is delamination of the laminate [1-2] and especially under the presence of various geometrical discontinuities like crack and notch which cause stress concentration as well as decrement in the value of natural frequency [3]. This decrement in the value of natural frequency becomes very dangerous because it increases the probability of resonance at a lower frequency in the case of the dynamic behaviour of the structure in different modes and also affects damping coefficient, mode shape, and natural frequency [4].

For eliminating the resonance problem, it is necessary to know the damping factors and natural frequency of the system and different mode shapes for strengthening the structure where as required [5].

2. Techniques of Vibration Analysis

There are two types of methods for investigation, first one is the analytical [6-7] and the second one is the experimental approach. The experimental approach gives an exact solution with various advantages like great exposer, high-level control, and clear-cut results to the researcher. There are some benefits with the analytical approach like less time consumption, no issue regarding machine set up and comparatively cheaper than experimental approach. Practically, it is not easy to get exact solution of all engineering complex problem, for this purpose, there are some methods which give approximate solution like F.E.M.[8] and F.E.M. has many software packages like ANSYS [9], ABAQUS, Solid Works, Fluent, FORTRON, etc.

3. Validation with Krawczuk & Ostachowicz

In the present examination results are first approved by model given by Krawczuk and Ostachowicz [10] for graphite-polyamide composite as shown in figure 1 and further examination is expanded for hybrid composite for different parameters of laminate like fibre volume fraction (V), ply angle (Φ) and with crack, the crack geometry is also taken as a parameter for present analysis which influences the natural frequencies of freely vibrated cantilever composite beam.

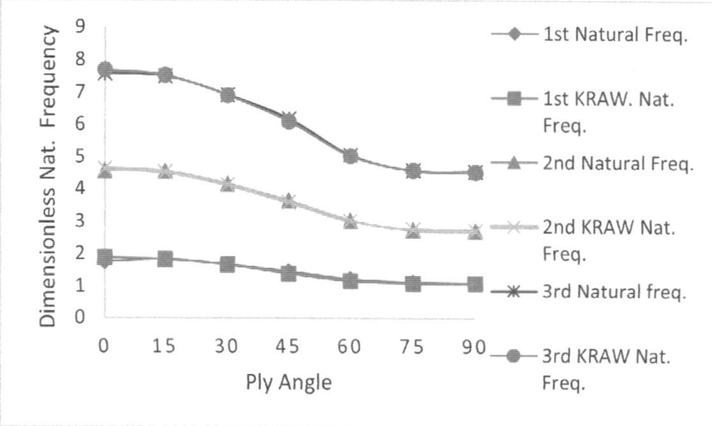

Figure 1. Comparison between present analysis and Krawczuk & Ostachowicz [10] for fibre volume fraction (V) =0.1

4. Materials and Methodology

4.1 Governing Equation

Governing equation for the free vibration condition of the beam is expressed by [11-12]

$$[K] - \omega^2[M]\{q\} = 0 \quad (1)$$

In the expression,
q = Degree of freedom, **K** = Stiffness matrix, **M** = Mass matrix

Table 1. Properties of boron epoxy, aramid epoxy and s-glass epoxy [13 - 15]

Mechanical Properties of Fibre (f) & Matrix (m)	Modulus of Elasticity (GPa)		Shear Modulus (GPa)		Poisson's Ratio		Mass Density (kg/m^3)	
	Em	Ef	Gm	Gf	νm	νf	ρm	ρf
Boron Fibre and Epoxy	3.25	393	1.25	163.75	0.3	0.2	1250	2700
Aramid Fibre and Epoxy	3.25	131	1.25	48.51	0.3	0.35	1250	1450
S-Glass Fibre and Epoxy	3.25	86.9	1.25	35.61	0.3	0.22	1250	2490

4.2 Model of Beam

In the present model of the beam the said crack is a transverse crack having V shape. The geometrical parameters of composite beam has taken as length (L), width (B) and height (H) are 1.0 m, 0.05 m, and 0.025 m respectively as shown in Figure 2. All the fibres are assumed to be oriented at an angle Φ.

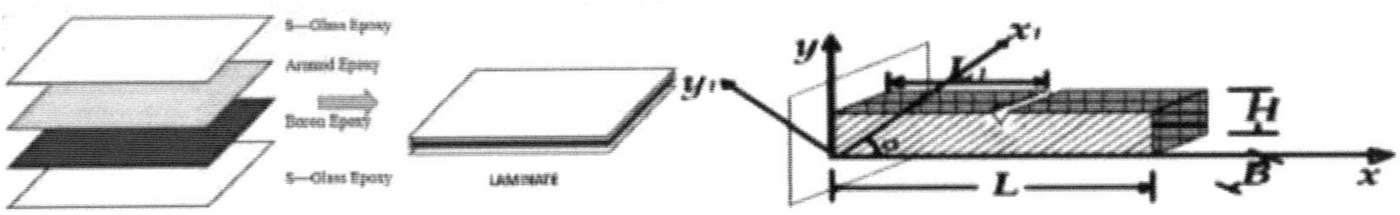

Figure 2. Schematic labelled design of laminate & laminated composite beam with transverse surface crack in clamped-free (C-F) configuration

4.3 Description of used element

In the present analysis, solid shell 190 (SOLSH 190) element is used as shown in figure 3. SOLSH 190 provides combine properties of the solid and shell elements.

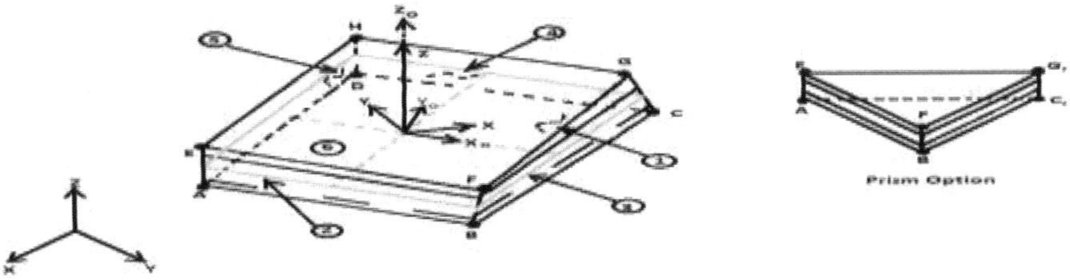

Figure 3. Geometry description of solid shell (SOLSH 190) element type

5. Results and Discussion

In the present investigation, the effect of various parameters is studied. Fibre volume fraction (V), relative crack depth (a/H), crack location (L1/L), support conditions, and ply angle of fibre (Φ) on transverse natural frequency studied and represented for stacking sequence **(0°/Φ°/0°/Φ°/0°/90°/(90-Φ)°/0°/(90- Φ)°/0°).**

5.1 Influence of Ply angle (Φ) on three least transverse natural frequencies for the cracked hybrid composite beam with relative crack depth (a/H) = 0.6, crack location (L1/L) = 0.5 and V=0.5

5.1.1 For (C-F) configuration

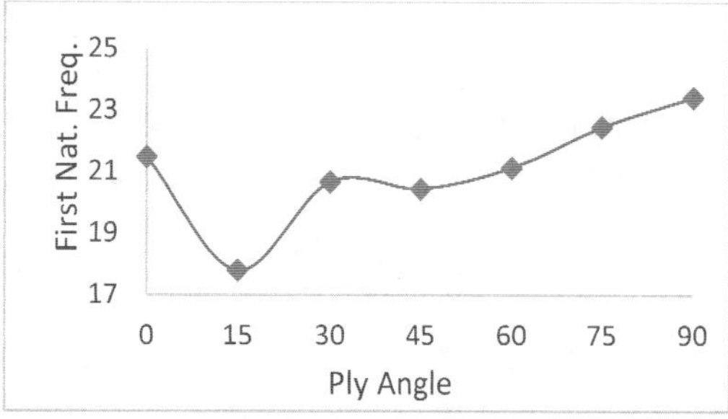

Figure 4. Ply angle (Φ) as a parameter of first natural frequency

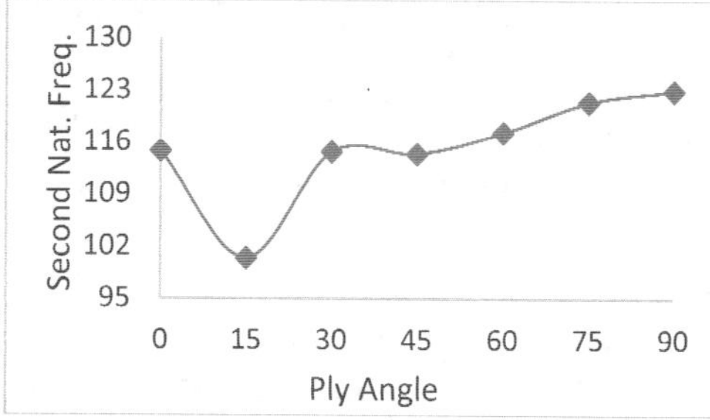

Figure 5. Ply angle (Φ) as a parameter of second natural frequency

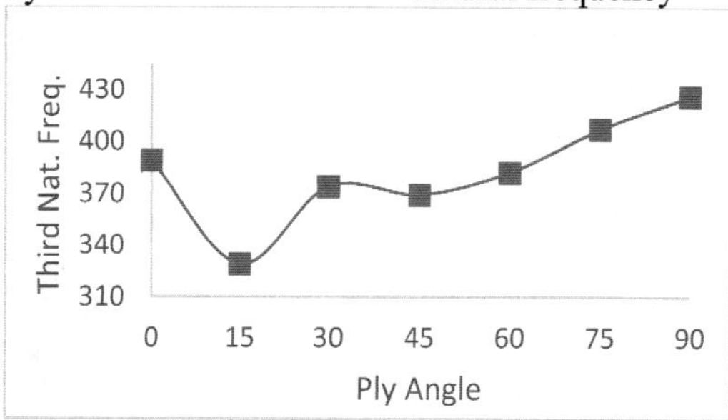

Figure 6. Ply angle (Φ) as a parameter of third natural frequency

5.1.2 For (C-C) configuration

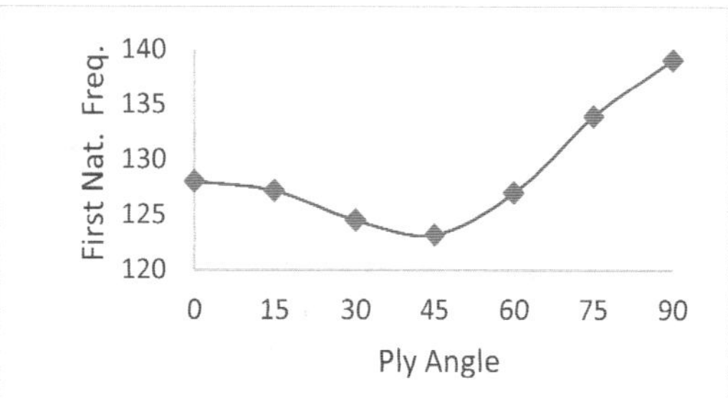

Figure 7. Ply angle (Φ) as a parameter of first natural frequency

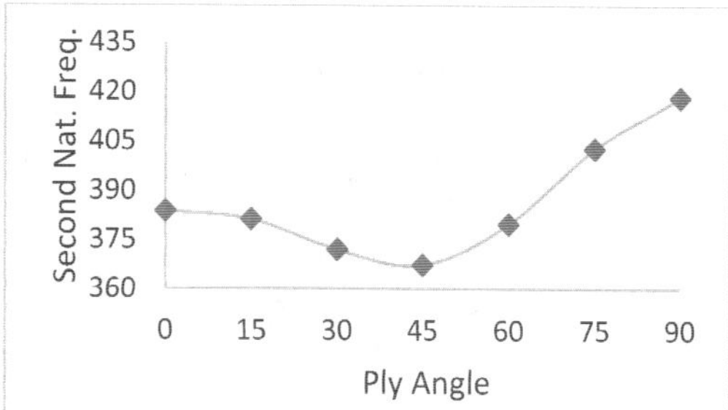

Figure 8. Ply angle (Φ) as a parameter of second natural frequency

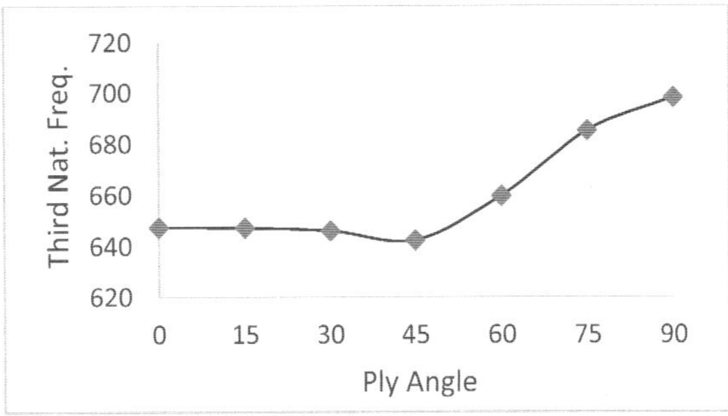

Figure 9. Ply angle (Φ) as a parameter of third natural frequency

From figure 4 to 6, it is cleared that minimum natural frequency is observed at 15° ply angle in the presence of transverse crack due to less stiffness of the beam for transverse vibration while the maximum natural frequency at 90° ply angle because of high stiffness of beam in C-F configuration. In C-C configuration the beam became less stiff at 45° and more stiff at 90° for which lowest and highest natural frequencies are observed respectively as shown in figure 7 to 9.

5.2 Influence of relative crack depth (a/H) on three least transverse natural frequencies for the cracked hybrid composite beam with Ply angle (Φ)=90°, crack location (L1/L) = 0.5, and fibre volume fraction (V)=0.5

5.2.1 For (C-F) configuration

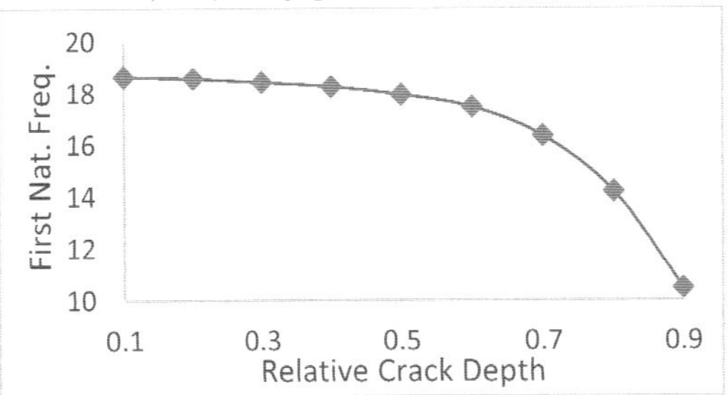

Figure 10. Relative crack depth as a function of first natural frequency

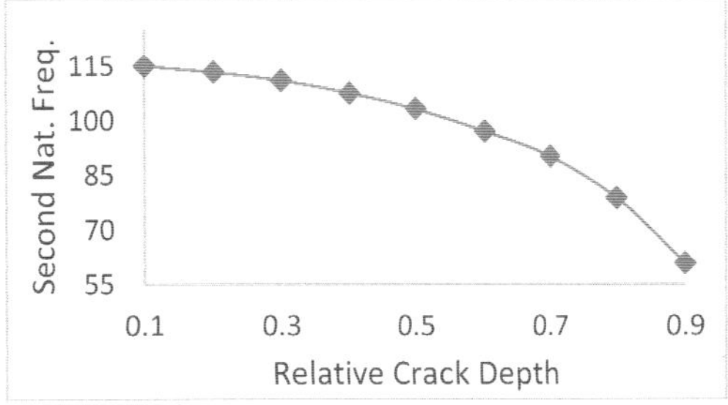

Figure 11. Relative crack depth as a function of second natural frequency

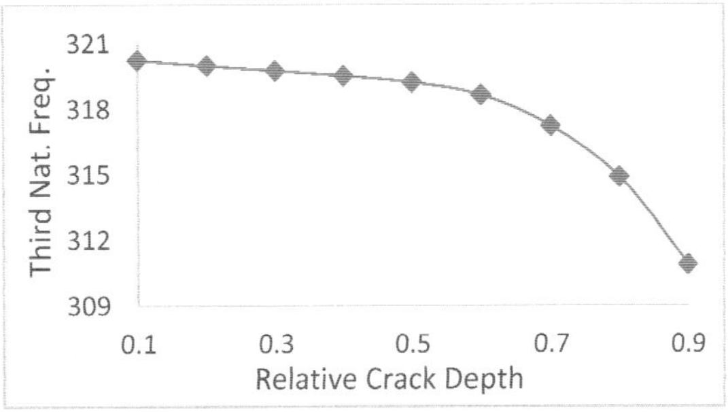

Figure 12. Relative crack depth as a function of third natural frequency

5.2.2 For (C-C) configuration

IOP Conf. Series: Materials Science and Engineering 1149 (2021) 012002 doi:10.1088/1757-899X/1149/1/012002

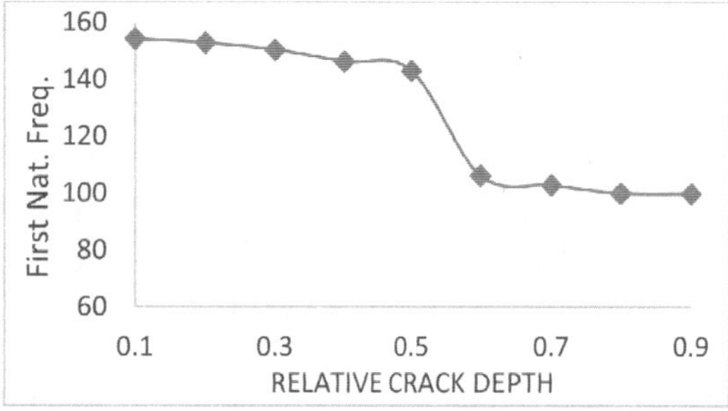

Figure 13. Relative crack depth as a function of first natural frequency

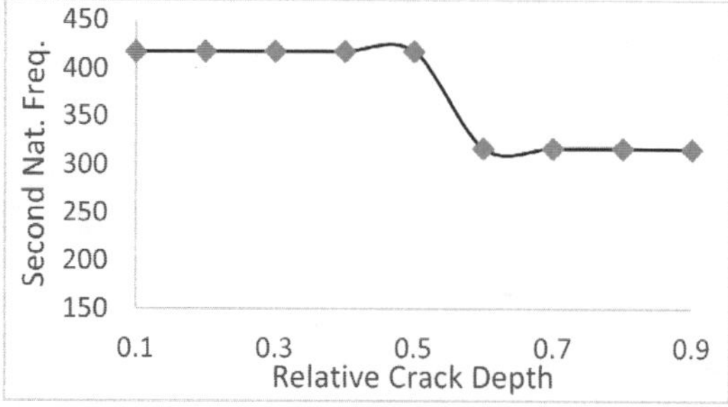

Figure 14. Relative crack depth as a function of second natural frequency

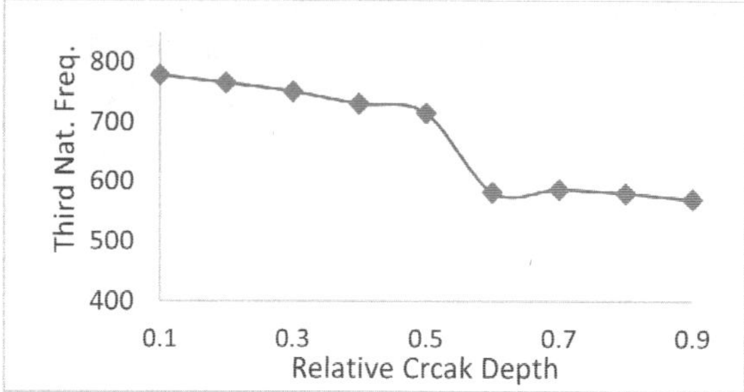

Figure 15. Relative crack depth as a function of third natural Frequency

All three natural frequencies are observed minimum and maximum at relative crack depth (a/H) = 0.9 and 0.1 respectively for both C-F and C-C configuration as shown in figure 10 to 15, because, since the size of crack increases the stiffness value of the beam decreases due to which natural frequency decreases.

5.3 Influence of relative crack location (L1/L) on three least transverse natural frequencies for the cracked hybrid composite beam with Ply angle (Φ) = 90°, crack depth (a/H) = 0.5 and fibre volume fraction (V)=0.5

5.3.1 For (C-F) configuration

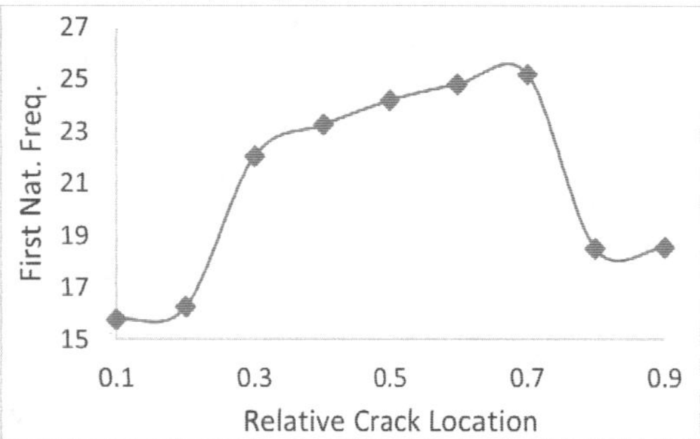

Figure 16. Relative crack location as a function of first natural frequency

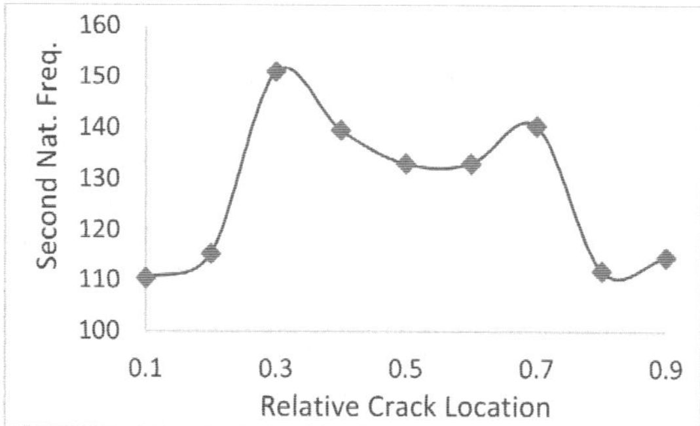

Figure 17. Relative crack location as a function of second natural frequency

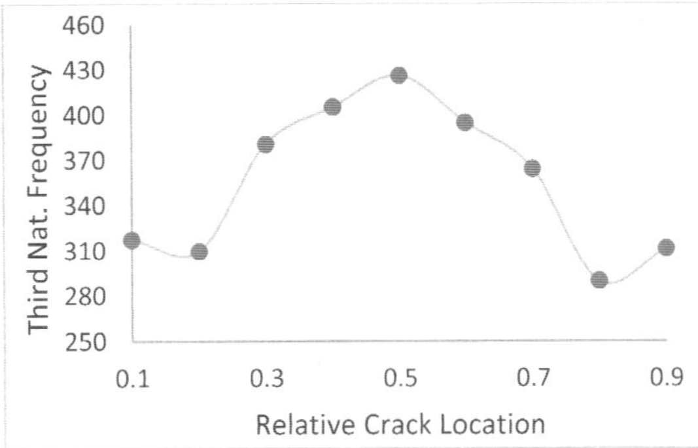

Figure 18. Relative crack location as a function of third natural frequency

5.3.2 For (C-C) configuration

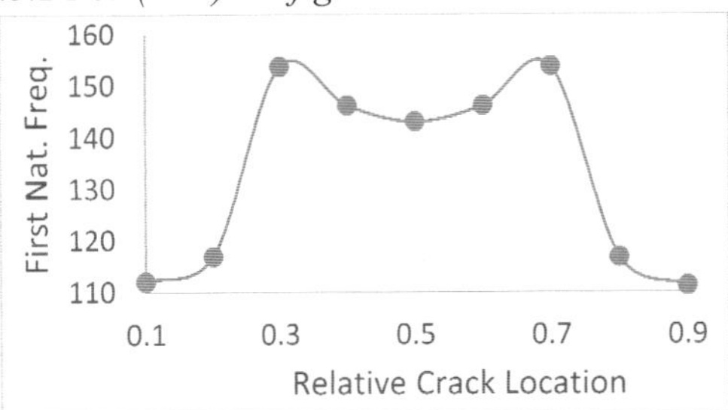

Figure 19. Relative crack location as a function of first natural frequency

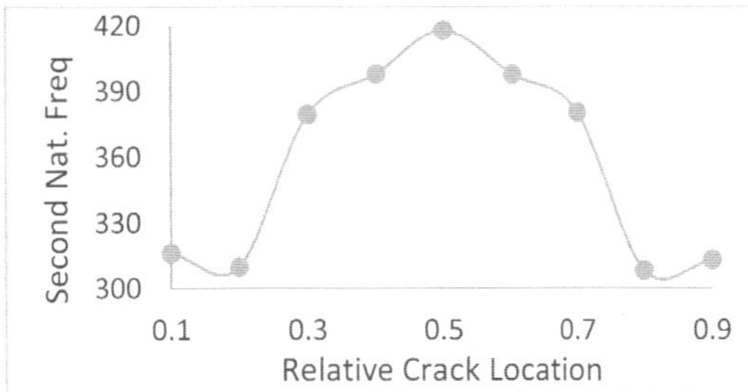

Figure 20. Relative crack location as a function of second natural Frequency

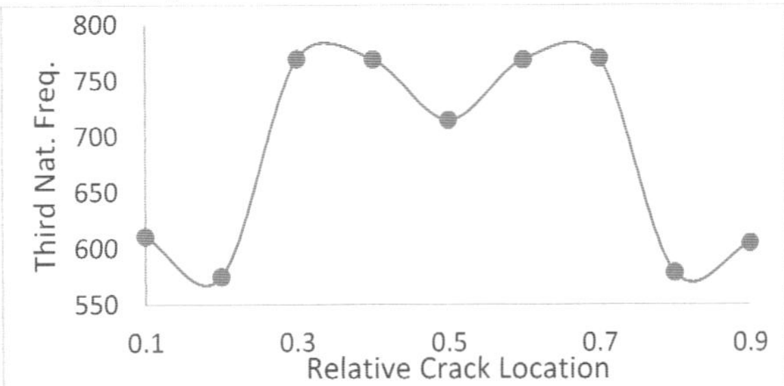

Figure 21. Relative crack location as a function of third natural frequency

In C-F configuration least natural frequencies are observed minimum when crack is developed nearer to fixed end as a result of high flexibility of the beam which causes high tendency of vibrations as shown in figures 16 to 18. All three least natural frequencies are observed maximum and minimum on two values of relative crack location because beam became symmetrical in C-C configuration as both ends are fixed so about crack location 0.5, there is a symmetry in trend as shown in figures 19 to 21.

6. Summary and Conclusion

1. It is cleared by the figure 4 to 6, for the cracked beam with the (0°/Φ°/0°/Φ°/0°/90°/(90-Φ)°/0°/(90-Φ)°/0°) stacking sequence for the crack geometry (a/H) = 0.6, (L_1/L) = 0.5, all three least transverse natural frequencies are minimum for the value of Φ equals to 15°, maximum at 90° for C-F configuration & by figure 7 to 9, minimum for the value of Φ equals to 45° and maximum at 90° for C-C configuration.
2. It is cleared by the figure 10 to 12, for the cracked beam with the (0°/Φ°/0°/Φ°/0°/90°/(90-Φ)°/0°/(90-Φ)°/0°) stacking sequence for the relative crack location (L_1/L) = 0.5, fibre volume fraction 0.5 and Φ equals to 90°, all three least transverse natural frequencies are minimum for the value of relative crack depth (a/H) equal to 0.9, maximum at crack depth (a/H) equal to 0.1 in C-F configuration & by figure 13 to 15, minimum for the value of relative crack depth (a/H) equal to 0.9 and maximum at crack depth (a/H) equal to 0.1 but there is sharp decrement is represented between 0.5 to 0.6 value of relative crack depth in the trend of all three natural frequencies in C-C configuration.
3. It is cleared by the figure 16, the lowest value of first natural frequency observed at relative crack location 0.1 and highest at 0.7 and there is a sudden increment in the value of first natural frequency, from 0.2 to 0.3 and again the trend is increasing linearly from 0.3 to 0.7 and again sudden decrement has seen from 0.7 to 0.8 for C-F configuration & by figure 17, lowest value of second natural frequency observed at relative crack location 0.1 and highest at 0.3 in C-F configuration.
4. It is cleared by the figure 18, the lowest value of the third natural frequency observed at relative crack location 0.8 and highest at 0.5 and there is an increment in the value of third natural frequency, from 0.2 to 0.3 and again trend is linearly increasing from 0.3 to 0.5 and linearly decreasing from 0.5 to 0.8 in C-F configuration.
5. It is cleared by the figure 19, the lowest value of first natural frequency observed at relative crack location 0.1 and 0.9 and highest at 0.3 and 0.7 for C-C configuration and trend is making mirror image about 0.5 & by figure 20, lowest value of second natural frequency observed at relative crack location 0.2 and 0.8 and highest at 0.5 for C-C configuration and the trend is making mirror image about 0.5.
6. It is cleared by the figure 21, the lowest value of third natural frequency observed at relative crack location 0.2 and 0.8 and highest at 0.3 and 0.7 for C-C configuration and trend is making mirror image about 0.5.

References

[1] **Danu, Sunita & Saini, Nitish & Khan, Yasar & Sharma, Harsh. (2019).** Composites for Vibro-Acoustics-A Review. *International Journal of Applied Engineering Research. 14. 6-11.*
[2] **Singh, Ashuvendra & Ahmad, Faraz & Saini, Nitish. (2019).** Finite Element Analysis Based Vibration behavior on Warren Truss Bridge.
[3] **Ali, A.T.Z. Aswan, A.J. Faizal, M. and Nisreen, A. 2009.** Free vibration Analysis and Dynamic Behaviour for Beams with cracks. *International Journal of Science Engineering and Technology.*
[4] **Saini, N.K., Panwar, M.S., Maurya, A. and Singh, P.A., 2019**. Design and Vibration Analysis of Go-kart Chassis. *International Journal of Applied Engineering Research, 14(9), pp.53-60.*
[5] **Saini, N.K. and Misra, A., (2017).** Vibration Analysis of Composite Cantilever Beam with Surface-Crack. *International Journal of Innovative Trends in Engineering, 31.*
[6] **Chondros, T.G. Dimarogonas, A.D. and Yao, J. 1998.** Longitudinal vibration of a bar with a breathing crack. *Engineering fracture mechanics,* 61(5):503-518.

[7] **Li, J. Huo, Q. Li, X. Kong, X. and Wu, W. 2014.** Vibration analyses of laminated composite beams using refined higher-order shear deformation theory. *International Journal of Mechanics and Materials in Design,* 10(1):43-52.

[8] **Krawczuk, M. 1994.** Rectangular shell finite element with an open crack. *Finite Elements in Analysis and Design,* 15(3):233-253.

[9] **Barbero, E.J. 2013.** Finite element analysis of composite materials using ANSYS®. CRC press.

[10] **Krawczuk, M. and Ostachowicz, W.M. 1995.** Modelling and vibration analysis of a cantilever composite beam with a transverse open crack. *Journal of Sound and Vibration,* 183(1):69-89.

[11] **Ganesan, R. and Zabihollah, A. 2007.** Vibration analysis of tapered composite beams using a higher-order finite element. *Part II: parametric study. Composite structures*, 77(3):319-330.

[12] **Krawczuk, M. and Ostachowicz, W.M. 1994.** A finite plate element for dynamic analysis of a cracked plate. *Computer methods in applied mechanics and engineering,* 115(1-2):67-78.

[13] **Saini, Nitish & Ahmad, Faraz & Danu, Sunita & Misra, Anadi. (2019).** Vibration Analysis of Fibre Metal Laminate Beam in ANSYS Parametric Design Language. *International Journal of Applied Engineering Research. 14. 149-155.*

[14] **Pandey, Ashutosh & Saini, Nitish & Ahmad, Faraz & Misra, Anadi. (2019).** Vibration Analysis of Stepped Laminated Composite Beam in ANSYS. *International Journal of Applied Engineering Research.* 14. 38-45.

[15] **Saini, N.K.,** Vibration analysis of laminated composite cantilever beam with transverse crack (Doctoral dissertation, GB Pant University of Agriculture and Technology, Pantnagar-263145 (Uttarakhand)).

Effect of flow velocity and pressure on carotid artery due to deposition of different shapes of plaque

Raman Yadav[1], Sharda Vashisth[2], Ranjit Varma[3] and Kaushal Kumar[4]

[1]Department of EECE, K.R. Mangalam University, Gurgaon, Haryana, India
[2]Department of EECE, The North Cap University, Gurgaon, Haryana, India
[3]Director, Delhi Technical Campus, Greater Noida, Uttar Pradesh, India
[4] Department of ME, K.R. Mangalam University, Gurgaon, Haryana, India
raman.y.90@gmail.com

Abstract: Atherosclerosis is the disease progressed because of the deposition of stenosisat the interior walls of arteries which results in reducing the size and causing blockage in the carotid artery leading to brain attack or stroke. Two-dimensional model of carotid artery is designed and analyzed by Computational Fluid Dynamics (CFD). Designing of general analytical model of artery followed by meshing of geometry, application of boundary conditions and analysis is done with ANSYS Fluent 16.2. The objective of this research is to analyze the consequence of flow velocity and pressure on the arteries having blockage to assess the diseases related to carotid artery. This paper provides analytical analysis of flow velocity and pressure on the artery due to deposition of different shapes of plaque. We consider plaque shape1, plaque shape2, cosine plaque and irregular plaque shapes. Hemodynamic variables like velocity and pressure differ with commute mostly in the shape of geometry. It's found that velocity is maximal at the constraint as compared to pre- and post-stenotic areas which shows that the artery having more stenosis area has maximum pressure and velocity on the wall of the artery.
Keywords: Atherosclerosis, Computational Fluid Dynamics (CFD), Contour, Stenosis, Bifurcation

1. Introduction

Atherosclerosis is the cardinal source of death and consequential impairment in the developed countries [1]. It is accountable for fatality and morbidness universally which is resulted because of debilitation of the immune system [2]. Cerebrovascular diseases are associated with the flow of blood in arteries. Atherosclerosis is the one of the vital arterial disease which is originated due to accumulation of plaque in arteries due to the accumulation of fat, cholesterol, carbohydrates, fibrous tissue and low-density lipoproteins [3, 4]. It is frequently referred to hardening of the arteries. Bifurcation and division of arteries are the most preferred sites for the progression of atherosclerosis. With the progression of the ailment, these plaques enlarge and lead to the impairment of flow velocity in the artery [5]. The symptom of atherosclerotic cerebrovascular disease is brain attacks or strokes. The Fluid dynamics of carotid artery gives us better understanding of flow of blood though artery and commencement and growth of atherosclerotic ailment [11, 19].

The role of the Carotid artery is to impart blood and other nutrients to the brain, face and neck of the human body. Here we have carotid artery, on both side of our neck. Each carotid artery can be divided into two division viz. internal carotid artery (ICA) and external carotid artery (ECA) [6, 17]. Internal carotid artery has larger diameter as compared to the external carotid artery as revealed in Figure 1. The purpose of the ICA is to impart blood to the brain and that of the ECA is to impart blood to the face, scalp and neck.

Laser Doppler anemometry has been widely used as experimental technique for the visualization and quantification of blood flow through arteries. B-mode ultrasound is another technique for the detection of early atherosclerosis which begins with intima-media thickening. Direct visualization of both artery wall and lumen is possible with it. This information contributes to significant function for the diagnosis and assessment of atherosclerotic ailment.

Computational Fluid Dynamics (CFD) [12] representation is used effectively for prediction of flow pattern within the artery, and for checking the initiation and progression of plaque in the arterial wall [8, 9, 15]. CFD, applied to 2-D arterial geometries having different plaque shapes, provide precious simulation of complex geometry. Computational methods are easy to simulate and gives result more accurate and cost effective [7, 10].

The shape and size of plaque are variable for patients [13, 14]. In this study, four different shapes of the stenosis are assumed, and its velocity and pressure contour are determined. The consequence of the stenosis shapes on the flow of blood throughout carotid artery is analyzed. The consequence of different shapes of plaques on the flow pattern and pressure of the artery at the bifurcation is also determined.

2. Methodology

2D idealize geometry of carotid artery is created by means of CFD Fluent for the simulation of blood flow. After that the geometry is meshed by inherent mesh tool in workbench. For meshing the geometries, tetrahedral elements are used. The anatomical geometrical representation of idealized carotid artery used in this study is shown in Fig. 2. Then boundary conditions are applied at inlet and outlets of the artery. The boundary condition functional for transient flow examination is inlet velocity according to the user defined function and the pressure for the stenosis artery at mutual outlet surface is 0 Pa. Carreau-Yasuda model is employed for the Non-Newtonian type of flow and for the Newtonian flow power law model is adopted.

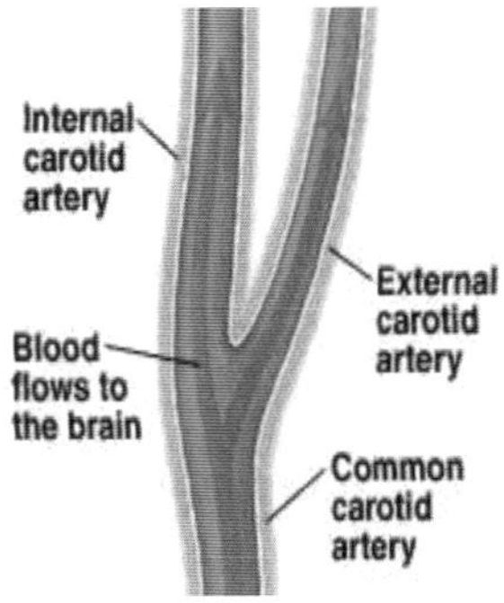

Figure 1. Anatomical model of healthy carotid artery

2.1. Software used

ANSYS Fluent® is economically accessible CFD software delivers unprecedented productivity, enabling simulation and highly reliable. It has capability to solve steady and transient simulations. It is communicating in programming language C which has wide variety to make it more versatile and flexible. On ANSYS workbench platform geometrical representation is designed, meshed and then relevant boundary conditions are decided. After identifying boundaries User Defined Functions (UDF) communicated in C are adjoined to the fluent, thus increases its adaptability. Variety of functions can be implemented using Fluent. Here finite volume discretization method is used to convert

Partial differential equation of conservation of mass and momentum into non-linear algebraic equations.

2.2. Governing Equations

The Navier- Stokes (1) and (2) represent the continuous fluid flow through artery. It consists of three equations; Continuity equation, Momentum equation and Energy equation which are obtained by applying conservation of fluid flow.

$$\nabla u = 0 \quad (1)$$

$$\rho \frac{\partial u}{\partial t} = -\nabla P + \mu \nabla^2 u \quad (2)$$

Where ρ be the density, u be the velocity vector, P be the pressure, μ be the coefficient of viscosity and t be the time.

2.3. Non-Newtonian Equations

For Non-Newtonian and Carreau model complex rheological behaviour of blood is approximated, here apparent viscosity is expressed as the function of shear rate [18] the governing equation is:

$$\mu = \mu_\infty + (\mu_0 - \mu_\infty)[1 + (\gamma\lambda)^2]^{n-\frac{1}{2}} \quad (3)$$

Parameters used for Carreau Non-Newtonian model are; time constant, λ=3.313s; power law index, n=0.3568; resting viscosity, μ_0=0.56P; infinite strain viscosity, μ_∞=0.0345 P; density of blood, ρ=1060 kg/m^3 [16]

3. Result

3.1. 2D idealized Carotid artery

The 2D geometry of carotid artery is created and meshed using ANSYS followed by calculations and further analysis. Fig. 2 shows the representation of healthy carotid artery along with its dimensions.

Blood flow refers to the movement of blood through the arteries. Flow of blood is inversely proportional to the area of the artery. Velocity of blood flow reduces as cross-sectional area increases. Velocity increases if the plaque is deposited in the artery which results in decrease of cross-sectional area of the artery. Pressure on the arterial wall may be defined as the measure of force that blood exerts on the wall of the artery as it moves blood through artery. Pressure and velocity of flow through artery is affected due to the deposition of the plaque in it. The stenosis shape results in varying the velocity and pressure of the blood on the wall. Velocity and pressure measurement of the blood flow through the artery due to presence of the stenosis are very important parameters to measure the severity of blockage which may lead to fatal situation if not diagnosed accurately.

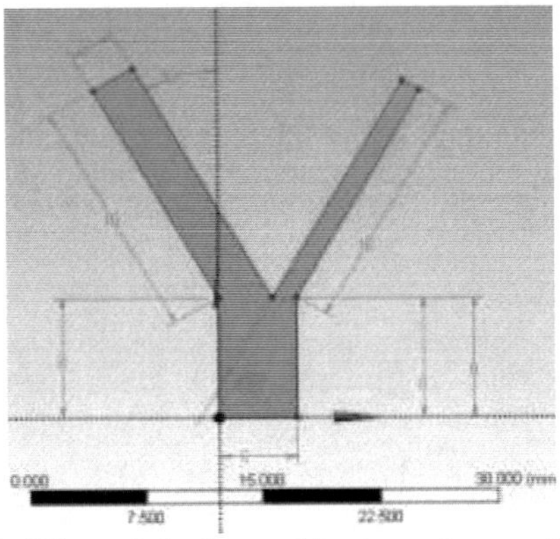

Figure 2. 2D model of carotid artery along with its dimensions

3.2. Velocity Contour

Shapes of the plaque vary from person to person. Fig. 3 shows the velocity contour of healthy artery. In our study we use different types of plaque shapes categorized as plaque shape 1, plaque shape 2, cosine plaque and irregular plaque as shown in Figures 4-7. Therefore, the velocity and pressure contour are calculated for all artery having different shapes of plaque.

From the results we can observe that velocity does not vary much before the blockage and remains nearly same in the proximal region. From the contours it is observed that velocity near the wall of the artery is nearly zero and the maximum velocity occurs at the region having maximum stenosis. The artery having maximum restriction results in increasing the flow velocity and the region having less percentage of restriction results in reducing flow velocity. This indicates that with increase in restriction, there is more possibility of deposition of plaque in arterial wall.

Figures 4-7 represent the velocity contours for different stenosis shapes taken in our study. It is noted that velocity contour of plaque shape 2 have maximum blood velocity at the stenotic region followed by plaque shape 1, irregular shape, cosine shape and healthy artery. Therefore, from the velocity contour figures it is revealed that the physiological effect of shape of stenosis is more in plaque shape 2. Area for fluid passage is also less in this case which in turn increases the chance of atherosclerosis and results in stroke. Velocity contour of the arteries exhibits how the different shapes of the stenosis influence the flow of blood through it. We have observed that the buildup of stenosis in the artery bring about blockage and the consequential reduction in region give rise to the flow of velocity of blood through it.

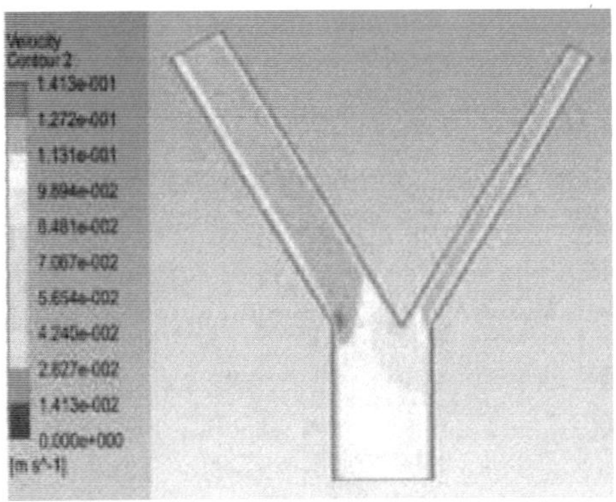

Figure 3. Flow velocity for healthy plaque

IOP Conf. Series: Materials Science and Engineering 1149 (2021) 012003 doi:10.1088/1757-899X/1149/1/012003

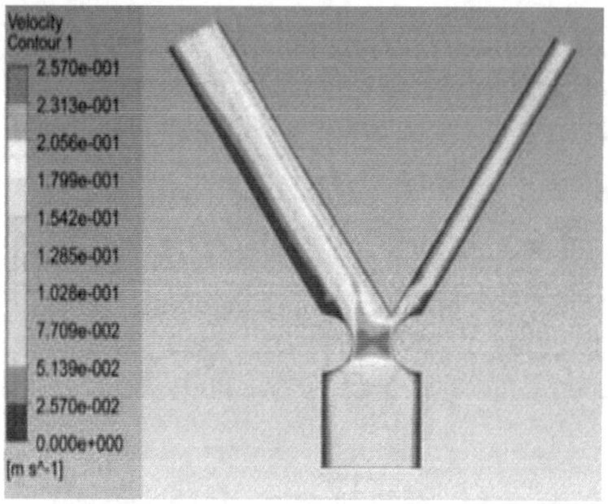

Figure 4. Flow velocity for plaque #1

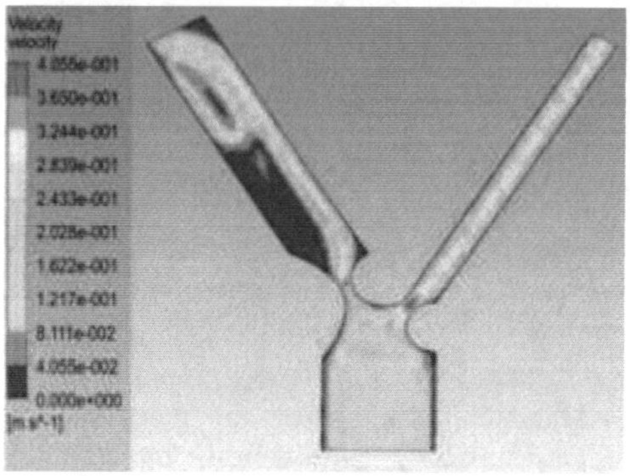

Figure 5. Flow velocity for plaque #2

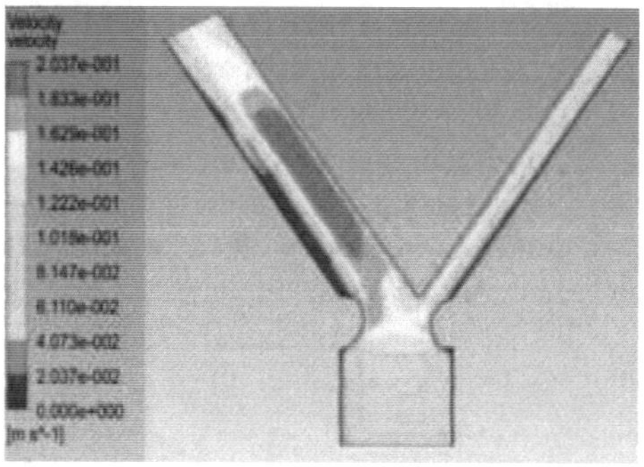

Figure 6. Flow velocity for Cosine plaque

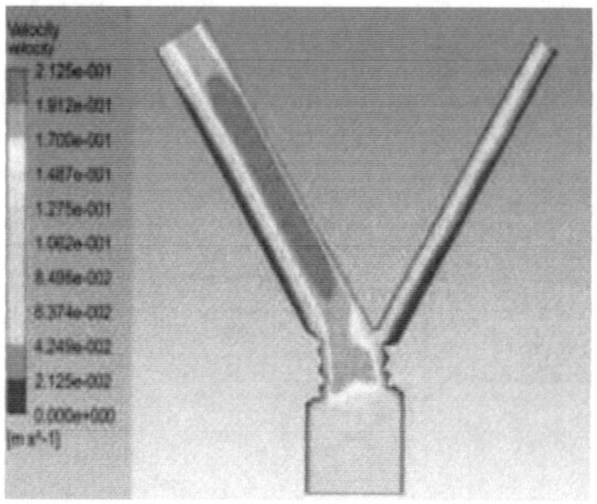

Figure 7. Flow velocity for Irregular plaque

3.3. Pressure Contour

Figures 8-11 shows the pressure contour of the artery having stenosis. Stenotic arteries in our study are arteries having plaque shape 1, plaque shape 2, cosine plaque and irregular plaque as discussed above. Pressure contour is calculated for all artery having different shapes of plaque. Through it we can measure the force exerted by the blood on the artery wall. As the percentage of restriction increases, the pressure on the artery also increases. Subsequently we can perceive that the maximal pressure takes place at the region having maximum stenosis. The artery having maximum restriction results in increasing the pressure through the artery whereas the region having less percentage of restriction results in reducing pressure through it. Maximum pressure through the artery was found at the mid of the stenosis and in the post stenotic region there is decrease in the pressure. Figs. 8-11 represent the pressure contours for different stenosis shapes taken in this study.

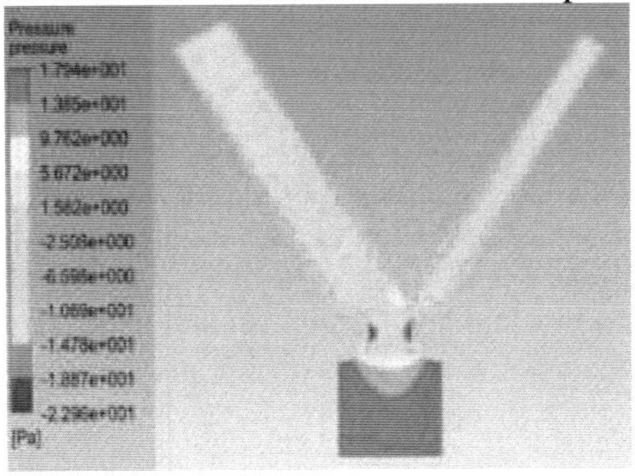

Figure 8. Pressure contour for Plaque #1

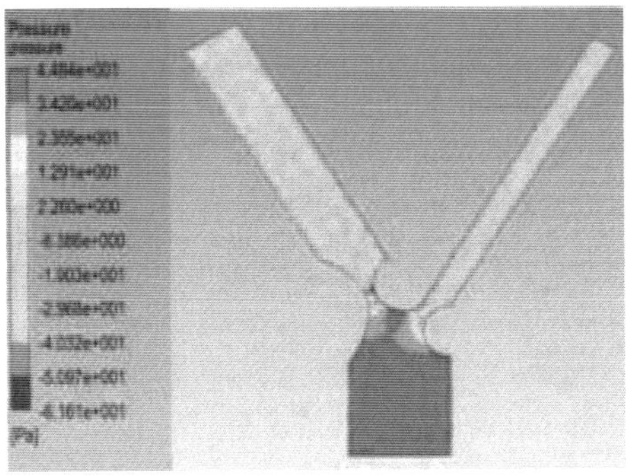

Figure 9. Pressure contour for Plaque #2

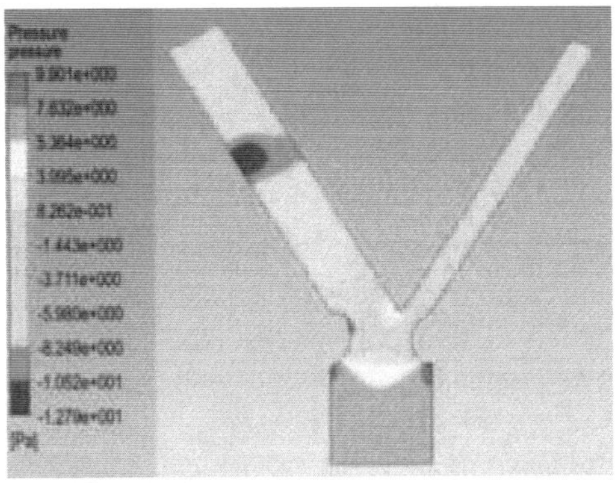

Figure 10. Pressure contour for Cosine Plaque

Figure 11. Pressure contour for Irregular Plaque

From figures it is noted that pressure contour of plaque shape 2 have maximum pressure at the stenotic region followed by plaque shape 1, irregular shape and cosine shape. Therefore, from the pressure contour figures it is revealed that the physiological effect of shape of stenosis is more in case of plaque shape 2. It is observed that the accumulation of stenosis in the artery leads to blockage and the considerable reduction in area bring about increase in the pressure of the blood through it.

4. Conclusion

The conclusion of this work can be outlined as observed: As stenosis in the artery increases, blood flow through it also increases. In this work, different shapes of stenosis are considered which shows that there is significant consequence of the shape of stenosis on the blood flow characteristics. Plaque shape 2 has maximum flow velocity followed by plaque shape 1, irregular and cosine plaque. Pressure through the artery in the presence of the plaque is also a diagnostic parameter which is also calculated for different plaque shapes and the pressure through the artery having plaque shape 2 is maximum then other arteries. From this we conclude that the artery having plaque shape 2 has maximum severity level and is more prone to rupture as compared to other plaque shapes.

References

[1] American Heart Association. http://www.americanheart.org

[2] ANSYS FLUENT, Computational Fluid Dynamics Software Package, Ver. 16.2, **(2015)**.

[3] Bhatia R, Vashisth S, Saini R, "Wall shear stress analysis in stenosed carotid arteries with different shapes of plaque", International Journal of Computer Applications, vol. 145 (4), **(2016)**, pp. 9-12.

[4] Chesler, N.C. and Enyinna, O.C., "Partical deposition in arteries ex vivo: Effect of pressure, flow and wave form", Journal of Biomechanical Engineering, **(2003)**, pp-389-394.

[5] Khan M, Vashisth S, Vijay R, Salhan AK, "Computer based real time systems for analyzing cardiovascular response to orthostatic stress", Journal of Biocybernetics and Biomedical Engineering, **(2015)**, pp. 232-239.

[6] Libby P, Ridker PM, Hansson GK, Leducq, "Transatlantic Network on inflammation in Atherosclerosis", Journal of the American College of Cardiology, 54(23), **(2009)**, pp. 2129-38.

IOP Conf. Series: Materials Science and Engineering 1149 (2021) 012003 doi:10.1088/1757-899X/1149/1/012003

[7] Marshall, I., Zhao, S., Papathanasopoulou, P., Hoskins, P., Xu, Y., "MRI and CFD studies of pulsatile flow in healthy and stenosed carotid bifurcation models", Journal of Biomechanics, **(2004)**, pp. 679-687.

[8] Misra J. C., Chakravaty S., "Flow in arteries in the presence of stenosis", Journal of Biomechanics, vol. 19, **(1986)**, pp. 907-918.

[9] Nanda Kumar, N., Sahu, K.C., Anand, M., "Pulsatile flow of a shear-thinning model for blood through a two-dimensional stenosed channel", Eur. J. Mech. Bfluids, **(2015)**, pp.-29-35.

[10]Perktold, K., Thurner, E., and Kenner, T., "Flow and stress characteristics in rigid walled and compliant carotid artery bifurcations models", Medical and biological Engineering and Computing, **(1994),** pp. 19.

[11]Raman Yadav, Sharda Vashisth, Ranjit Verma, "Computer based Methods to Analyze Carotid Stenosis", IEEE International Conference on Computing and Communication Technologies for Smart Nation (IC3TSN), **(2017)**, pp. 270-272.

[12]Staniloae CS, Ambrose JA., :Identification of Vulnerable Atherosclerotic Plaques. Exerpt Rev Cardiovasc Therapy", **(2003)**, pp. 353-365.

[13]Vashisth S, Khan M, Vijay R, Salhan AK, "Acquisition and analysis of human carotid pulse waveforms during tilt table maneuvers", International Journal of Applied Biomedical Engineering, 6(1), **(2013)**, pp. 32-39.

[14]Vashisth S, Khan M, Vijay R, Salhan AK, Sondhi S, "Real time acquisition, processing and analysis of human carotid pulse waveforms using MATLAB", Nirma University International Conference on Engineering (NUiCONE), **(2011)**, pp.1-5.

[15]Woollard KJ, Geissmann F., "Monocytes in atherosclerosis: subset and functions", Nature reviews Cardiology, 7(2), **(2010)**, pp. 77-86.

[16]Y. B. Fan, W.T. Jiang, Y.W. Zou, J.C. Li, J. Chen, X.Y. Deng, "Numerical simulation of pulsatile non-Newtonian flow in the carotid artery bifurcation", Acta MechanicaSinica, **(2009)**, pp. 249-255.

[17]Young D. F., Tsai F. Y., "Flow characteristics in models of arterial stenosis: I steady flow", Journal of Biomechanics, vol. 6, **(1973)**, pp. 395-410.

[18]Z. Ding, K. Wang, J. Li, X. Cong., "Flow field and oscillatory shear stress in a tuning fork shaped model of the average human carotid bifurcation", Journal of Biomechanics, **(2001)**, pp. 1555-1562.

[19]Z.H. Sun, B. Mwipatayi, T. Chaichana, C. Ng, "Hemodynamic effect of calcified plaque on blood flow in carotid artery disease: A preliminary study", ICBBE, **(2009)**.

Pipe Stress Analysis of Pump System in Process Plant

Mr. Akhilesh Kumar Jha[1*], Dr. S Kennedy[2] and Mr. Kamlesh Kumar Jha[3]

[1]School of Mechanical Engineering, Galgotias University, Greater Noida, India.

***akhilesh.rajpur@gmail.com**

Abstract- Every process and LNG plants operate many pumps and compressors in the processing unit. Performing the pump stress Analysis is immensely critical. The Pump's high vibration on the process plant will seriously affect the pipe and equipment's smooth operation. The resonance effect between pipe and equipment will create a severe additional desecration. Before installation or formal process, we need to perform static analysis, and dynamic analysis (modal, forced response analysis) is mandatory for the pump system. The pump vendor must perform a Pulsation analysis on the piping system for reciprocating machinery subject to pulsation flow. In this study, flexibility analysis of pump piping system as per process piping code B31.3 using CAESAR-II programming followed by placing of diverse pipe supports. The support type and position of the supports as per the output get from static and dynamic examination. By the software analysis of CAESAR II, the efficiency of pipe flexibility and expansion of the piping system's natural frequency is authenticating by include more oriented and load-bearing support to the elbow and Valve, which prevents the resulting of resonance successfully. The stress analysis, modal analysis, pulsation analysis, and vibration analysis method proposed in this paper will increase the lifespan and operation reliability of pipe and equipment.

Keywords: Acoustic study, ASME B31.3, CAESAR-II, Frequencies, Modal analysis, Pulsation analysis.

1. Introduction

A pump is one of the most critical pieces of equipment in the process plant. Pumps are mainly being used to transfer the service fluids or gases from one location to another. In the case of upstream oil and gas field applications, pumps play an essential role in pumping the wells' oil or gas and transferring them to the onshore storages or pipelines. The standard type of pumps is a centrifugal pump, reciprocating pumps, screw pumps, etc., which are used universally. In each process or LNG plants have stand by pump to assure continuous operations in case of maintenance of the first pump. High-temperature service pump piping is the most critical system to perform the stress analysis. The stress analysis is the most important stage of the pipe design to restrict failure and cause mishap. It is associated with high degree of responsibility [1] Apart from the stresses caused by the pipe dead weight, fluid density and isolation piping systems are also subjected to thermal changes, operating and design pressure and occasional condition like water hammer, Wind, Seismic etc. Generally, the piping stress analysis's greatest challenge is to provide the is to provide the system sufficient flexibility to absorb the thermal expansions. In the present day, pipe stress analysis plays much critical role to fix the piping layout. It is still one the challenging task for engineers to work in this field. The piping shall be routed initially, considering flexibility. Then it will be easier for the stress engineer to do the stress analysis and avoid back and fro communications within the piping department. This paper aims to present a stress analysis of pump piping systems to limit the nozzle loads within the allowable limits. [2- 4]

2. Literature Review

Some researchers worked on many parameters related to pump piping. It includes finalization of nozzle allowable, modification of pump piping, etc. Payal Sharma et al [5] have mentioned that the design of the piping system is governed by the international code and standard ASME B31.3. The author also highlights about the basic concept of flexibility and SIF. The Appendix-P of ASME B31.3 is the guideline provide for doing stress analysis of the piping system. The author has presented the stress criteria background and explains why Appendix P is formulated based on confusing logic that may lead to the piping system's unsafe design. The researchers' studies mainly focus on overcoming the low nozzle allowable provided by manufacturers [6]. L.C. Peng [7] had raised concerns about the change in equipment reliability due to connecting piping. So, the piping shall be designed so that minimum loads shall be transferred to the connecting equipment so that the pumping efficacy can increase the equipment's reliability. Simizu [8] studied the nozzle load analysis for process pumps. They have found that the Pump may require to withstand nozzle loads under extreme conditions, which often exceeds the criteria stated in API610. They have gone into detail of shaft end displacement of centerlines mounted Pump under nozzle loads. A research has stated that the current permissible loads for piping loads on rotating equipment nozzles specified by manufacturers are too low. To accommodate the design specification of connecting piping system, the pump manufacturer must increase the allowable loads. Literature work does not adequately specify the design modification. The Pump allowable can be kept within limits by systematic improvement in pump piping and to satisfy the latest edition requirement of API 610 [9]. Therefore, there is a requirement of modification in piping routing and supporting to reduce the nozzle loads of pumps.

3. Experimental analysis

Piping system are very much critical system in different process parameter such as pressure, pipe material specification, temperature, surge, seismic loads, wind loads etc., acting concurrently. Mainly two types of analysis are usually preferred in oil and gas industry a) static examination and b) dynamic examination. The static loads are due to thermal, pressure, equipment displacement, fluid weight, etc. The dynamic loads are seismic, wind, water hammer, surge, two phases, etc., which usually create upset. Apart from the method of flexibility analysis and mechanical stress analysis traditional method: 1) Approximation process 2) Exact analytical methods 3) Modal tests.

3.1. Inputs for piping stress analysis

Stress Isometric from piping designer, P&ID & Line Designation Table (LDT from process department, Equipment GAD and other vendor data from the mechanical group.PFD /datasheet if required from the process department, Specification of piping (PMS) from the piping material group, All control valves and other valve data from Instrumentation. Project-specific nozzle allowable standard, overall plot plan, and area plot plan for finding HPP elevation and equipment orientation from piping layout group.

3.2. Governing codes and standards for pipe stress analysis

ASME B31.3: Process piping Code
Centrifugal Pumps: API 610
Centrifugal Compressors: API 617

3.3. Stress analysis of pump piping system

The analysis of pump piping consists of suction and discharge piping. The static analysis and modal analysis of suction and discharge lines of Pump existing in the refinery are considered in the current paper.

One of the piping unit's refinery systems consists of 2 pumps with two suction and discharge nozzles. The pump piping loads exerted on the pump discharge nozzles have exceeded the pump manufacturer's limit. It is required to bring the nozzle loads within the limit by stress analysis (static) of the pump piping system. The static analysis of the complete system is performed in finite element analysis software (CAESAR-II). It includes finalizing the piping system routing and supports. The pump nozzle loads are within allowable limits, suggesting the changes or modifications in the piping system such that the system is optimal in terms of cost.

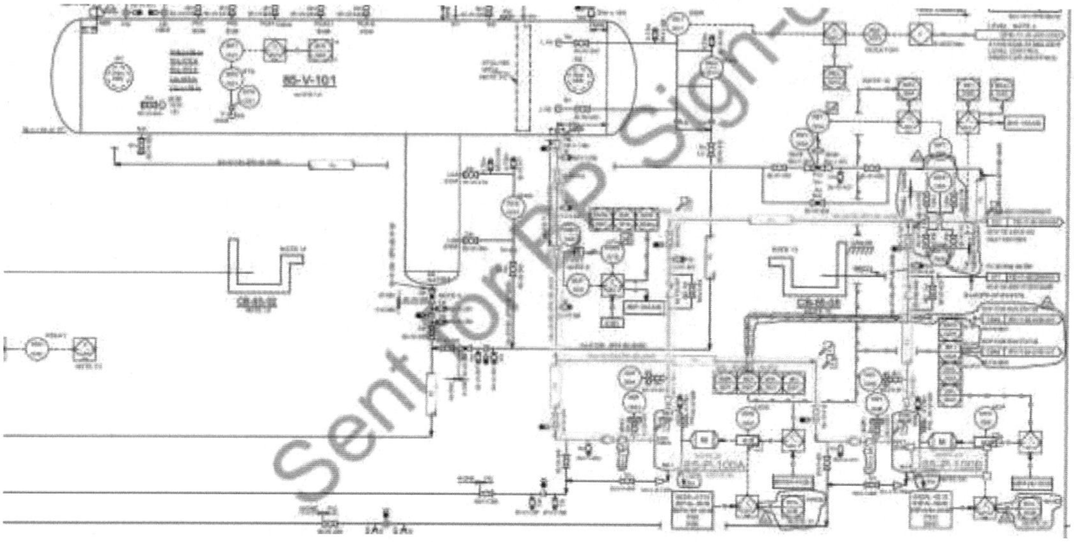

Figure 1. Process and instrumentation diagram for pump system. The yellow highlighted pipeline is pump suction and discharge line.

3.4. Optimization of piping

The design and cost comparison among the cases mentioned above shall be the ideal approach to present the design's optimization. However, these are not being used in consultancies due to many drawbacks like time, cost, etc. For this project, the optimization is achieved mainly by material optimization and engineered support, resulting in cost optimization. Even if we save on a few of the fittings or spring support, it will save a large amount of commodity, welding, handling, test & inspection cost.

If we consider a reduction in the number of supports or the simple supports rather than specially designed supports, it will be a cost saver. The standard support is usually fabricated and welded at the site. However, in special supports, these are required unique design, manufacturing, and installations. Thus, the first choice should be simple supports, and in extreme cases where simple supports are not practical or not acceptable considering design analysis, SPS shall be used. The problem cases are being analyzed in terms of the cost of material used in the piping. The best and optimum is design, which is safe in terms of allowable limits and cost-effective [10].

3.5. Stress evaluation

The standard-purpose FEA codes will not consider the supports' geometric properties, valve data, and engineered support. Constructing the analytical models for each load case scenario is of utmost critical. Moreover, the evaluation of allowable stresses for each load case is very strenuous. So, CAESAR-II program was used in this research, which is used in all the process industry. From the below figure, software modelling is done as per stress isometric. Procedure of the piping modeling (1) piping system part; (2) Node and coordinates of the piping system; (3) Piping parameters" (like length, diameter, and schedule); (4) Provide process and insulation. Firstly, complete the equipment modelling and then start modelling the piping based on piping isometric drawing. Always make a closed system to get the correct results. Typically, pump lines are connected to heat exchanger, horizontal or vertical vessel and tank. So, it will create a closed secure network. Then run the Caesar-ii piping model to check code stresses, displacements, loads on nozzle and support, natural frequencies etc.

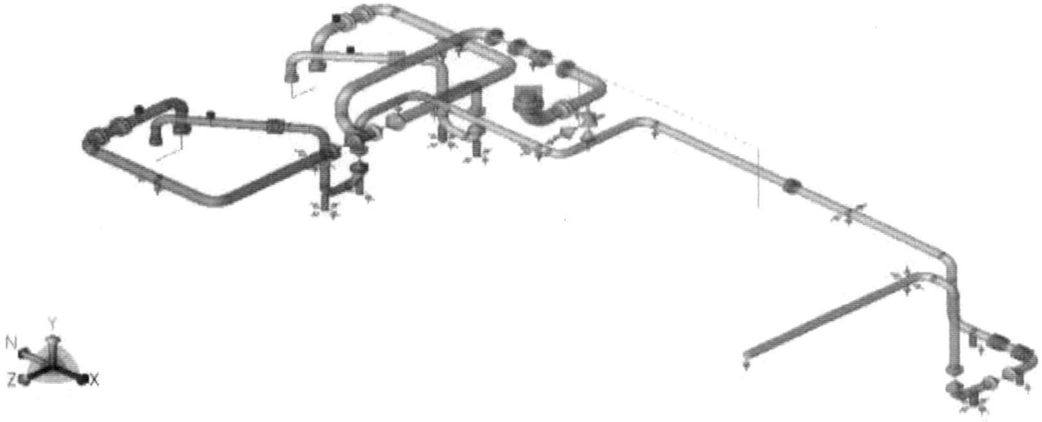

Figure 2. The geometry of operating-Stand by Temperature profile for a two-pump system isometric view. Pump-A is in working condition and pump-B will be in stands by pump.

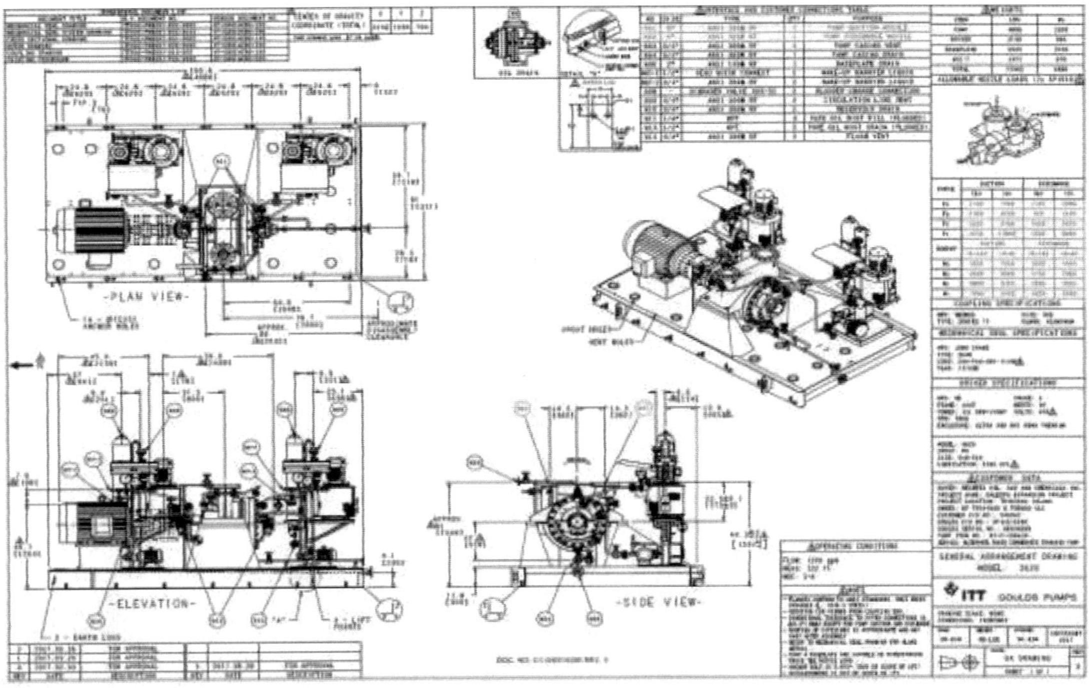

Figure 3. General arrangement drawing for pump. All the calculation related details (nozzle allowable, anchor location, pump type, size and rating of pump mentioned).

3.6. Load cases

For each stress calculation the following load cases are required as minimum for pump piping system. So, considering T1 as operating temperature, T2 as maximum stress and T3 as minimum stress, the following load cases are required. The load cases combination is as per ASME B31.3, process piping specification.

Loads Defined in Input
W - Weight
T1 - Thermal Case #1
T2 - Thermal Case #2
T3 - Thermal Case #3
T4 - Thermal Case #4
T5 - Thermal Case #5
P1 - Pressure Case #1
HP - Hydro. Pressure
H - Hanger Loads
U1 - Unif Load Case #1
U2 - Unif Load Case #2
WIN1 - Wind Load Case #1
WIN2 - Wind Load Case #2
WIN3 - Wind Load Case #3
WIN4 - Wind Load Case #4
WW - Water Filled Weight
WNC - Weight No Contents

Drag a column header and drop it here to group by that column

	Definition	Name	Stress Type
L1	W+H	W+H (HGR)	HGR
L2	W+T1+P1+H	W+T1+P1+H (HGR)	HGR
L3	WW+HP+H	WW+HP+H (HYD)	HYD
L4	W+T1+P1+H	W+T1+P1+H (OPE)	OPE
L5	W+T2+P1+H	W+T2+P1+H (OPE)	OPE
L6	W+T3+P1+H	W+T3+P1+H (OPE)	OPE
L7	W+T4+P1+H	W+T4+P1+H (OPE)	OPE
L8	W+T5+P1+H	W+T5+P1+H (OPE)	OPE
L9	W+P1+H	W+P1+H (SUS)	SUS
L10	WNC+H	WNC+H (SUS)	SUS
L11	W+T1+P1+H+U1	W+T1+P1+H+U1 (OCC)	OCC
L12	W+T1+P1+H-U1	W+T1+P1+H-U1 (OCC)	OCC
L13	W+T1+P1+H+U2	W+T1+P1+H+U2 (OCC)	OCC
L14	W+T1+P1+H-U2	W+T1+P1+H-U2 (OCC)	OCC
L15	W+T1+P1+H+WIN1	W+T1+P1+H+WIN1 (OCC)	OCC
L16	W+T1+P1+H+WIN2	W+T1+P1+H+WIN2 (OCC)	OCC
L17	W+T1+P1+H+WIN3	W+T1+P1+H+WIN3 (OCC)	OCC
L18	W+T1+P1+H+WIN4	W+T1+P1+H+WIN4 (OCC)	OCC
L19	L11-L4	L11-L4 (OCC)	OCC
L20	L12-L4	L12-L4 (OCC)	OCC
L21	L13-L4	L13-L4 (OCC)	OCC
L22	L14-L4	L14-L4 (OCC)	OCC
L23	L15-L4	L15-L4 (OCC)	OCC
L24	L16-L4	L16-L4 (OCC)	OCC
L25	L17-L4	L17-L4 (OCC)	OCC
L26	L18-L4	L18-L4 (OCC)	OCC
L27	L19+L9	L19+L9 (OCC)	OCC
L28	L20+L9	L20+L9 (OCC)	OCC
L29	L21+L9	L21+L9 (OCC)	OCC
L30	L22+L9	L22+L9 (OCC)	OCC
L31	L23+L9	L23+L9 (OCC)	OCC
L32	L24+L9	L24+L9 (OCC)	OCC
L33	L25+L9	L25+L9 (OCC)	OCC
L34	L26+L9	L26+L9 (OCC)	OCC
L35	L4-L9	L4-L9 (EXP)	EXP
L36	L5-L9	L5-L9 (EXP)	EXP
L37	L6-L9	L6-L9 (EXP)	EXP
L38	L7-L9	L7-L9 (EXP)	EXP

Figure 4. Load cases table for stress summary, restraint load, displacement, and nozzle check for sustained, operating, expansion and occasional cases.

IOP Conf. Series: Materials Science and Engineering 1149 (2021) 012004 doi:10.1088/1757-899X/1149/1/012004

4. Results and discussion

4.1 Output results for code compliance report

Stress check for all the multiple load cases combine in a single report using the code compliance report.

Load Case	From Node	Code Stress lb./sq.in.	Allowable Stress lb./sq.in.	To Node	Code Stress lb./sq.in.	Allowable Stress lb./sq.in.	Piping Code

```
LOAD CASE DEFINITION KEY

CASE 7 (HYD) WW+HP+H
CASE 14 (SUS) WNC+H
CASE 23 (OCC) L23=L15-L8
CASE 31 (OCC) L31=L23+L13
CASE 32 (OCC) L32=L24+L13
CASE 33 (OCC) L33=L25+L13
CASE 34 (OCC) L34=L26+L13
CASE 35 (OCC) L35=L27+L13
CASE 36 (OCC) L36=L28+L13
CASE 37 (OCC) L37=L29+L13
CASE 38 (OCC) L38=L30+L13
CASE 39 (EXP) L39=L8-L13
CASE 40 (EXP) L40=L9-L13
CASE 41 (EXP) L41=L10-L13
CASE 42 (EXP) L42=L11-L13
CASE 43 (EXP) L43=L12-L13
CASE 44 (EXP) L44=L8-L9
CASE 45 (EXP) L45=L8-L10
CASE 46 (EXP) L46=L9-L10


Piping Code: B31.3       = B31.3 -2016, Jan 31, 2017

***  CODE COMPLIANCE EVALUATION PASSED  ***

Highest Stresses: (lb./sq.in.)
Ratio (%):                 52.2     @Node  5600  LOADCASE: 33 (OCC) L33=L25+L13
Code Stress:            13887.9     Allowable Stress:    26600.0
Axial Stress:            2699.0     @Node  5620  LOADCASE: 34 (OCC) L34=L26+L13
Bending Stress:         14241.4     @Node  5510  LOADCASE: 46 (EXP) L46=L9-L10
Torsion Stress:          2752.2     @Node  6760  LOADCASE: 34 (OCC) L34=L26+L13
Hoop Stress:             5135.0     @Node    80  LOADCASE: 31 (OCC) L31=L23+L13
Max Stress Intensity:   22877.7     @Node  5510  LOADCASE: 46 (EXP) L46=L9-L10
```

Figure 5. Code compliance report.

4.2. Nozzle load check

The below load table shows that the nozzle loads are well within the API 610 code allowable limits. For rotary equipment, nozzle load qualification on design and upset condition is not recommended.

NOZZLE CHECK REPORT: Nozzle Loads Screening
Various Load Cases

Node	fa lb.	fb lb.	fc lb.	Forces Check	ma ft.lb.	mb ft.lb.	mc ft.lb.	Moments Check

LOAD CASE DEFINITION KEY

CASE 8 (OPE) W+T1+P1+H
CASE 10 (OPE) W+T3+P1+H
CASE 11 (OPE) W+T4+P1+H
CASE 12 (OPE) W+T5+P1+H
CASE 13 (SUS) W+P1+H
CASE 14 (SUS) WNC+H

Node	fa lb.	fb lb.	fc lb.	Forces Check	ma ft.lb.	mb ft.lb.	mc ft.lb.	Moments Check
20	Absolute Method							
Limits	1700	2200	1400		5200.0	3800.0	2600.0	
8(OPE)	-58	-31	-197	0.141	-1485.9	270.8	349.6	0.286
10(OPE)	55	102	35	0.047	122.7	8.5	106.3	0.041
11(OPE)	-61	-52	-187	0.133	-1331.2	261.0	300.8	0.256
12(OPE)	35	-8	22	0.021	-180.1	-4.1	-129.8	0.050
13(SUS)	44	12	7	0.026	29.3	58.8	-103.0	0.040
14(SUS)	40	8	5	0.023	25.4	55.2	-60.9	0.023
1020	Absolute Method							
Limits	1700	2200	1400		5200.0	3800.0	2600.0	
8(OPE)	267	50	-473	0.338	1816.3	673.9	-394.6	0.349
10(OPE)	493	-83	55	0.290	-146.3	-90.0	50.3	0.028
11(OPE)	471	-6	5	0.277	-2.7	-5.6	205.9	0.079
12(OPE)	260	42	-470	0.336	1773.5	671.7	-428.5	0.341
13(SUS)	477	-17	9	0.280	-10.9	-9.2	183.2	0.070
14(SUS)	282	-7	6	0.166	-16.7	17.4	74.3	0.029

Figure 6. Nozzle loads check actual vs allowable.

4.3. Modal analysis

The Modal analysis of the inlet and outlet piping system performed independently through CAESAR II programming. In this research we observed that the minimum natural frequency of the system more than 10 Hz, which is much higher than the industry recommended EI guidelines frequency 7Hz. So, the resonance effect between pump and piping would not happen.

NATURAL FREQUENCY REPORT

MODE	(Hz) FREQUENCY	(Radians/Sec) FREQUENCY	(Sec) PERIOD
1	10.196	64.064	0.098
2	11.400	71.631	0.088
3	11.925	74.929	0.084
4	13.391	84.136	0.075
5	15.044	94.523	0.066
6	15.401	96.767	0.065
7	17.699	111.209	0.056
8	18.254	114.693	0.055
9	19.494	122.481	0.051
10	20.442	128.441	0.049
11	20.683	129.955	0.048
12	21.342	134.097	0.047
13	21.496	135.064	0.047
14	21.970	138.041	0.046
15	24.244	152.331	0.041

Figure 7. Natural frequency report of the system.

4.4. Recommendation to reduce piping stresses and optimization piping layout

Increase the fundamental natural frequency by reducing the support span, avoid SBC near turbulence source (e.g., closed valves, short radius or mitered bends, tees, and reducers), and add protection to SBC.Adequacy of Rigidity, Skirt support: Filter /scrubber/volume bottles (no legs support unless Rigidity braced) Axial stop+ guiding near pressure throttling piping, e.g., Orifice plate, Valve. [11]

5. Conclusion

The systematic study of pump piping systems is performed using FEA software CAESAR-II and process piping code ASME B 31.3-2016 for flexibility analysis. [12]

The insightful and FEA software results is observed. The pipe stress analysis is performed for the piping system is checked manually based on the process design code B31.3-2016, and the system is stress analyzed with the help of FEA software.[13]

The outcomes are breaking down and tracked down that the Piping framework is protected. In this study elaborate the procedure and special consideration to be followed for pump piping analysis. [14]

It concludes that provided with enough flexibility, this is considered in the initial stage of piping routing. Some of the means to increase flexibility are 45 deg connections; spring supports, axial stop used other than anchor supports, and overall flexibility up to the header piping.[15]

As per ASME B 31.3 and API 610 pump nozzles are within the code allowable stress, Nozzle loads, restraint loads, natural frequencies, sagging, leakage check and support loads are well within the standard.

6. References

[1] Se-Yun Hawang, Min-Seok Kim and Jang-Hyun lee, "Thermal stress analysis of process piping system installed on LNG vessel subject to hull design loads," MDPI, Journal of marine science and engineering, 2020.
[2] L. Peng, T. Peng, Pipe Stress Engineering, Houston, Texas, USA, ASME Press, 2009.
[3] ASME B31.3, Process Piping. ASME, American Society of Mechanical Engineers, 2016.
[4] S.D. Sajish, S. Jaladeen, P. Selvaraj, "Stress Indices for Non-radial Branch Connections for Piping," Science Direct, Elsevier, Procedia Engineering 55, 2013.
[5] Payal Sharma, Mohit Tiwari, Kamal Sharma, "Design and Analysis of a Process Plant Piping System," International Journal of Current Engineering and Technology, April 2014.
[6] Mohinder L. Nayyar, Piping Handbook, 7th edition, Tata McGraw-Hill, 2000.
[7] L.C. Peng," Understanding Piping Code Stress Evaluation and ASME B 31.3 Appendix P," 2013.
[8] T. Shimizu and H. Teshiba, "Analysis of nozzle load for process pumps," 1985.
[9] API Standard 610 11th Edition, pp. 30-35, September 2010.
[10] C.B. Rajeev, G. Ramesh, M.V. Kuppusamy, "Optimization of Integrity Testing of the Piping system in a Nuclear Fuel Cycle facility," Science Direct, Elsevier, vol 86, pp 539-545, 2014.
[11] L.C. Peng, "Quick check on Piping Flexibility," Peng Engineering, Texas.
[12] S. U. Bhave and P. Sonawane, "Analysis of pump piping based on piping configurations," IJIRST, Volume 2, Issue 09, 2016.
[13] L.C. Peng, "Equipment Reliability Improvement through Reduced Pipe Stress," 1993.
[14] N. Prabhu Kishore and S. Prabhu, "Modeling and stress analysis of pump piping,"
IOP conference series, Material science and engineering 455, 2018.
[15] S. Kannappan, Introduction to Pipe Stress Analysis, Knoxville, Tennesse, John Wiley & Sons, 1985.

Simulation and kinematic analysis of KUKA KR5 Arc robot

Aparna Jha[1*], Manoj Soni[1] and Mohd Suhaib[2]

Indira Gandhi Delhi Technical University for Women, New Delhi, India
Jamia Millia Islamia, New Delhi, India
[*]**jhaaparna1006@gmail.com**

Abstract. Nowadays robots have become an indispensable part of manufacturing systems, for gripping needs manipulators are needed. Trade sector is touching industry 4.0 and asks for a much-advanced automation process. In which commercial robots play an essential role in automating the methods such as material handling, palletizing, painting, assembly lines, and numerous applications. With the increase in demand and need to have more research on engineering robots, artificial intelligence, and machine learning. A systematic design and analysis study is needed to get the best output for a specific robot. In India, there is scarcity of Indian based robot manufacturers to satisfy the rising need. Hence, there is more influence of foreign robot manufacturer in Indian markets. Due to the deficient information in robotics, robot part and resources are unavailable. More and more open source and easily access information regarding industrial robot is needed. So, this work emphasis on the developing and designing six degrees of freedom robotic arm and its kinematic analysis particularly forward and inverse kinematics. This paper gives an idea of the Denavit Hartenberg method for forward and inverse kinematics of a 6 axis KUKA KR5 Arc robot and uses RoboAnalyzer simulation software. The Robo Analyzer is simulation software that is used for design, kinematics, and dynamic analysis.

1. Introduction

In a new era of engineering automation, Robotics has to turn out to be fundamental and economical manpower in the field of research and mass manufacturing. In robotic research, robotic kinematic parameters and end position of end effectors play a very crucial role in its control strategy [1]. In the studies of robotics, kinematic study is the footing of robot technology study and the prominent characteristics to know robot pathway planning as well as its motion regulator [2]. The Denavit Hartenberg process is most considered to define the relation among forward and inverse solutions to study the kinematics in the robot kinematic model. To perform the kinematic analysis and establishing a correlation b/w the end effector and its characteristics, the transformation matrix given by Denavit Hartenberg is used [3].

RoboAnayzer is simulation software comprising of option putting the Denavit Hartenberg features of the robotic arm. This software gives an option to select from one to seven degrees of freedom [4]. An operator may also make a customized robot after changing kinds of joints and sizes. Kinematic transformations are proved using Roboanalyser software simulation [5]. When any value is put in the (DH) Denavit Hartenberg table, initial as well as the final manipulator joint dimensions may be altered for the forward kinematic investigation. This simulation software also has an integral simulation representation of robots from numerous creators such as MTAB, FANUC, ABB, KUKA, etc. Any handler may differ numeric values inside the input column and trace the path covered by the manipulator.

1.1. Characteristics of Roboanalyze :
This software is utilized to get the kinematic analysis and dynamic analyses of robotic manipulators. Mainly it is used to get the following results:

- DH characteristics simulation,
- Forward Kinematics,
- Forward Dynamics
- Inverse Kinematics,
- Inverse Dynamics
- Path Planning

2. Denavit-Hartenberg Illustration of the Kuka 3r Arc Robot

Countless industrial robots are made up of quite a lot of rigid links associated with joints. The degrees of freedom show the no. of self-determining input desired to rightly put all the fixed links of any robot about the stationary or any fixed link [3]. All rigid links being coupled through such joints. These joints may be the (P) prismatic joints displaying translatory motion and (R) revolute joint displaying revolution motion [4]. For calculating the outcome of the joint adjustable on a location as well as the orientation of the robotic arm, it is crucial to make a relationship among base as well as end-effectors of the manipulator. Figure 1 shows the graphic user interface, in which the upper window shows the build of the robot and bottom boxes of the window shows D-H parameters.

Numerous industrial robots are made up of several rigid links fixed to each other by joint. The degrees of freedom give a sum of the autonomous inputs required to accurately position all of the links of the robotic manipulator about ground or the immovable link [6]. All rigid links are coupled through joints. All these linkage joints may be the (P) prismatic joint that shows the translation motion as well as the (R) revolute joint that shows revolution motion [7]. For calculating any result of the joint adjustable on the position along with the alignment of the robot, where it is crucial to create a relation among coordinate planes fixed to its base position and the end-effectors of the robot [8].

- (α) Twist angle: angle among orthogonal projections of joint axes onto a plane normal to common normal
- (θ) Joint angle: angle among orthogonal projections of common normal to plane normal to joint axes.
- (d) Joint offset: lengths of intersections of common normal on joint axis
- (a) Link length: a measure of the path among common normal to the axis.
  α, θ - varies if the joint is revolute and prismatic respectively [9]

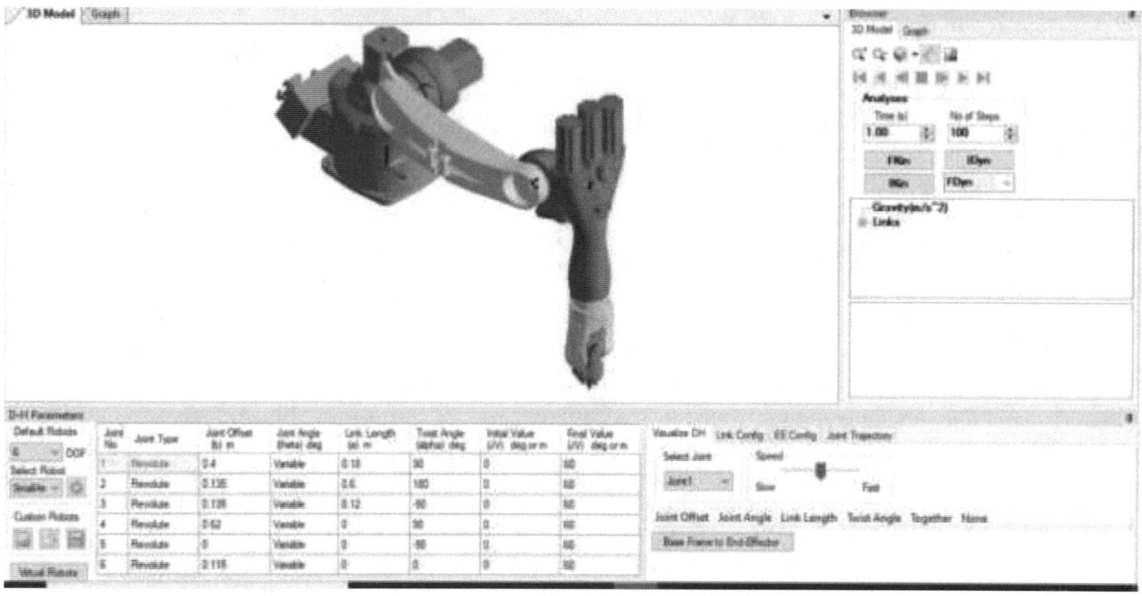

Figure 1. Model of KUKA KR5 Arc

3. Analysis of Forward Kinematics

For calculating forward kinematics analysis, measures of joint angles are entered, position and orientation of end-effectors of the robot are obtained. According to the D-H formulation, a homogenous matrix transformation is achieved in the form of a 4x4 matrix using the orientation of the base of the robotic arm to the end-effectors [10]. In the homogenous transforming matrix, the DH parameters from the base of the link to end-effectors link must be attached, further, it is multiplied to calculate the position of end-effectors in the KUKA 3R Arc robot in 3-D space, with the below equation:

$$ {}^{b}T_{ee} = {}^{0}T_{1} * {}^{1}T_{2} * {}^{2}T_{3} * {}^{3}T_{4} * {}^{4}T_{5} * {}^{5}T_{6} \quad \text{...... (1)} $$

which can further be represented as

$$ {}^{b}T_{ee} = \begin{bmatrix} & & & x \\ & R & & y \\ & & & z \\ 0 & 0 & 0 & 1 \end{bmatrix} \quad \text{...... (2)} $$

In which, R (3×3) denotes the positioning of end-effectors utilizing the direction cosine procedure as well as $[x,y,z]^T$ signifies the position vectors regarding its reference coordinate frame. Transformation homogenous matrix process could be pictured in RoboAnalyzer software by putting input value in the DH table which is given in said software. In figure 2 an illustration is shown, (position of end-effectors).

IOP Conf. Series: Materials Science and Engineering 1149 (2021) 012005 doi:10.1088/1757-899X/1149/1/012005

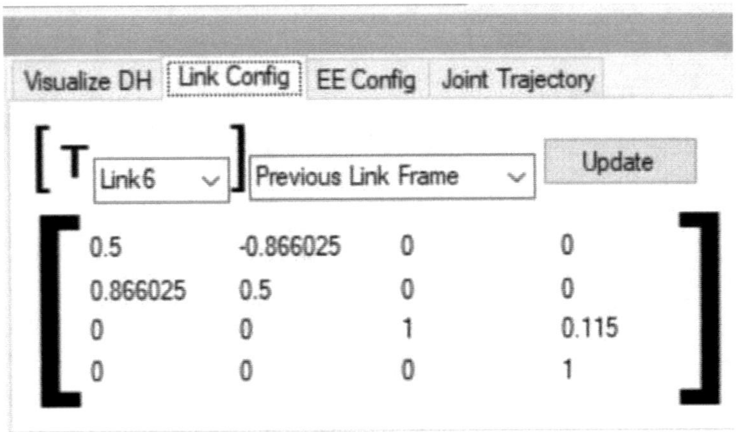

Figure 2. Transformation homogenous Matrix in the software.

In the graph, Transformation homogenous matrix could be proved as well as pictured. The final location of end-effectors in the $[x,y,x]^T$ 3D space is shown in Figure 3, where the (x)- axis in the graph denotes the estimated time of simulation, and here (y)- axis denotes the position value of the end-effectors.

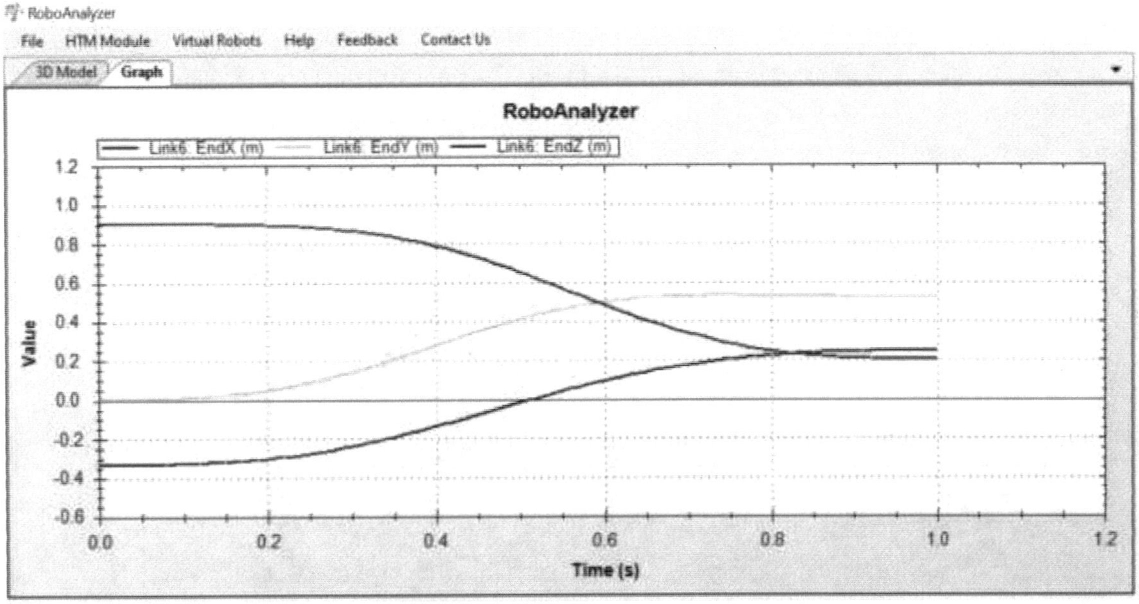

Figure 3. Graph showing position of end-effector of robot.

3.1. Robot Motion Visualization

It is very difficult to visualize the motion of the robot without the support of any software tool. 3D software for Robot visualization has played a very important role by enabling viewers to visualize the path traced by the robot in 3D space as well as the path followed by end-effectors. The said software permits an operator to enter essential D-H characteristics. As soon as all inputs are provided, the forward kinematics processes are being executed in the simulation of the robot. As given in figure 4, this also gives the location of the end-effectors in the matrix.

IOP Conf. Series: Materials Science and Engineering 1149 (2021) 012005 doi:10.1088/1757-899X/1149/1/012005

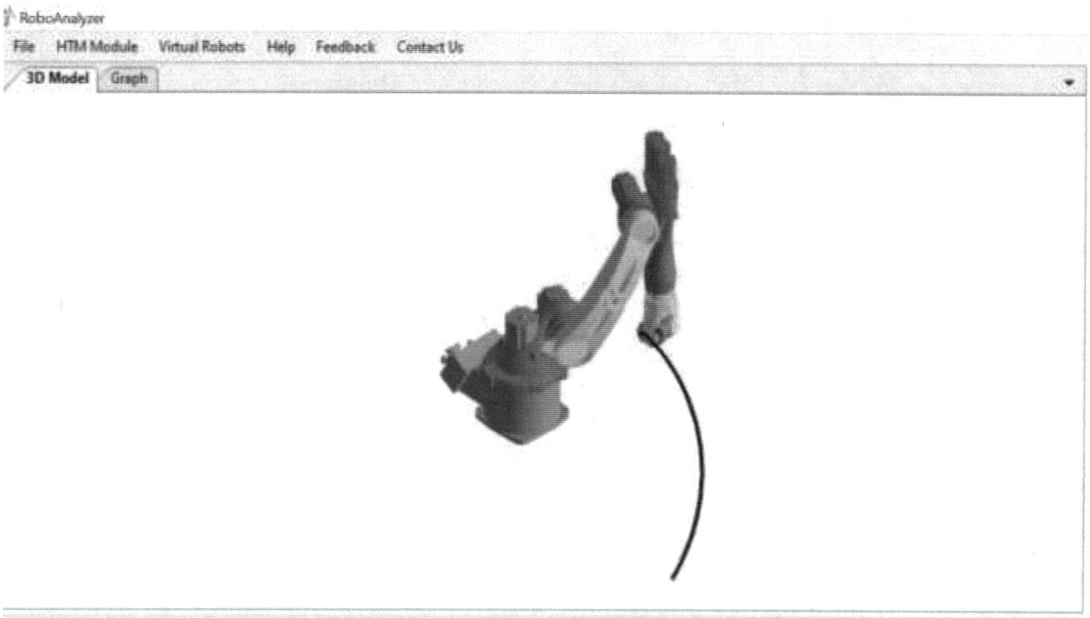

Figure 4. Path drawing and visualization of KUKA KR5 Arc.

The RoboAnalyzer software also comprises of Virtual Robot Module (VRM) that allows an operator the usage of a teach pendant-style user interface for the operation of any one of the already loaded industrial robots, which figure 5 also shows.

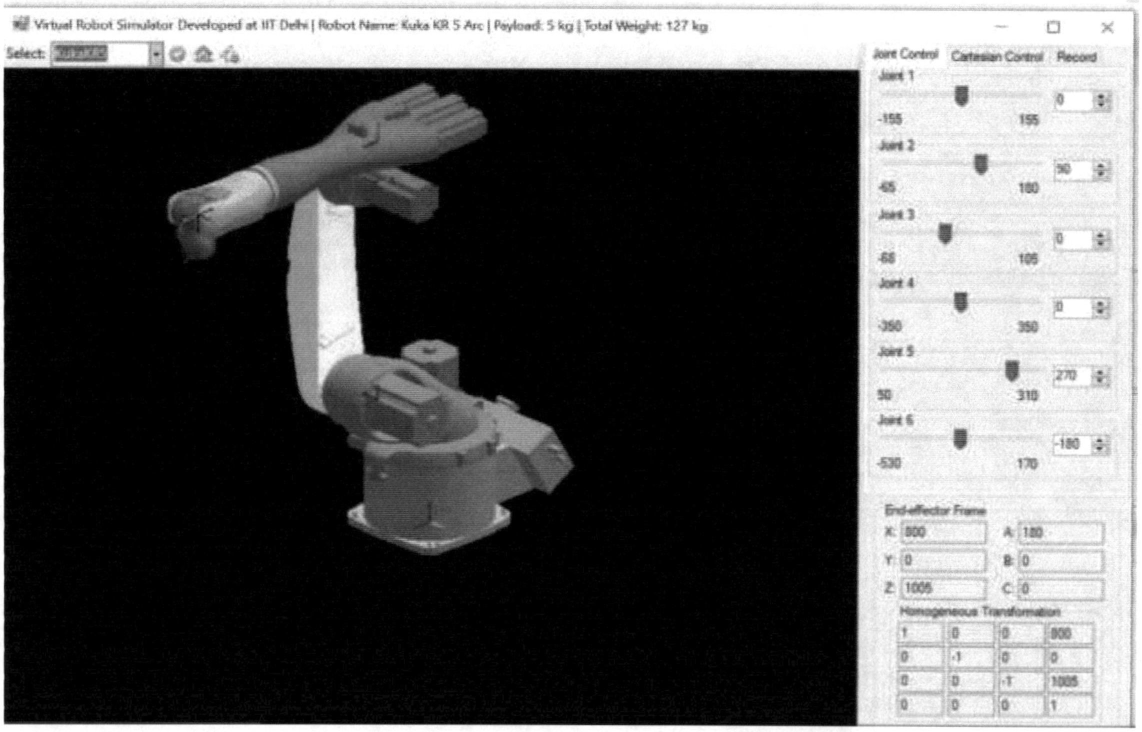

Figure 5. The virtual model of the KUKA KR5 Arc in RoboAnalyzer is displayed.

IOP Conf. Series: Materials Science and Engineering 1149 (2021) 012005 doi:10.1088/1757-899X/1149/1/012005

4. Analysis of Inverse Kinematics

Inverse kinematics demarks with defining a joint angle for any given position and orientation of end-effectors of a robotic manipulator in the 3-D space [11]. Analysis of Inverse kinematics deals with computational equations to attain solutions for any specified position of the robot. With the increase in numbers of the degrees of freedom, there is also an increase in the total number of solutions of inverse kinematics. RoboAnalyzer software has an inbuilt inverse kinematics module in which an operator may select a robot from the set menu to achieve inverse kinematic analysis. This software displays several solutions in situation of inverse kinematic analysis. Figure 6 show the illustration of the inverse kinematics analysis.

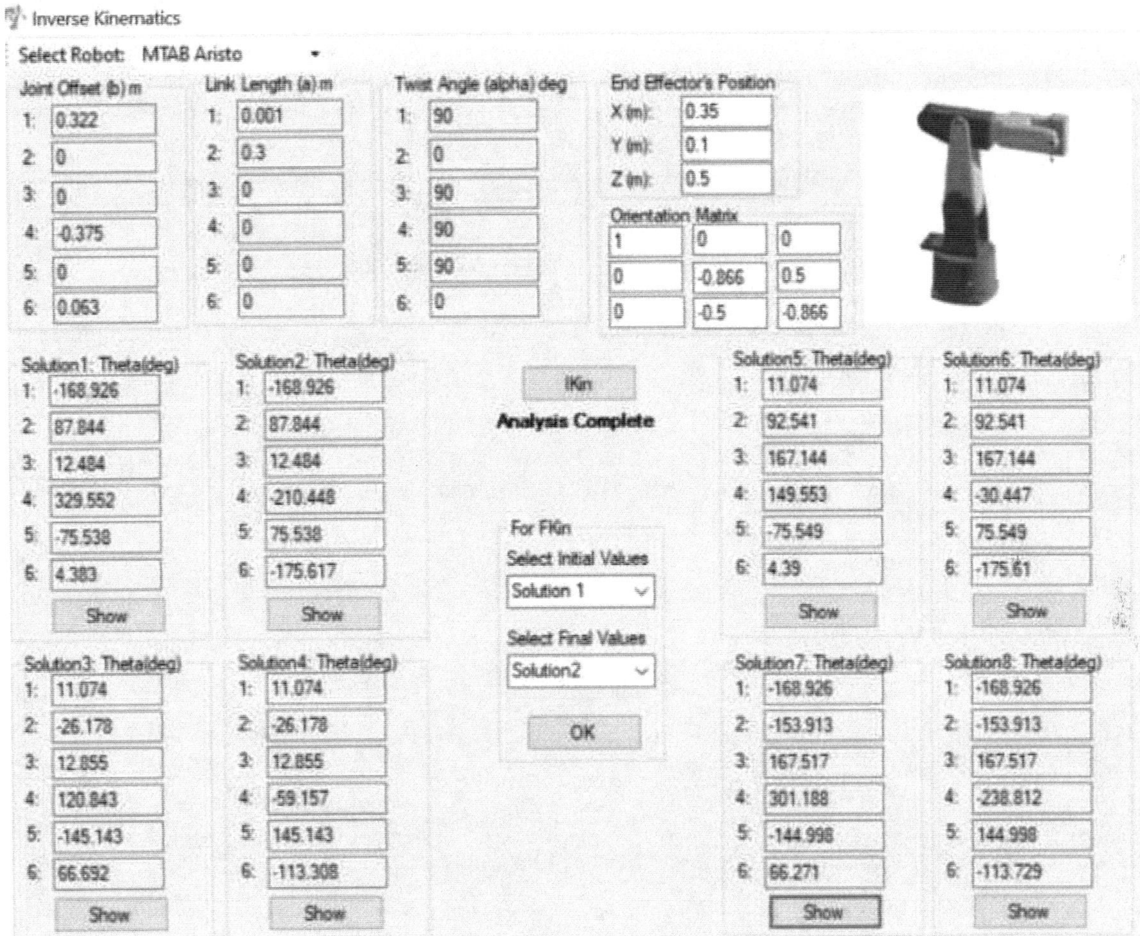

Figure 6. Solutions of Inverse Kinematics.

5. Conclusion

The KUKA 3R Arc robot is selected and the motion study is calculated by the means of robot picturing software. Kinematic studies of the said robot can act as the laid foundation for the proposed designs of robot for further use in industries like electronics, warehouses, automobiles, etc. With the help of such analysis, it's easier to understand the kinematic structure and verification can be done to check the correctness. This paper can be helpful for analysis of the kinematic studies of the robotic gripper having different dimensions of the link lengths, forward and inverse analysis of kinematic model can be achieved using D-H model in this software (Roboanalyzer).

References

[1] Luo, J. J., & Hu, G. Q. (2005). Study on the simulation of robot motion based on MATLAB. JOURNAL-XIAMEN UNIVERSITY NATURAL SCIENCE, 44(5), 640.
[2] Sun, X., Xie, Z., Jian, K., & Zhang, J. (2012). Dynamics analysis and simulation of 6-PSS flexible parallel robot. Nongye Jixie Xuebao(Transactions of the Chinese Society of Agricultural Machinery), 43(7).
[3] S K SAHA" Introduction to Robotics", CHINA MACHINE PRESS, 2009: 76 -145.
[4] Ramish, S. B. H., & Kanwa, F. (2016). Design of a 3 DoF robotic arm. In The sixth international conference on Innovative Computing Technologir (INTECH 2016).
[5] Aravinthkumar, T., Suresh, M., & Vinod, B. (2021). Kinematic Analysis of 6 DOF Articulated Robotic Arm. International Research Journal of Multidisciplinary Technovation, 3(1), 1-5.
[6] Gupta V, Chittawadigi RG, Saha SK. Robo analyzer: Robot visualization software for robot technicians. ACM Int Conf Proceeding Ser. 2017; Part F1320:1–5.
[7] Flanders, M., & Kavanagh, R. C. (2015). Build-A-Robot: Using virtual reality to visualize the Denavit–Hartenberg parameters. Computer Applications in Engineering Education, 23(6), 846-853.
[8] S.K. Saha, Introduction to Robotics, Tata McGraw Hill, 2008
[9] Singh, A., Kanchan, R. P., & Singh, S. (2016). A Review Paper on Analysis and Simulation of Kinematics of 3R Robot with the Help of RoboAnalyzer. International Journal of Engineering Research, 5(04).
[10] Khan, W. A., Zhuang, H., & Angeles, J. (2007). RVS4W: A visualization tool for robot design. Proceedings of the Canadian Engineering Education Association (CEEA).
[11] Huang, G. S., Tung, C. K., Lin, H. C., & Hsiao, S. H. (2011, May). Inverse kinematics analysis trajectory planning for a robot arm. In 2011 8th Asian Control Conference (ASCC) (pp. 965-970). IEEE.

IOP Conf. Series: Materials Science and Engineering 1149 (2021) 012006 doi:10.1088/1757-899X/1149/1/012006

Study of Fabrication and Mechanical Characterization of Magnesium Metal Matrix Composites

A Nirala[1,2*], S Soren[1], Navneet Kumar[2], Raj Kumar[2], AK Shrivastava[2], H Prasad[2], Sanjeev Kumar Sharma[3] and J K Yadav[2]

[1]Department of Fuel, Minerals and Metallurgical Engineering, Indian Institute of Technology (ISM) Dhanbad, India
[2]Galgotias College of Engineering and Technology, Greater Noida 201310, India
[3]Department of Mechanical Engineering, Amity University Uttar Pradesh, Noida, India
* **niralaiitk@gmail.com**

Abstract. Magnesium matrix composites are used widely because of light weight, higher mechanical, and physical properties and due to that, it is used potentially in numerous applications of aviation and defense organizations. Improving specific strength, stiffness, dampening behavior, wear behavior, shrinkages, and fatigue properties, in comparison with traditional engineering materials, is greatly affected by the integration of reinforcement into the magnesium metal matrix. The present research paper describes the results and merits of various reinforcements in magnesium and its alloy ant it addresses important processes, such as interfacial transformation, agglomeration, fiber-matrix interaction, and the difficulties with the distribution of reinforcement particles. The effect of strengthening on the microstructural and mechanical properties was often discussed in detail, such as Tensile properties, hardness and wear properties of the composites. Furthermore, the development and invention of new magnesium metal matrix composites are challenges for future study. The study of nanocomposites of magnesium matrix is a future research aim for the automotive/aviation sector.

1. Introduction

In particular, scientific research as well as commercial use, magnesium and its alloys have gained widespread interest as energy conservation and efficiency demands and it is increased because of lighter in weight. In automotive and aerospace applications, these properties play an important role in lowering fuel consumption and greenhouse emissions [1-3]. However, instead of their reduced creep properties low strength, and low wearing resistance at high temperatures inhibited use of magnesium alloys[5]. Reinforcements are then necessary to enhance the characteristics of the base metal (Magnesium) or alloys. Magnesium metal matrix composite may be a suitable option for aluminum metal matrix composite. The lightest materials (Mg matrix Composite), along with advantages at high temperature is high modules of elasticity, high resiliency, high creep tolerance and wear resistance. Consequently, their ductility was diminished and their general use was restricted[6,7]. A well judged choice of the reinforcement (Morphology and composition) will accomplish the desired properties. The reinforcement materials should be unwavering and non-reactive at the specified working temperature as well.

Verities of techniques have been practiced to produce magnesium matrix composite such as stir casting[8-16], squeeze stir casting[17-20], and powder metallurgy technique[21-22]. Stir casting was considered an easy to adjust and most used process. among these techniques. The near-net shaping of composites by traditional foundry processes is also an advantage of this process. The stir casting method for the mixing of reinforcing particles with the melt has taken place under the safety atmosphere of CO_2/SF_6 or Ar inert for preventing the burning of molten magnesium. These protective gases can be immersed in a stir casting in the melt due to the vortices, resulting in increased porosity of the composite obtained. However, because of the oxidation environment above the melt surface, the procedure would not remove the oxide formation of elevated magnesium. The Present scenario of research and manufacturing of magnesium matrix composites is also minimal relative to aluminum matrix composites. One of the major factors may be that Mg matrix composites are difficult to extract due to their high chemical activity.

It's not large enough to achieve those special properties in its purest form, on the whole, a high strength-to-weight ratio is allocated with various elements. In different nonferrous metals, magnesium is used widely as an alloying element. Magnesium alloys are also used where lightweight is of prime significance in structural and non-structural applications. Several researchers have been studied and their findings were carried out on composites with Mg metal matrix.

2.Literature Survey

2.1. Reinforcement of SiC in Mg Matrix

The wear behavior of composites (MMC), which were strengthened by carbide particulate silicon particles (SiCp) during dry sliding, and was investigated by Lim et al. [4]. The composites have marginally superior resistance to wear at minimum load, but the effect on the wear of the SiC partial reinforcement is not as conclusive at higher loads. Wear became the dominant and allowed the magnesium matrix to become distorted at the interface. Wang et al. [23] performed an ultrasonographical analysis on the novel stir casting of hardened magnesium matrix composites by micro-SiC Particles. In contrast to conventional casting, the ultrasonic operation increased the mechanical behaviors of composites manufactured in various parameters substantially. Adding SiC particles means grain size is refined. Grain composite sizes declined as particle contents were increased. As the particle content grew, the modulus of elasticity and ultimate tensile strength was significantly strengthened .Kandil et al.[24] have fabricated and investigated SiC reinforced Mg alloy (AZ91) matrix composites and found that morphological and mechanical properties are enhanced as expected. The reinforcements of SiC had been preheated for 2 hours at 1100°C to deter oxidation. Graphite crucible is used to melt matrix materials in an inert atmosphere at 725°C. The preheated SiC reinforcements have been mixed to the vortex of the molten matrix and stirred for one minute at a temperature of 700°C. Homogeneity or dispersion of the reinforcement in the matrix phase is examined through SEM(Scanning Electron Microscopy) and energy dispersion X-ray analysis. With the rise in SiC particles, the hardness is increased but the % deformation is minimized.A two-step stir casting technique was used by Lianxi et al. to fabricate SiCw/ZK51A magnesium matrix composite. The two-step stir casting technique is the latest fabrication technique and it showed improved strength and modulus of the composite than unreinforced alloy. The SiC reinforcement is a stable reinforcement of ZK51A alloy and it is confirmed by the absence of interfacial reaction between matrix and reinforcement [25].

Rzycho´n et al [26]. produced magnesium alloy matrix composites reinforced with SiC particles and studied their mechanical properties. They have reinforced 45µm sized SiC in 0.3, 2, 5, and 10% volume fractions. To eliminating the oxide layer of the matrix it is preheated in an inert atmosphere in the furnace at 450°C and temperature raised to 720°C to melt the matrix phase and hold for 20 minutes and obtained a homogeneous temperature to entire molten metal. Preheated reinforcements have been added to the molten matrix as predecided content of the reinforcements. Agitation of the slurry has been done for 10 minutes by stainless steel stirrer (stirrer speed 150 rpm) to get a homogeneous mixture. Now at the end of the process, a uniformly distributed slurry of composites is poured in cylindrical graphite mold of 4 cm diameter at 720°C. The microstructural examination has been performed by Scanning electron microscopy (SEM) and optical microscopy to demonstrate a consistent reinforcement distribution in the matrix. The composite hardness is improved but the final strength and strain are decreased by increasing SiC content.

2.2 Al_2O_3 Reinforced Mg-MMC

The magnesium alloy matrix (AM100) of alumina (Al_2O_3) reinforced composites had been made by Jayalakshmi et al. using squeeze-casting methodologies. There were reinforcements of 15%, 20%, and 25% Al_2O_3 volume fraction added to the matrix. They studied the highest hardness (165 BHN) for 25% reinforced and it about twice as much as a base alloy. Tensile properties have been investigated at room temperature, 100, 150, and 200°C. The unreinforced base alloy's overall tensile strength is highly susceptible to temperature and simultaneously raising temperature reduces the strength of the composites. Temperature influences the strength of the composites, and it has been found that strength at test temperature is 1/3rd of room temperature strength. Present research findings also explained % elongation was enhanced by raising the temperature [27].

The authors explored the tensile behavior and microstructure of the fibers-ceramic particles in the squeeze casting of the Magnesium AM60 hybrid-composite. Composite AM60 based hybrid incorporating alumina fibers enhancing microstructure and tensile behavior. Microscopy of the optical and scanning electron consistently demonstrates strengthening without agglomeration in an alloy matrix. The composite reinforced with alumina has examined their mechanical properties and found improved modulus of elasticity, tensile strength, and hardness [28].

Al_2O3 reinforced Mg alloy (AZ31B) composites produced by the disintegrated melt deposition technique, subsequently machined followed by hot extruded by authors and they have investigated tribological properties of Mg alloy (AZ31B) composites. The wear properties have been studied at 10 m/s speed and 10 N and 30 N load. At low speeds, composite wear values are greater than Mg alloy and it is increased by increasing the nano-alumina reinforcement. The lower output strengths of the composites are largely responsible [29].

2.3 TiC Reinforced Mg-MMC

T wear behavior of TiC reinforced Mg MMC's are manufactured using the powder metallurgical (P/M) process, and studied by Narayanasamy et al. They did mix 0, 5, 10, 15, and 20 wt. % of reinforcement in magnesium matrix. In order to explain the distribution of TiC particles in the magnesium matrix, microstructural analysis of as-sintered specimens is carried out. Furthermore, the impact of TiC is to be examined on the hardness and wear properties. Inhomogeneous and agglomerated TiC particles were found on the surface of Mg–20TiC composites when the TiC content was greater than 15%. When the TiC content is increased, the density of sintered composites increases and the hardness increases up to 15% volume fraction of TiC. When the TiC content is increased from 15% to 20%, there is a slight improvement in hardness, and the wear loss of composites is nearly equal. As compared to pure Mg, Mg composites made with TiC with a different volume fraction have better microhardness, less wear, and a higher friction coefficient[30].

Table.1. The finding of the present study along with reinforcement

S.No.	Reinforcement	Finding	Reference
1	SiCp	Wear properties enhanced	C. Y. H. Lim et al. [4]

2	Micro-SiC	Reinforcements improved the ultimate tensile strength, yield strength, and elastic modulus.	X. J. Wang et al. [23]
3	SiC	The strain decreases by increasing the SiC particles' volume fraction.	A. Kandil et al. [24]
4	SiCw	In comparison to the unreinforced matrix alloy, modulus and mechanical strength increase considerably.	H Lianxi et al.[25]
5	SiC	Improved hardness value, ultimate tensile strength but decreases % elongation with the increase of SiC.	T. Rzycho´n et al. [26]
6	Al_2O_3	Tensile strength decreases and strain improved at test temperature compared to room temperature	Jayalakshmi et al.[27]
7	Al_2O_3	In contrast to matrix alloy, the mechanical properties such as elastic module, tensile strength, and hardness increased.	Zhang et al. [28]
8	Al_2O_3	With the rise in nano-alumina, wear resistance has improved.	Nguyen et al. [29]
9	TiC	Density and wear properties enhanced by TiC	P. Narayanasamy et al. [30]
10	B_4C	Enhanced wear properties by addition of B4C	Q.C. Jiang et al. [31]
11	Graphite particle	By increasing the reinforcements strain value increases dramatically, but when it reaches 15%, it decreases.	Aatthisugan et al. [32]

2.4 Mg-MMC's Reinforced with B_4C
The research analyzed B4C particulates made of powder metallurgy (P/M) process, which have been reinforced with different fractions of 10,15% and 20%. The authors studied precipitation phenomena in magnesium matrix composites of B4C reinforcement. They inferred that the ceramic reinforcement promotes Mg17Al12 nucleation at the reinforcement-matrix interface by favoring the heterogeneous precipitation process. The inclusion of B4C particulates increases the composites' hardness greatly. When compared to a 10% B4C/Mg composite, the wear rate of a 20% B4C/Mg composite is smaller. The P/M process was used to successfully fabricate magnesium matrix composites that were reinforced with 10, 15, and 20% B4C particulates. The hardness and wear resistance of the composite increased after B4C particulate was incorporated into the matrix[31].

2.5 Graphite Particulate Reinforced Mg-MMC
Aatthisugan et al. [32] stated that Mg matrix composites reinforced by graphite particles (particle size of 50 μm reinforced 5-20%) and stir casting methods have been employed to fabricate the composites. A dynamic mechanical analyzer has been used to calculate the damping capacities of the composites. The findings show that graphite particles have a significant impact on the damping capacities of the

graphite reinforced composites. The strain of the composites increases initially up to 15 volume % graphite reinforcement but decreases when graphite is more than 15 volume %.

2.6 Fe_2O_3 Reinforced Hybrid Mg-MMC

According to Rodríguez, C. Gómez et al., nano-alumina or nano-iron oxide was applied to the Magnesia matrix up to 5% by weight. X-ray diffraction (XRD) and scanning electron microscopy (SEM) was used to investigate the crystalline phases and microstructure properties of specimens sintered at 1600 °C for 4 hours in an electric furnace. Density and porosity are used to describe the physical properties. A cold crushing strength (CCS) test was used to investigate the mechanical properties. Also, the chemical activity of slag attacks was investigated. The presence of nano-iron oxide in the formation of magnesioferrite spinel induced by the magnesium matrix was thus found which enhanced the sintering operation. The bonding structure was also affected by nano-iron oxide due to a direct bonding enhancement. The inclusion of nano-alumina in the magnesia matrix, on the other hand, caused magnesium-aluminate spinel formation, resulting in lower properties than those obtained by adding nano-iron oxide[33].

3. Conclusion

Many conflicts must be overcome to enhance the engineering use of magnesium matrix composites, such as manufacturing methods, reinforcement effects, and mechanical properties reinforcement effect on microstructure and mechanical properties. From the previous works, the key results were drawn between Magnesium metal matrix and reinforcements as shown in Table.1 and it showed the effects of reinforcements by addition of reinforcements. The current study emphasized that reinforcements such as SiC, B4C, and Al2O3 improved mechanical properties such as tensile strength, yield strength, strain, and hardness. However, the addition of TiC, nano, and micro-sized graphite particles improved the wear properties. The application of nanoparticles and their impact on mechanical/physical/chemical properties may be affected unexpectedly, and future research will be required to predict such possibilities.

References

[1] S. Nimityongskul, M. Jones, H. Choi, R. Lakes, S. Kou and X. Li // *Mater. Sci. Eng. A* 527 (2010) 2104.

[2] B.L. Mordike and T. Ebert // *Mater. Sci. Eng. A* 302 (2004) 37.

[3] M. Habibnejad-Korayem, R. Mahmudi and W.J. Poole // *Mater. Sci. Eng. A* 519 (2009)198

[4] C.Y.H Lim, S.C Lim, M Gupta//*Wear*,Volume 255, Issues 1–6,2003,Pages 629-637,ISSN 0043-1648, https://doi.org/10.1016/S0043-1648(03)00121-2

[5] P. Poddar, V.C. Srivastava, P.K. De and K.L. Sahoo // *Mater. Sci. Eng. A* (2007) 357.

[6] M.Y. Zheng, W.C. Zhang, K. Wu and C.K. Yao // *J. Mater. Sci.* 38 (2003) 2647.

[7] V. Viswanathan, T. Laha, K. Balani, A. Agarwal and S. Seal // *Mater. Sci. Eng. R* 54(2006)121

[8] V. Laurent, P. Jarry, G. Regazzoni, and D. Apelian: //*J. Mater. Sci.*, 1992, 27, pp. 4447-59.

[9] A. Martin and J. Llorca: //*Mater. Sci. Eng.*, 1995, A201, pp. 77-87.

[10] R.A. Saravanan and M.K. Surappa://*Mater. Sci. Eng.*, 2000, A276, pp. 108-16.

[11] S.C. Sharma, B. Anand, and M. Krishna: // *Wear*, 2000, 241, pp. 33-40.

[12] A. Bochenek and K.N. Braszczynska: //*Mater. Sci. Eng.*, 2000, 290A, pp. 122-27.

[13] Y. Cai, M.J. Tan, G.J. Shen, and H.Q. Su: // *Mater. Sci. Eng.*, 2000, 282A, pp. 232-39.

[14] H. Hu: // *Scripta Mater.*, 1998, 39, pp. 1015-22.

[15] T. Imai, S.W. Lim, D. Jiang, Y. Nishida, and T. Imura: //*Mater. Sci. Forum*, 1999, 304/306, pp. 315-20.

[16] A. Luo//*Metall. Mater. Trans*., 1995, 26A, pp. 2445-55.

[17] C. Mayencourt and R. Schaller//*Mater. Sci. Eng*.. 2002, A325, pp. 286-91.

[18] M. Yoshida, S. Takeuchi, J. Pan, G. Sasaki, N. Fuyama, T. Fujii, and H. Fukunaga//*Adv. Compos. Mater*., 1999, 8, pp. 258-68.

[19] L. Hu and E. Wang: // *Mater. Sci. Eng*., 2000, 278A, pp. 267-71.

[20] M.Y. Zheng, K. Wu, and C.K. Yao//*Mater. Sci. Eng*., 2001, 318A, pp. 50-56.

[21] H. Ferkel and B.L. Mordike: // *Mater. Sci. Eng*., 2001, 298, pp. 193-99.

[22] B.W. Chua, L. Lu, and M.O. Lai//*Compos. Struct*., 1999, 47, pp. 595-601.

[23] X.J. Wang, N.Z. Wang, L.Y. Wang, X.S. Hu, K. Wu, Y.Q. Wang, Y.D. Huang//*Materials & Design*, Volume 57,2014, Pages 638-645, ISSN 0261-3069, https://doi.org/10.1016/j.matdes.2014.01.022

[24] Kandil, A// *Journal of Engineering Sciences*, 40, 1, 255–270, 2012. https://journals.ekb.eg/article_112726.html

[25] Lianxi, H. and Erde, W.//*Materials Science and Engineering, A*, 267–278, 2000. https://www.mdpi.com/2075-4701/8/6/431/pdf

[26] Rzycho´n, T., Dybowski, B., Gryc, A., Dudek, M.//*Archieves of Foundry Engineering*, 15, 1, 99–102, 2015.

[27] S. Jayalakshmi, S.V. Kailas, S. Seshan//*Composites: Part A* 33 (2002) 1135–1140. http://library.nmlindia.org/FullText/CA330281135.pdf

[28] Zhang, Xuezhi & Zhang, Qiang & Hu, Henry//*Materials Science and Engineering:* A. 607. (2014) 269–276. 10.1016/j.msea.2014.03.069..

[29] Nguyen, Q. B., Sim, Y. H. M., Gupta, M., & Lim, C. Y. H.//*Tribology International*, *82*(Part B), (2015) 464-471. https://doi.org/10.1016/j.triboint.2014.02.024

[30] P. Narayanasamy, N. Selvakumar, P. Balasundar//*Materials Today: Proceedings*, Volume 5, Issue 2, Part 2,2018, Pages 6570-6578, ISSN 2214-7853

[31] Q.C. Jiang, H.Y. Wang, B.X. Ma, Y. Wang, F. Zhao,//*Journal of Alloys and Compounds*,Volume 386, Issues 1–2,2005, Pages 177-181, ISSN 0925-8388 http://doi.org/10.22214/ijraset.2017.1004

[32] Aatthisugan, I., Rose, A. R., & Jebadurai, D. S// *Journal of magnesium and alloys*, *5*(1), (2017) 20-25 https://doi.org/10.1016/j.jma.2016.12.004

[33] Rodríguez, C. G., Roy, T. D., Shaji, S., Rodríguez, G. C., Quiñonez, L. G., Rodríguez, E., ... & Aguilar-Martínez, J. A. //*Ceramics International*, *41*(6), (2015) 7751-7758.

Security and Privacy Issues in Location Dependent Services for Mobile Communication: A Synergistic Review

Chitrangada Chaubey[1*], Swapnil Raj[2] ans Suresh Kaswan[3]

[1,2,3] School of Engineering and Information Technology, Sanskriti University,Mathura, India.
***chitrangada111@gmail.com**

Abstract. With a growing rate of mobile usage in the current networking era, mobile devices have been adopted widely. With mobile proliferation, location awareness is gaining huge insights in mobile and communication technologies for which, information subsystem has to be maintained on geo-located systems. Location Dependent Services (LDS) fetch relevant information from the geo-located information subsystems and deliver it to the user based on their location preferences and choices. Despite having the desirable features and services, LDS does not provide satisfactory protection to the user's geographic location. Preservation of location has become one of the utmost concerns in mobile network communication. In this paper, we will go through the security & privacy challenges for LDS in mobile networks. Precisely, this paper covers the cryptographic approaches and technologies to enhance privacy to promote location privacy in mobile network communication. The different approaches covered in the review are first compared and then the further research areas have been identified as well as concluded.

1. Introduction

With the increasing usage rate of wireless appliances and the headway in sensing technology, it has become very easy to locate or spot any individual's precise location anytime and anywhere. Hence, as a result, a new service of application—LDS is being coined [1-4]. Precisely, location-dependent service is a mobile computing application that is associated with the user's location: this location is shared through requesting location access [5,6]. It felicitates the users to make use of the device's geographical position for different services LDS allows the users to access the latest location information about their nearby entities and also permits the businesses to share the recent updates against client's query [7-9].

The location information of the user consists of X-Y-Z coordinates which are generated by either of Location Determination Technology (LDT), like EOTD, GPS, Cell-ID, etc [10-12]. Location-based information is the basic section of the smartphone experience that promotes the mobile computing applications which are being used for Geographic location-based social networking, route navigation while traveling, land searches in real estate, retail and digital marketing, advertisements and product promotions, finding friends, restaurants, gym, petrol pump, medical shop, etc within the range [13,14]. With the proliferation of location-dependent services, LDS users are facing some crucial issues of their detailed and sensitive data getting accessed by someone [15]. As our location redirects our personality more than what we have mentioned in our resume and portfolio, so this problem of getting location accessed is becoming a serious privacy threat in our lives. By accessing a user's location anyone can misuse it in many ways so this should be considered and resolved by discovering privacy measures [16]. If the user wants to access this service along with confidence, that their private data is completely protected[17,18]. Considering security & privacy threats, because of the location data exchange, some privacy & security threats to clients exist too [19].

This paper has considered and extracted the location-dependent services, their components, and different privacy and security issues involved with them. Also, a detailed case study has been done and the findings have been listed. Figure 1 represents the various application areas of Location Dependent Services [17-19].

IOP Conf. Series: Materials Science and Engineering 1149 (2021) 012007 doi:10.1088/1757-899X/1149/1/012007

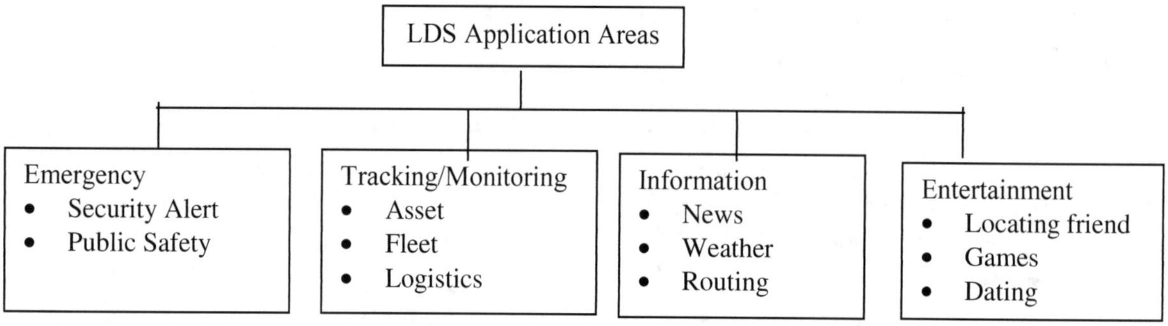

Figure.1 Application areas of Location Dependent Services

1.1 Components of LDS

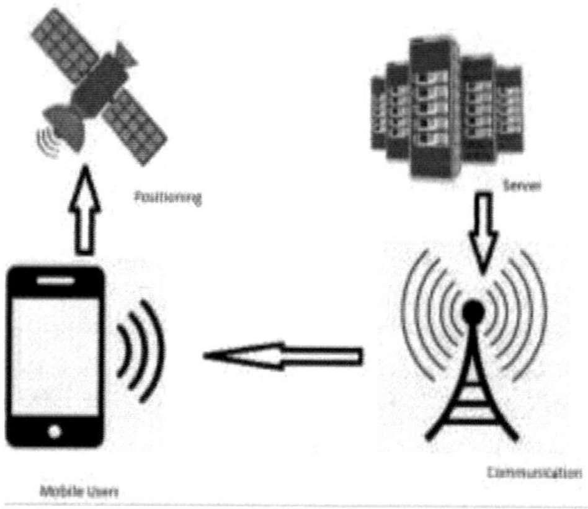

Figure.2 Typical Architecture of LDS

Figure 2 shows the typical architecture of LDS[20-22]which requires four basic components as:

- Mobile Device and User: A device through which the user submits the service information. This requested service can be interpreted by either graphs, pictures, text, speech, and so on. These devices(gadgets) include Laptops, Mobile Phones, and PDAs [23].
- Positioning: GPS falls under the most popular positioning system, which tracks the client's location and conveys the same as asked by the user [24,25].
- Communication: It works as a channel for network units. It is used to exchange the data & resource request of a client from a mobile terminal to the LDS provider and afterward the information services from the service contributor to the user or client [26,27].
- Server: It acts as a content Provider to hold and convey geographic data of the client.

Figure 3 represents the basic areas of research for the Location Dependent Services [28].

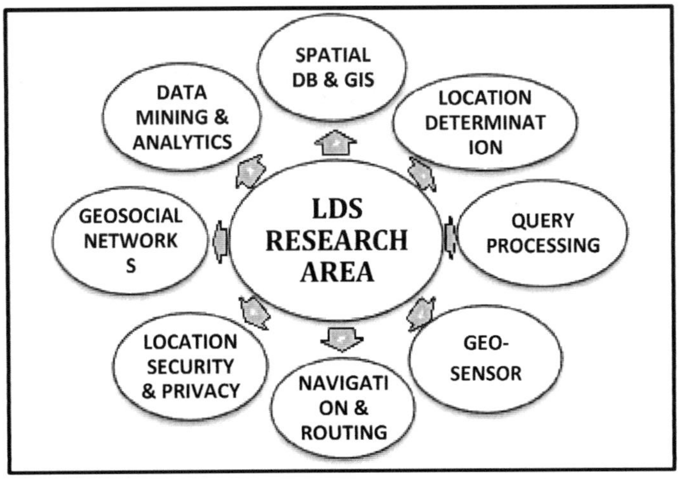

Figure.3 Basic Research Areas of LDS

2. Security & Privacy Requirements in Location Dependent Services

Considering Location Dependent Services, most of the security & privacy intimidations related to LDS have still not been appropriately addressed. Internet of Things and cryptography (apart from being an emerging concept) sis being expected that various usage areas like transportation, healthcare, and manufacturing, will get benefit from these two. [29-31]. The following are some properties with definitions necessary for privacy & security in mobile communication networks.

2.1 Security Concerns

1. *Authentication:* Only authenticated users are supposed to grant permission when they request for joining the LDS i.e. when the Service Provider (SP) is requested for location information by the user through a connecting station [32,33].
2. *Traceability:* Any user who mishandles the network is supposed to get traceable and reported by an authority. This helps in reducing the impact of the misbehaving user all across the network [34,35].
3. *Confidentiality:* It assures that the message delivery is done to authorized parties only. It can be achieved through a secure key management system's encryption techniques [36].
4. *Efficiency:* A lightweight management scheme should be used to ensure effective operation.

2.2 Privacy Concerns

1. *Anonymity:* The ability of a user to access something without disclosing the identity to the third-parties [37].
2. *Short-term linkability:* A receiving vehicle (VANET) must be capable of verifying that the multiple messages it is receiving within a small-time frame [38].
3. *Pseudonymity:* It gives assurance of resource access by a user without sharing its identity, but can still be answerable for that use [39].
4. *Location Privacy:* The location of the user must be shielded from unapproved access [40].

3. Literature Survey

Below are some issues related to location preservation and security in Mobile Networks, that have been reviewed in the current paper.

Table 1. Location privacy addressed by various Privacy Enhancement schemes in mobile networks.

Category	References	Privacy Metric	Location Privacy attack Reported	Issues Reported
Mix	[32]	Anonymity lot	Correlation	Makes it hard to track the users along with their Orientation.
Mix	[29]	Tracking of Location	Tracing success	Used an optimized algorithm for deployment of mix-zone strategy to improve user's location privacy.
Mix	[40]	Size of anonymity lot& Entropy	Timing & Transition	Used a variety of placement algorithms for mix-zone and Constructions to improve user's privacy.
Obfuscation	[35]	Obfuscation lot size	Easy Tracking	Provides an overview of obfuscated LDS to Intuitively raise privacy preferences.
Obfuscation	[36]	Imprecise range of LDS query	Inference	A random query was used to hide the user's identity in the query with cloaked enabled information evaluation.
Cloaking	[22]	Size of Anonymity lot	tracking of Location	The intention of the LDS data is altered to reach enumerated anonymity limitations in the region.
Cloaking	[39]	Anonymity lot size	N/A	A set of peers is formed by the mobile node before requesting for LDS using either multi-hop or single routing.
Cloaking	[1]	Anonymity lot size	N/A	TTP application, Casper which enables the users to register a particular privacy profile, through which the user's exact location is blurred by anonymizer into a cloaked area.
Cloaking	[30]	Entropy	Trail analysis	Used an optimal technique of mobility-aware cloaking to restrict the tracing.
Cloaking	[3]	Periodic Queries & cloaked spatial region	Query sampling and tracking	Used a spatial information-based cloaking technique to differentiate between the boundary of location access and query privacy.
Cloaking	[2]	Anonymity lot size	Inference	A decentralized system, Prive (for query depersonalization), which assures it under any of the user spread.
Cloaking	[13]	Anonymity lot size	Correlation	Users' privacy profiles are defined by cluster isolated algorithms.
Cloaking	[11]	Anonymity lot size	Inference	A prohibition-based linear computation is used for spatial query processing.

Table 2. Location privacy addressed by various cryptographic schemes in mobile networks.

Category	Strategy	Privacy Metric	Location Privacy attack Reported	Issues Reported
PIR	Nearest neighbor [20]	K-degree of anonymity	Correlation	Approximation of Nearest neighbor using Hilbert ordering
PIR	Nearest neighbor [31]	Communication cost & overhead	Disclosure of User's identity	User and database both are protected by location generalization into coarse-grained regions.
PIR	Oblivious transfer & PIR [32], [33]	Performance-based On communication efficiency	Disclosure of User's identity	Privacy is preserved using oblivious transfer for location data queries.
PIR	*k* nearest neighbors [40]	Computation complexity & overhead	Data & Location privacy	LBS grant to retrieve location as per user's request.
PIR	*k* nearest neighbors [14]	Communication overhead & complexity	Location privacy and Query processing	As per the user-cantered request, Location information is retrieved.

Location proof	Variation of received signals [26]	GPS Variations	Protection against location constituting	Location data are being updated according to continuous variation in GPS traced data with aerial satellites.
Location proof	Correlation [4]	Formal proof	Privacy and Integrity	Cross-party verifiers location proof based on triangulation.
Location proof	Trust-based provisioning [18]	The system is based on a PKI infrastructure	Privacy preservation against location forging and data modification.	Location and Contents are geo-tagged from trusted LDS for later revival.
Key managemen t	Trail Analysis[3]	Re-keying efficiency	Location Encryption	Hierarchy-based encryption membership-based location information.
Key managemen t	Anonymity [12]	Key Management and Message delivery overhead	Location Decryption	As per the keys provided, the group members are supposed to access the location for further of their use.
Key managemen t	Pseudo proximity [11]	Response Time & Throughput	DoS attacks	Make use of spatial & temporal approval to construct secure ranking based on key graphs

4. Conclusion

This paper has presented a broad review of location privacy in LDS. Security and privacy in mobile networks have been discussed along with the shortcomings and technologies used in previous work and methodologies. A codification has been introduced for recent publication's comparison along with the privacy metrics concerns and adversary models. This paper has subdivided the current research work into two categories that are cryptographic approaches and privacy enhancement for location privacy in mobile networks. Also, various categories of location privacy issues and challenges have been identified and discussed along with enhancement suggestions.

References

[1] ZongdaWu1, Guiling Li, Shigen Shen1, Xinze Lian, Enhong Chen, Guandong Xu, 2020 'Constructing dummy query sequences to protectlocation privacy and query privacy in location-based services', *Springer Science+Business Media,* LLC.

[2] Fabian Burmeister, Paul Drews, Ingrid Sachirmer, 2021 'Modeling the C(o)urse of Privacy-critical Location-based Services – Exposing Dark Side Archetypes of Location Tracking', *Proceedings of the 54th Hawaii International Conference on System Sciences.*

[3] Abdur R. Shahid, Niki Pissinou, S.S. Iyengar, Kia Makki, 2020 'Delay-aware privacy-preserving location-based services under spatiotemporal constraints', *Int J Commun Syst.*

[4] Y. Cao, S. Yang, G. Min, X. Zhang, H. Song, O. Kaiwartya, andN. Aslam, 2017 'A cost-efficient communication framework for battery- switch-based electric vehicle charging', *IEEE Communications Maga- zine,* **55**, pp. 162–169.

[5] E. C. Eze, S. Zhang, and E. Liu, 2014 'Vehicular ad hoc networks (vanets): Current state, challenges, potentials and way forward', *20th International Conference on Automation and Computing*, pp. 176–181.

[6] F. Yang, S. Wang, J. Li, Z. Liu, and Q. Sun, 2014 'An overview of internet ofvehicles', *ChinaCommunications*, **11**, pp.1–15.

[7] J. Lin, W. Yu, X.Yang, Q.Yang, X.Fu, and W.Zhao, 2017 'A real- time end-route route guidance decision scheme for transportation-based cyberphysical systems', *IEEE Transactions on Vehicular Technology*, **66**, pp.2551–2566.

[8] Y. Cao, Y. Miao, G. Min, T. Wang, Z. Zhao, and H. Song, 2016 'Vehicular-publish/subscribe (v-p/s) communication enabled on-the- move ev-charging management',*IEEE Communications Magazine,* **54,** pp.84–92.

[9] J.-H. Song, V. W. Wong, and V. C. Leung, 2008 'Secure Location VerificationforVehicularAd-hocNetworks', *IEEE Global Telecommunications Conference*, pp.1–5.

[10] O. Kaiwartya, A. H. Abdullah, Y. Cao, A. Altameem, M. Prasad, C. T. Lin, and X. Liu, 2016'Internet of vehicles: Motivation, layered architecture, network model, challenges, and future aspects',*IEEE Access*, **4**, pp. 5356–5373.

[11] S. Bao, W. Hathal, H. Cruickshank, Z. Sun, P. Asuquo, and A. Lei, 2017 'A lightweight authentication and privacy-preserving scheme for vanets usingteslaandbloomfilters', ICTExpress.

[12] Y. Leng and L. Zhao, 2011 'Novel design of intelligent internet-of-vehicles management system based on cloud-computing and internet-of-things',*Proceedings of 2011 International Conference on Electronic Me-chanical Engineering and Information Technology*, **6**, pp. 3190–3193.

[13] J. Lim, H. Yu, K. Kim, M. Kim, and S. B. Lee, 2017 'Preserving location privacy of connected vehicles with highly accurate location updates',*IEEE Communications Letters*, **21**, pp. 540–543.

[14] A. Wasef, R. Lu, X. Lin, and X. Shen, 2010 'Complementing public key infrastructure to secure vehicular ad hoc networks [security and privacy in emerging wireless networks]',*IEEE Wireless Communications*, **17**, pp. 22–28.

[15] M. Srivatsa, A. Iyengar, J. Yin, and L. Liu, 2009 'Scalable key management algorithms for location-based services', *IEEE/ACM Transactions on Networking*, **17**, pp. 1399–1412.

[16] K. Rabieh, M. M. E. A. Mahmoud, and M. Younis, 2017 'Privacy-preserving route reporting schemes for traffic management systems', *IEEE Transactions on Vehicular Technology,* **66**, pp. 2703–2713.

[17] A. Wasef and X. Shen, 2013 'Emap: Expedite message authentication protocol for vehicular ad hoc networks', *IEEE Transactions on Mobile Computing*, **12**, pp. 78–89.

[18] X. Yi, R. Paulet, E. Bertino, and V. Varadharajan, 2014 'Practical k nearest neighbour queries with location privacy', *IEEE 30th International Conference on Data Engineering*, pp. 640–651.

[19] X. Yi, R. Paulet, E. Bertino, and V. Varadharajan, 2016 'Practical approximate k nearest neighbour queries with location and query privacy', *IEEE Transactions on Knowledge and Data Engineering*, **28**, pp. 1546– 1559.

[20] H. Hu and J. Xu,2009 'Non-exposure location anonymity', *IEEE 25th International Conference on Data Engineering,* pp. 1120–1131.

[21] T. Xu and Y. Cai, 2007 'Location anonymity in continuous location-based services', *Proceedings of the 15th Annual ACM International Symposium on Advances in Geographic Information Systems,* **39**:1–39:8.

[22] W. Luo and U. Hengartner, 2010 'Veriplace: A privacy-aware location proof architecture', *Proceedings of the 18th SIGSPATIAL International Conference on Advances in Geographic Information Systems*, pp. 23–32.

[23] T. Peng, Q. Liu, and G. Wang, 2017 'Enhanced location privacy preserving scheme in location-based services',*IEEE Systems Journal*, **11**, pp. 219–230.

[24] T. H. You, W. C. Peng, and W. C. Lee, 2007 'Protecting moving trajectories with dummies', *International Conference on Mobile Data Management*, pp. 278–282.

[25] X. Lin, X. Sun, P. H. Ho, and X. Shen, 2007 'Gsis: A secure and privacy preserving protocol for vehicular communications', *IEEE Transactions on Vehicular Technology*, **56**, pp. 3442–3456.

[26] L. Buttyán, T. Holczer, and I. Vajda, 2007 'On the effectiveness of changing pseudonyms to provide location privacy in vanets', *Proceedings of the 4th European Conference on Security and Privacy in Ad-hoc and Sensor Networks*, pp. 129–141, Springer-Verlag.

[27] J. Freudiger, M. Raya, M. F. P. Papadimitratos, and J.-P. Hubaux, 2007 'Mixzones for location privacy in vehicular networks', WiN-ITS Vancover, British Columbia, Canada.

[28] C. Zhang, R. Lu, P. H. Ho, and A. Chen, 2008 'A location privacy preserving authentication scheme in vehicular networks', *IEEE Wireless Communications and Networking Conference*, pp. 2543–2548.

[29] S. Guo, D. Zeng, and Y. Xiang, 2014 'Chameleon hashing for secure and privacy-preserving vehicular communications', *IEEE Transactions on Parallel and Distributed Systems*, **25,** pp. 2794–2803.

[30] C. Bettini, S. Mascetti, X. S. Wang, and S. Jajodia, 2007 'Anonymity in location-based services: Towards a general framework', *International Conference on Mobile Data Management*, pp. 69–76.

[31] R. Shokri, J. Freudiger, M. Jadliwala, and J.-P. Hubaux, 2009 'A distortion-based metric for location privacy', *Proceedings of the 8th ACM Workshop on Privacy in the Electronic Society*, WPES '09, (New York, NY, USA), pp. 21–30.

[32] J. Krumm, 2007 'Inference attacks on location tracks', *Proceedings of the 5th International Conference on Pervasive Computing,* PERVASIVE'07, pp. 127–143, Springer-Verlag.

[33] G. Calandriello, P. Papadimitratos, J.-P. Hubaux, and A. Lioy, 2007 'Efficient and robust pseudonymous authentication in vanet', *Proceedings of the Fourth ACM International Workshop on Vehicular Ad Hoc Networks*, VANET '07, (New York, NY, USA), pp. 19–28.

[34] R. Lu, X. Lin, H. Zhu, P. H. Ho, and X. Shen, 2008 'Ecpp: Efficient conditional privacy preservation protocol for secure vehicular communications', *IEEE INFOCOM 2008 - The 27th Conference on Computer Communications*.

[35] X. Lin, R. Lu, X. Liang, and X. Shen, 2011 'Stap: A social-tier assisted packet forwarding protocol for achieving receiver-location privacy preservation in vanets', *Proceedings IEEE INFOCOM,* pp. 2147–2155.

[36] R. Lu, X. Lin, T. H. Luan, X. Liang, and X. Shen, 2011 'Anonymity analysis on social spot-based pseudonym changing for location privacy in vanets', *IEEE International Conference on Communications (ICC),* pp. 1–5.

[37] R. Lu, X. Lin, T. H. Luan, X. Liang, and X. Shen, 2012 'Pseudonym changing at social spots: An effective strategy for location privacy in vanets', *IEEE Transactions on Vehicular Technology*, **61**, pp. 86– 96.

[38] R. Yu, J. Kang, X. Huang, S. Xie, Y. Zhang, and S. Gjessing, 2016 'Mixgroup: Accumulative pseudonym exchanging for location privacy enhancement in vehicular social networks',*IEEE Transactions on Dependable and Secure Computing*, **13,** pp. 93–105.

[39] K. Sampigethaya, M. Li, L. Huang, and R. Poovendran, 2007 'Amoeba: Robust location privacy scheme for vanet', *IEEE Journal on Selected Areas in Communications*, **25**, pp. 1569–1589.

[40] S. Eichler, 2007 'Strategies for pseudonym changes in vehicular ad hoc networks depending on node mobility', *IEEE Intelligent Vehicles Symposium*, pp. 541–546.

Dimensional Synthesis of Watt-I Metal Mechanism based on Motion Generation

Khalid Nafees

Department of Mechanical Engineering, Inderprastha Engineering College, Ghaziabad, India
khalidnafees@gmail.com

Abstract. In industrial automation, there is always increasing demand of generating desired motion with accuracy and precision. It is possible when the mechanism generates motion for large number of precision points. Therefore, in present work, the dimensional synthesis of a 6-bar, Watt-I mechanism has been carried for twenty coupler positions. The single degree of freedom mechanism has revolute pairs at each of its joint and is able to generate required motion. All the links of the mechanism are made of metal. The dyad-triad design equations are formed in terms of complex number algebra. To solve these equations, a MATLAB code is generated and utilized to yield the dimensional parameters i.e. lengths and angular orientations of all links of the mechanism. The complete synthesis methodology is modeled and demonstrated on a numerical problem. To verify the outcomes, SAM 7.0 application software has been incorporated and applied on the given example.

1. Introduction

Inventing useful machine has always been keen interest of mankind. A machine consists of single mechanism or a group of mechanisms that work together to accomplish some useful work. Dimensional synthesis of such mechanisms is required to develop new advanced machines for modern applications. It includes assurance of correct link length along with its orientation to satisfy the required kinematic task. The number of links and their configuration varies depending upon the intricacy of the kinematic task to be performed. The configuration of the mechanism may be of 4-bar, 6-bar, 8-bar, slider crank mechanism or any other configuration that fulfills the requirements of kinematic task viz., path generation, motion generation, function generation to be performed. Dimensional synthesis for any such kinematic task involves finding out approximate/exact length of all the links of the mechanism that constitute and configure the mechanism. Also, it involves determination of orientations of all the links of the mechanism at its home position.

Dimensional synthesis can be carried out using graphical or analytical methods. But, graphical methods are restricted due to drawing accuracy. On the other hand, the analytical methods are in general simple and easy to apply. When they are used in combination with high-end programming softwares, they yield dimensions of all design parameters of the mechanism with improved accuracy. Sometimes, any optimization technique is also incorporated to determine the optimal design parameters of the mechanism to fulfill the sophisticated requirement of the task to be performed with minimum error. E.g., the optimal synthesis of a 6-bar linkage having rotational constraints was carried out in which coupler point generates the desired path [1]. The aim of synthesis is to get the required path as close to the specified path as possible. The author(s) considered that the specified path as a conjunction of a circular arc along with rectilinear segment. Many researchers have evaluated dimensional synthesis of various mechanisms for different kinematic tasks as discussed under subsequent section of related work.

2. Related work

Till starting of 20th century, only graphical methods were more prevalent to carry out dimensional synthesis of mechanisms. Rose [2] carried out synthesis of 5-bar mechanism using loop equations. Subsequently, Rawat [3] performed synthesis of variable topology mechanisms. Sun et al. [4] proposed a synthesis methodology of spherical 4-bar mechanism for motion generation. The author (s) performed harmonic analysis and observed that the output parameters are dependent on the rotation angle operator of the rigid body. Farhang et al. [5] discussed the motion generation based synthesis of those mechanisms which have input cranks with relatively small length. The author concluded that the coupler tracing point generate only elliptical curve for such mechanisms. Soh and Ying [6] projected design procedure for synthesis of 6-bar mechanism and 8-bar slider mechanisms to generate complex motion. The author (s) demonstrated the design process on wheelchair that can renovate its configuration as walking support for rehabilitation purposes. Shen et al. [7] formulated a computer code to generate prescribed coupler positions of fixed and moving pivot loci for 4-bar mechanism based on maximum driver static torques and coupler load. Schreiber et al. [8] combined circle point search method with homotopy method to dimensionally synthesis the planar Stephenson mechanisms for generating given motion. Martin et al. [9] suggested a MathCAD computer code and demonstrated the synthesis of a planar 4-bar crank rocker mechanism. The author (s) also considered feasible transmission angles while synthesizing the mechanisms. Balli and Chand [10-11] carried out the dimensional synthesis of variable topology 5-bar mechanism for motion between extreme positions. Their work was confined to single coupler tracing point. So, Nafees and Mohammad [12-13] extended the dimensional synthesis work of 5-bar mechanism for double offset tracing points. Balli and Chand [14-15], carried out synthesis of planar 7-link mechanisms that transfer motion between two extreme positions. Gadad et al. [16-17] synthesized variable topology 7-bar mechanisms for transmitting motion between extreme positions. Furthermore, Daivagna and Balli [18-20] performed synthesis of 5-bar and 7-bar variable topology mechanism to transfer motion between two dead-centre positions. Recently, Qaiyum and Mohammad [21] performed PSO based optimal synthesis of Stephenson-III mechanism.

Researchers have carried out synthesis, but they lack in performing synthesis for large number of precision points. To fulfill the requirements of high accuracy and precision in industrial automation, it is necessary that mechanisms should generate motion for greater number of precision points. Therefore, in present work, dimensional synthesis of a 6-bar (Watt-I) mechanism has been performed with 20 precision points. The design equations have been derived and represented in terms of complex number mathematics. Finally, a numerical problem has been demonstrated and verified to elaborate the synthesis methodology for given mechanism.

3. Arrangement of planar Watt-I linkage

The Watt-I is a planar kinematic six-bar linkage comprising of binary and ternary links with revolute pair at each of its joints. There are 3 binary, 2 ternary and 1 binary offset links (Refer **Figure 1**).

The connection of various links of the Watt-I linkage is shown in Fig. 1. The binary link 1 is fixed at pivots O_1 and O_2. The crank link 2 rotates about fixed point O_1 and the input motion is imparted to the linkage through this link. The ternary link 3 connects with crank link 2, binary offset link 4 and ternary link 6 at points, A, B and F respectively. The binary link 4 has an offset at point C which is called as coupler tracing point. The dimensional synthesis of the whole linkage is based on precision points and orientation of the binary offset link 4. Another ternary link 6 connects with fixed link 1, binary link 5 and ternary link 3 at points O_2, E and F respectively. The ternary link 6 is also called oscillating link as it oscillates about fixed point O_2. The binary link 5 makes revolute pair with link 4 and link 6 at points D and E respectively.

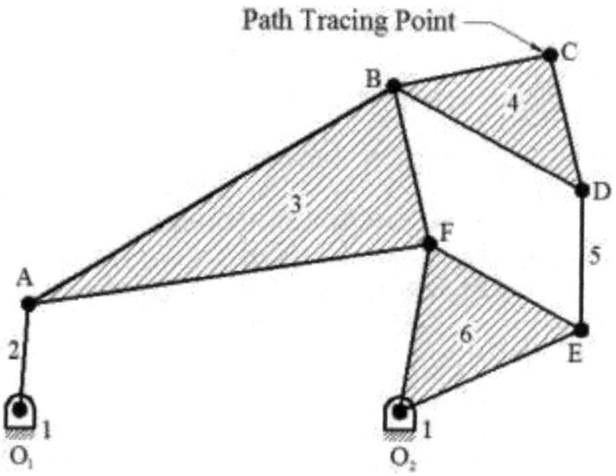

Figure 1 Arrangement of Planar Watt-I Linkage

4. Arrangement of planar Watt-I linkage for prime position

Consider the home position of the linkage as shown in **Figure 2**. The initial position of each revolute joint is expressed by $O_1A_0B_0C_0D_0E_0F_0O_2$. With the rotation of crank link from O_1A_0 to O_1A_j by an angle θ, the position of each revolute joint of the linkage is also shifted. The linkage acquires a prime position which is expressed by $O_1A_jB_jC_jD_jE_jF_jO_2$ (Refer **Figure 2**). All the remaining links i.e., ternary link ABF, ternary link O_2EF, binary link DE and binary offset link BCD have rotated by orientation angles of α_j, β_j, ϕ_j and γ_j respectively. The coupler tracing point C is also shifted from position C_0 to C_j thereby the displacement of this point is expressed by δ_j. For motion generation, there is requirement of 20 prescribed precision points (C_j) along with 20 orientations of binary offset links, which yields 20 prime position of the given linkage.

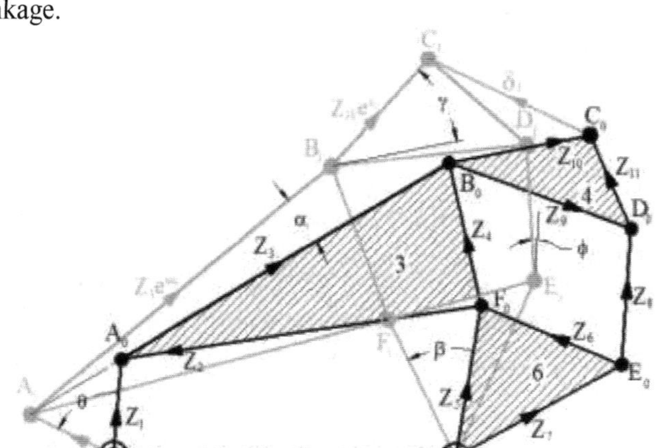

Figure 2 Displacement of Planar Watt-I Linkage from Initial to Prime Position By δ_j

5. Formation of design equations

By considering the various independent vector loops, the equations are derived as explained below:

Consider the independent vector loop $O_1A_jB_jC_jC_0B_0A_0O_1$ (Refer **Figure 2**), the design equation [22] is formed as

$$Z_1e^{i\theta_j} + Z_3e^{i\alpha_j} + Z_{10}e^{i\gamma_j} - \delta_j - Z_{10} - Z_3 - Z_1 = 0$$

$$Z_1(e^{i\theta_j} - 1) + Z_3(e^{i\alpha_j} - 1) + Z_{10}(e^{i\gamma_j} - 1) = \delta_j \tag{1}$$

Consider the independent vector loop $O_2E_jD_jC_jC_0D_0E_0O_2$ (Refer **Figure 2**), the design equation [22] is formed as

$$Z_7e^{i\beta_j}+Z_8e^{i\varphi_j}+Z_{11}e^{i\gamma_j}-\delta_j-Z_{11}-Z_8-Z_7=0$$

$$Z_7(e^{i\beta_j}-1)+Z_8(e^{i\varphi_j}-1)+Z_{11}(e^{i\gamma_j}-1)=\delta_j \quad (2)$$

Consider the independent vector loop $E_0F_0B_0D_0E_0$ (Refer **Figure 2**), the design equation [22] is formed as

$$Z_6+Z_4+Z_9-Z_8=0 \quad (3a)$$

Consider the independent vector loop $E_jF_jB_jD_jE_j$ (Refer **Figure 2**), the design equation [22] is formed as

$$Z_6e^{i\beta_j}+Z_4e^{i\alpha_j}+Z_9e^{i\gamma_j}-Z_8e^{i\varphi_j}=0 \quad (3b)$$

On subtracting equation (3a) from (3b), we get

$$Z_6(e^{i\beta_j}-1)+Z_4(e^{i\alpha_j}-1)+Z_9(e^{i\gamma_j}-1)-Z_8(e^{i\varphi_j}-1)=0 \quad (3)$$

For 20 precision points, j varies from 1 to 20 in design equations (1) to (3).

Consider closed loop $F_0A_0B_0F_0$, the unknown vector Z_2 is determined as (Refer **Figure 2**)

$$Z_2=Z_4-Z_3 \quad (4)$$

Consider closed loop $O_2F_0E_0O_2$, the unknown vector Z_5 is determined as (Refer **Figure 2**)

$$Z_5=Z_6+Z_7 \quad (5)$$

Consider closed loop $O_1O_2F_0A_0O_1$, the unknown vector Z_{12} is determined as (Refer **Figure 2**)

$$Z_{12}=Z_1-Z_2-Z_5 \quad (6)$$

6. Algorithm to obtain solution of design equations

As the number of unknown parameters is less than number of equations, the solution of equations (1) to (6) cannot be obtained manually. Therefore, a MATLAB computer code is developed to determine the solution of these equations. The algorithm for this program involves the following steps:

Step1. Input the value of prescribed parameters that comprises of coupler point orientation of binary offset link BCD (γ_j) and coordinates of coupler point C (P_j) for 20 precision points.

Step2. Determine displacement vector (δ_j) for each coupler point C by subtracting coordinates of its prime position from initial position based on all precision points.

Step3. Input the values of assumed parameters and freely chosen parameters.

Step4. For j = 1, 2,......20, calculate values $e^{i\theta_j}$, $e^{i\alpha_j}$, $e^{i\beta_j}$, $e^{i\gamma_j}$ and $e^{i\varphi_j}$.

Step5. Determine value of lengths Z_1, Z_3 and Z_{10} in vector form using equation (1).

Step6. Determine value of lengths Z_7, Z_8 and Z_{11} in vector form using equation (2).

Step7. Determine value of lengths Z_6, Z_4 and Z_9 in vector form using equation (3).

Step8. Determine value of length Z_2 in vector form using equation (4).

Step9. Determine value of length Z_5 in vector form using equation (5).

Step10. Determine value of length Z_{12} in vector form using equation (6).

7. Numerical problem

7.1. Problem Statement

It is required to carry out to dimensionally synthesize a planar kinematic six-bar Watt-I linkage for motion generation. The mechanism is prescribed with 20 motion generation positions corresponding to coupler orientations as given in Table 1 (Refer **Figure 3**).

7.2. Prescribed Parameters

The prescribed parameters include coupler point motion of binary offset link BCD (γ_j) and displacement vector (δ_j). The displacement vector (δ_j) is obtained by subtracting coordinates of the prime position from initial position at equal intervals (i.e. θ_j {0, 2π}, where j = 1, 2,...20).

Table 1 Prescribed Parameters

Prescribed Point (P_j)	Abscissa (X_j)	Ordinate (Y_j)	Prescribed Coupler Orientation (γ_j) rad.
P_1	0.950	0.941	5.500
P_2	0.886	0.935	8.995
P_3	0.838	0.925	10.428
P_4	0.809	0.915	9.740
P_5	0.801	0.906	6.875
P_6	0.816	0.897	1.776
P_7	0.857	0.888	-5.615
P_8	0.927	0.875	-15.470
P_9	1.027	0.851	-27.674
P_{10}	1.152	0.805	-41.081
P_{11}	1.280	0.732	-52.540
P_{12}	1.378	0.655	-58.098
P_{13}	1.423	0.621	-57.067
P_{14}	1.426	0.645	-51.566
P_{15}	1.400	0.703	-43.602
P_{16}	1.351	0.771	-34.435
P_{17}	1.283	0.836	-24.866
P_{18}	1.202	0.887	-15.584
P_{19}	1.114	0.921	-7.162
P_{20}	1.028	0.938	0.000

7.3. Assumed Parameters

The assumed parameters include orientation of crank O_1A (θ_j), ternary link ABF (α_j), ternary link O_2EF (β_j) and binary link DE (ϕ_j) and the corresponding range of these parameters are $\{-2\pi/45, 5\pi/36\}$, $\{-5\pi/12, \pi/18\}$ and $\{-\pi/18, \pi/6\}$ respectively [23].

7.4. Prescribed Parameters

The design parameters include dimensional lengths of fixed link O_1O_2, binary links (O_1A and DE), ternary links (ABF and O_2EF) and binary offset link BCD. Each side of these links is considered as design parameter and represented in vector notation as Z_1, Z_2, Z_3, Z_4, Z_5, Z_6, Z_7, Z_8, Z_9, Z_{10}, Z_{11} and Z_{12} (Refer **Figure 2**).

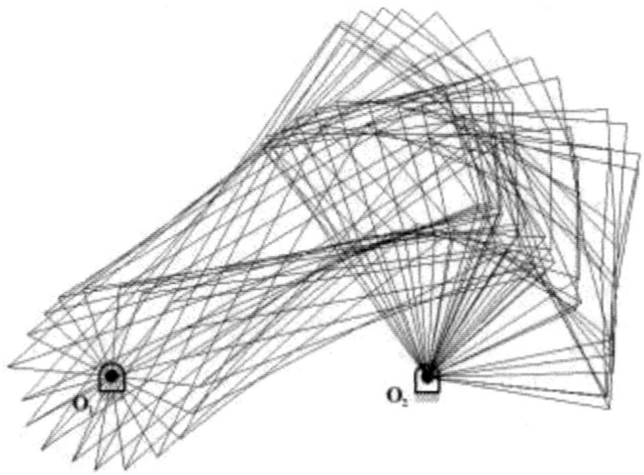

Figure 3 Representation of 20 Prime Positions of Planar Watt-I Linkage

IOP Conf. Series: Materials Science and Engineering 1149 (2021) 012008 doi:10.1088/1757-899X/1149/1/012008

8. Result and discussion

The dimensional length and angular positions of each link is determined by solving equations (1) to (6) with the help of MATLAB coding are:

$Z_1 = 0.20 \angle 157^o$; $Z_2 = 0.74 \angle 15^o$; $Z_3 = 0.77 \angle 37^o$; $Z_4 = 0.29 \angle 111^o$;
$Z_5 = 0.31 \angle 119^o$; $Z_6 = 0.31 \angle 12^o$; $Z_7 = 0.40 \angle 246^o$; $Z_8 = 0.25 \angle 95^o$;
$Z_9 = 0.39 \angle 7^o$; $Z_{10} = 0.29 \angle 46^o$; $Z_{11} = 0.25 \angle 138^o$; $Z_{12} = 0.70 \angle 0.27^o$

These final dimensional length with angular orientations of each link of given Watt-I mechanism for motion generations based on 20 precision points are graphically shown in **Figure 4**.

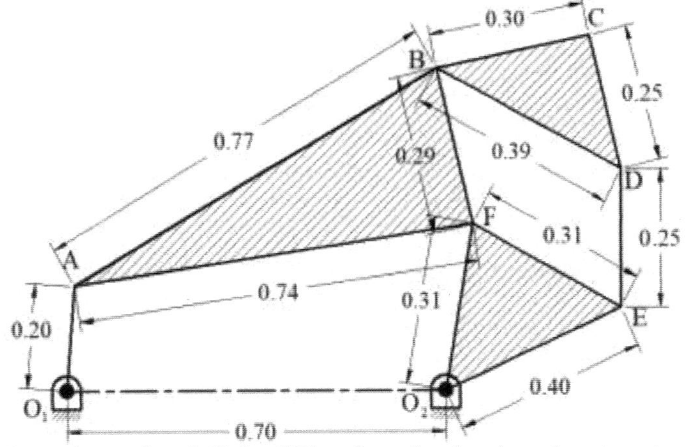

Figure 4 Final dimensions of each link of Watt-I mechanism based on twenty precision points motion generation

9. Conclusions

In this paper, the state-of-art is summarised for dimensional synthesis of a 6-bar, Watt-I mechanism that generates required motion for 20 precision points. For this purpose, various design equations are derived and formed by considering various independent vector loops. The parameters have been identified as prescribed parameters, assumed parameters and design parameters. The SAM 7.0 software has been used to verify the final dimensional parameters determined using MATLAB coding. The advantage of present work is that it will assist in the design of mechanisms that will fulfil the accuracy and precision requirement of automation industry. These useful mechanisms will be effective in transforming intricate motions for a greater number of precision positions. The existing methodologies have low accuracy and poor precision due to limited number of precision points. Therefore, comparing with other methods, this work overcomes the limitation of existing methodologies and provides benefit of improved accuracy and precision.

Nomenclature

i : A subscript used for representing link length parameters and varies from 1 to 12.
j : A subscript used for representing precision point parameters and varies from 1 to 20.
Z : Vector symbol representing side of each link.
δ_j : Displacement of coupler point from initial position C_0 to prime position C_j
θ_j : Angular movement of binary crank link O_1A from initial position O_1A_0 to prime position O_1A_j
α_j : Angular movement of ternary link ABF from initial position $A_0B_0F_0$ to prime position $A_jB_jF_j$
β_j : Angular movement of ternary link O_2EF from initial position $O_2E_0F_0$ to prime position $O_2E_jF_j$
γ_j : Angular movement of binary offset link BCD from initial position $B_0C_0D_0$ to prime position $B_jC_jD_j$
ϕ_j : Angular movement of binary link DE from initial position D_0E_0 to prime position D_jE_j

References

[1] Bulatović R R and Đorđević S R 2012 Optimal synthesis of a path generator six-bar linkage *J. Mech. Sci. Technol.* **26 (12)** 4027–4040.
[2] Rose S 1961 Five-bar loop synthesis. *Mach. Des* **12** 189-195.
[3] Rawat Y R 1966 Synthesis of variable topology mechanisms. *Dissertation*, IIT Bombay, India.
[4] Sun J, Chen L and Chu J 2016 Motion generation of spherical four-bar mechanism using harmonic characteristic parameters *Mech. and Mach. Theory* **95** 76-92.
[5] Farhang K, Midha A and Hall A S 1988 Synthesis of Harmonic Motion Generating Linkages - Part II: Path and Motion Generation *J. of Mech. Design* **110 (1)** 22-27.
[6] Soh GS, Ying F 2015 Motion Generation of Planar Six- and Eight-Bar Slider Mechanisms as Constrained Robotic Systems *J. Mech. Robot* **3 (3)** 1-8.
[7] Shen Q, Al-Smadi Y M, Martin P J, Russell K, Sodhi R S 2009 An extension of mechanism design optimization for motion generation *Mech. and Mach. Theory* **44 (9)** 1759-1767.
[8] Schreiber H, Meer K, Schmitt BJ 2002 Dimensional synthesis of planar Stephenson mechanisms for motion generation using circle point search and homotopy methods *Mech. and Mach.Theory* **37 (7)** 717-737.
[9] Martin P J, Russell K, Sodhi R S 2007 On mechanism design optimization for motion generation *Mech. and Mach. Theory* **42 (10)** 1251-1263.
[10] Balli S S, Chand S 2002 Five-bar motion and path generators with variable topology for motion between extreme positions *Mech. and Mach. Theory* **37** 1435-1445.
[11] Balli S S, Chand S 2001 Synthesis of a five-bar mechanism with variable topology for motion between extreme positions *Mech. and Mach. Theory* **36** 1147-1156.
[12] Nafees K, Mohammad A 2016 Synthesis of a planar five-bar mechanism consisting of two binary links having two offset tracing points for motion between two extreme positions *J. Mech. Robot* **8 (4)** 1-3.
[13] Nafees K, Mohammad A 2017 Dimensional synthesis of a planar five-bar mechanism for motion between two extreme positions *Australian Journal of Mechanical Engineering* 1-8.
[14] Chand S, Balli S S 2001 Synthesis of seven-link mechanism with variable topology. *In:* Proceedings of the CSME-MDE-2001 Conference, Concordia University, Montreal, Canada.
[15] Balli S S, Chand S 2003 Synthesis of a planar seven-link mechanism with variable topology for motion between two dead-center positions *Mech. and Mach. Theory* **38** 1271-1287.
[16] Gadad G M, Ramakrishan H V, Srinath M S, Balli S S 2012 Dyad synthesis of planar seven-link variable topology mechanism for motion between two dead-centre positions *Journal of Mechanical and Civil Engineering* **3(3)** 21-29.
[17] Gadad G M, Daivagna U M, Balli S S 2005 Triad and dyad synthesis of planar seven-link mechanisms with variable topology *Proceedings of the 12th National Conference on Machines and Mechanisms (NaCoMM'05)* 67–73.
[18] Daivagna U M, Balli S S 2007 FSP Synthesis of an off-set five-bar slider mechanism with variable topology *Proceedings of the 13th National Conference on Mechanisms and Machines (NaCoMM'07)* IISc Bengalore 345-350.
[19] Daivagna U M, Balli S S 2011 Synthesis of five-bar slider mechanism with variable topology for finitely separated positions *Adv. Mech. Eng.* 1-10.
[20] Daivagna U M, Balli S S 2010 Synthesis of a seven-bar slider mechanism with variable topology for motion between two dead-center positions. In: Proceedings of the World Congress on Engineering, London, U.K. 2010.
[21] Qaiyum A, Mohammad A 2020 Optimal synthesis of stephenson-III linkage using particle swarm optimization *International Journal of Mechanical and Production Engineering Research and Development* **10(3)** 15989-15998.
[22] Sandoor G N, Erdman A G 1984 Mechanism Design: Analysis and Synthesis. Vol I. Prentice-Hall of India Private Limited, New Delhi.
[23] Artas Engineering Software, Available at https://www.artas.nl/en (Accessed on Feb 2021)

IOP Conf. Series: Materials Science and Engineering 1149 (2021) 012009 doi:10.1088/1757-899X/1149/1/012009

Significant complications declining usage of Electrical Discharge Machining process

Gurpreet Singh Matharou[1] and Basanta Kumar Bhuyan[1]

[1] Department of Automobile Engineering, Manav Rachna International Institute of Research and Studies,Faridabad,India
gsmatharou1@gmail.com

Abstract. In the present paper, an effort has been laid down to develop a framework for the micro EDM (μ EDM) in today's industrial environment. The μ EDM has shown considerable involvement in ceramic industries where workpiece needs to be machined in micro sizes. μ-EDM owing to stronger surface quality and higher precision is a futuristic machining process for building up micro tools and alternative micro components. However, there are several complications that demand to be discussed before μ-EDM can be dealt with as a decisive technique and it can carry out its performance. This investigation includes of the EDM system scheduling and electrode wear issues. Particular consideration has been offered to aspects and methods critically varying the efficiency involving positioning approaches during grinding of electrode and new μ-EDM operations.

1. Introduction

The EDM process is noticeable since 1770-74 when Dr. Joseph Priestly (an English Scientist) invented the electrical flow erosive effects. In 1943 EDM process was discovered by Dr. B.R. Lazarenko. They had created Makarenko circuit which is of interest in current scenario also [1]. There was a complication with inadequate electrode which had been amended in 1960's by treatment of pulse and solid-state generators in extension to revolving systems. In 1980's first CNC based EDM was started in USA [2]. Nowadays EDM uses approaches based on non-traditional energy origins such as electrical (electrons and ions), sound and light, mechanical and synthetic etc [3]. With technology improvements and advancement in materials, fresh and harder to structure materials such as composite components, ceramics, and alternative metallic objects finds space in aerospace, industrial and fundamental systems. Machining of such tough substances becomes useless with classical practices of machining and thus EDM process for material removal becomes significant [4]. Electrical discharge machining (EDM) as a non-traditional material removal process employed in aerospace, surgical and automotive region in charge to develop dies, punches, moulds, precision parts etc [5]. This process can handle all electrically conductive components regardless of their firmness, hardness, shape and proportion. EDM has showed usage in mould and tools manufacture setups such as drilling of pieces and in producing turbine parts for airline industries, e.g., in tool creating setups the material discharged from a metal slab to create drill sections or in the specific cutting of substances such as drilling narrow holes or openings that are lesser than 0.05-0.06 inches [6]. Another convenience of EDM over conventional machining methods is that with EDM, it is viable to cut pre-hardened steel in complicated outlines or cavities without the inclination to mitigate the pre-seasoned steel by conventional heat treatment methods, and later re-harden it [7]. It has likewise demonstrated profitable and compelling in the machining of hard, high quality, and temperature opposition of conductive material [8]. It assumes an excellent job in the production of least-cost items with increasingly dependable quality affirmation. Micro EDM is an important type of EDM and has shown lots of applications in ceramics industry [9].

Many researchers had worked on different combinations of workpiece-electrodes, different dielectrics combinations, different machining input variables to ascertain the effects on different responses. Mukhopadhyay et al [10] had attempted micro drill operation on μ-EDM of Monel K-500 sheet using a Cu electrode with deionized water as a dielectric, to access the effects of current (I_p), pulse on time (T_{on}), working time (WT), Al powder concentration on circularity. Response surface methodology (RSM) used for model development. The optimum circularity value of 0.976 suggested for input parameter values I_p=1.2A, WT=1.8 sec, and Al concentration of 3.33 gm/liter. Mallick et al [11] had attempted micro electrochemical discharge machining (μECDM) on glass (silica) using a stainless-steel electrode (straight and curved) with NaOH and KOH electrolyte solution. The study was an attempt to find out the effect of voltage (v), electrolyte concentration (EC) and tool shapes (straight and curved) on material removal rate (MRR), the width of cut (WOC), heat affected zone (H.A.Z), and surface roughness (R_a). They reported an increase in the H.A.Z, MRR, and WOC with an increase in V for both types of electrolytes and tool shapes. Lowest HAZ reported at 50V, 10%wt KOH electrolyte with curve tool. They suggested optimum condition of V= 50V, 20 % wt. of KoH, and advocated straight shape electrode for micro channel cutting on glass (silica) material. Tiwary et al [12] had studied the μ-EDM operation of Ti-6AL-4V with a brass electrode, to access the effect of I_p, powder type (Cu, Ni, Co) mixed in de-ionized water and powder concentration (2,4,6 g/L) on MRR, electrode wear rate (EWR), taper, and overcut(oc). Dielectric with powder mixed in de-ionized water offers the best MRR, and Normal de-ionized water dielectric provides better EWR, taper, and OC. They suggested cobalt mixed de-ionized water at 4 g/L with I_p 1.5A to be an optimized setting for machining the titanium composite using PCA and overall principal component index (OPCI). Jahan et al [13] investigated during μ-EDM of micro holes on NiTi Shape memory alloy (SMA) and Ti-6Al-4V, the R_a at the center point and edges are reported to be dissimilar, they suggested the phenomenon was due to an advanced concentration of spark at the center in relation to the sides. The electrode reported to turn out to be hemispherical and tapered, resulting in a higher spark at the center. Due to the formation of oxide layers, microhardness has reported to increase in both the materials. Better R_a has been reported in NiTi (SMA). Singh et al [14] had attempted μ-EDM of two different Al_2O_3 composites with conducting fillers in the form of multi-walled carbon nanotubes (MWCNT) in 2.5 and 5 vol. % of concentration. High MRR has been reported in composite with 5 vol. % of MWCNT. A 60% enhancement in MRR (both composites) at tool rotation of 750 rpm. R_a has reported to reduce by 20%. With an increase in tool rotation speed, less recast layer deposition at edges had been reported. Better machining characteristics of brass tool reported, in comparison with copper and WC. The research study focuses on the leading issues of μ-EDM machining and linked causes that needs to be discussed for producing a dependable process that can lead to accurate results. Also, the paper comprises of the μ-EDM investigation area and therefore will offer research subsequent trends.

2. Micro EDM

In μ-EDM, the workpiece (anode) and tool (cathode) are separated by dielectric fluid (kerosene or deionized water) and was fed with a pulsed voltage. **Figure** 1 illustrates the simplified depiction of μ-EDM. The workpiece and tool are brought closer until the dielectric in the channels plunges down and allows current to move through it causing sparks. The spark intensity can be altered by changing the input parameters of the machine [15]. The μ-EDM offers extensive prospects to generate miniature topographies, micro components and the fabrication of μ electro-mechanical arrangements [16].

IOP Conf. Series: Materials Science and Engineering 1149 (2021) 012009 doi:10.1088/1757-899X/1149/1/012009

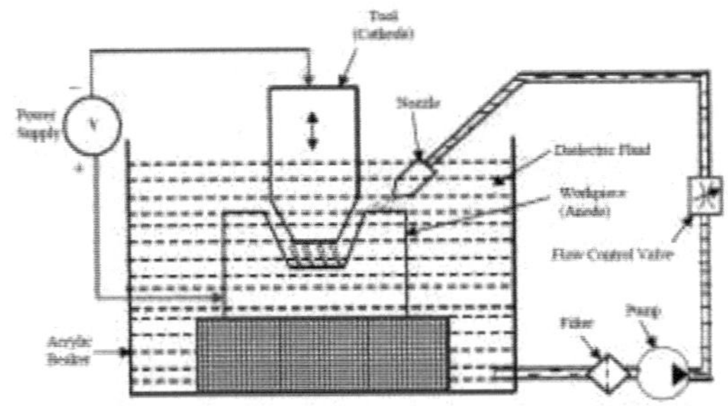

Figure 1 μ-EDM Schematic Representation

The μ-EDM technique employed for manufacture of micro pieces can be segregated under four groups [17]. Wire μ-EDM where an electrode diameter-wire is diminished to about 0.02mm to decrease a conductive work piece. μ-EDM drilling utilizes micro electrodes to drill work piece short pits of diameter 0.035-0.45mm. Die sinking μ-EDM [5] where electrode image is replicated on the work piece with a little overcut. μ-EDM milling where micro electrodes are employed to develop three dimensional cavities by seeking a transfer approach as witnessed in traditional milling [18].

2.1 Process parameters

The specifications in micro-EDM are employed to deal with the behavior measures of the micro-EDM fabricating process. Process parameters are the controllable machining inputs that regulates the environment in which machining is being carried out. Pulse on time (T_{on}) is the period of flow of discharge and is the stage through which the electrode is heated by the strong plasma channel. Pulse off time (T_{off}) shows the period during which nil flow of discharge exists and the dielectric is given time to clear the melted plasma away from the workpiece and restore its insulating properties. Spark gap (Sg) is the distance amongst the electrode and workpiece for respective EDM cycle. Duty factor (τ) is a percentage of the pulse interval with reference to the total cycle time. Current (I_p) denotes electrical charges travelling between the tool and workpiece. I_p has a direct influence to the MRR. Voltage (v) can be read across the tool electrode/workpiece gap before the spark current starts to flow. Straight polarity in which the electrode is connected to negative terminal & work-piece to the positive terminal. Reverse polarity is just the reverse [19].

2.2 Response variables

These criteria project the various actions of μ-EDM results. Material removal rate (MRR) & electrode wear rate (EWR) shows the volume of material removed from workpiece & electrode respectively, with reference to time and usually measured in mm^3/min. The surface roughness (R_a) displays the surface finish in μm. It is generally measured using profilometer. The overcut (OC) shows the difference between the hole/cavity size and the electrode size [19].

3. Shortcomings of Micro EDM

The μ-EDM offers contactless machining expertise which is applicable for brittle substances machining. The machining process gives very less impact on the workpiece and likewise supports in producing proper microstructures. It offers tolerances of the range of +/- 0.005 [20]. The hardness of the workpiece makes no influence on the machining. It has demonstrated usage in exceedingly hard materials such as nimonic, stellite, tungsten carbide, etc. It likewise presents potential of machining complex shapes under thin-walled environment of micro-level without leaving burrs in it. Fragile and subtle parts can be machined due to a no force process. Despite the capacities achieved even though these processes are not utilized extensively because of the accessibility process and due to non-accuracy of machine tools [21]. The research will concentrate on some issues limiting the operation of μ-EDM and will present positive explanations. This section presents the above discussed operations of μ-EDM and recognizes complex operations. A recent process to handle model is required for product miniaturization. Several issues have turned up because of variation in micro machining requirements.

3.1 Formation of work piece and electrode

For development of minor craters preparation for threading of wire, work piece and electrode formation are a serious trouble. These cracks might have micro diameters with various aspect ratio and relies on the file to be fabricated. In preparation they are made either by Traditional drilling or EDM drilling. The holes positioning efficiency with endorsement to the degree expected should be bigger so as to facilitate the practice of automatic threading and to restrict limited circuiting after wiring process. The holes threading on specialized systems of micro wire is exceedingly crucial. In trial of micro die sinking, it demands at least two to three electrodes developed either by EDM or micro milling. There could be chances of misrepresentation and irregular shapes of electrodes geometry. The electrode improvements not merely are costly & time dominating but likewise may produce distinct result since three electrodes may not have identical properties [5]. Trajectory EDM employs a simple shaped pipe, electrode or shaft of diameter between 0.15 and 0.42mm. The conducting electrode is dissipated against an electrode sacrificed in a process specified as grinding of EDM [22].

3.2 Handling of electrodes and its parts

There are several issues with electrode operation and the relevant parts when the direction is to cut down the diameter of the wire. Initially already occurring wire instruments were bought to accept slight diameter wires but they demand reasonable time for arrangement of machine [23]. There is a massive inconvenience for wire establishment as the spool place to nozzle of interweaving is substantial. The brakes productive forces cannot be employed owing to fine wires. This leads to periodic fractures in wire and requires periodic manual intervention [24]. In die sinking of μ-EDM drill or milling, varied designs and machineries can be adopted to uphold handling smaller parts and electrodes. Therefore, sub structures are combined into structures of micro EDM for fabrication on the apparatus and maintaining desired micro electrodes shapes and sizes. WEDG (wire electro discharge grinders) includes identical subsystems are ceramic guides & treating units [25].

3.3 Process Planning

In μ-EDM process strategy is of ultimate relevance and should be thoroughly set down dealing with the miniature size of elements and machined surface tolerances. During the machining process and during preparation stage several mistakes exist pointing to failed results. These mistakes may be because of imperfection equipment and stochastic sparking process [5]. Several papers aim ways of upgrading operation of EDM like the MRR, R_a and reduction in EWR. For μ-EDM the criteria of process are nevertheless at the improvement stage and their impacts on achievement measures needs to be formulated. The optimization of specifications is based on investigation of transform to establish the impact of every process variable on the desired machining features [26]. There is reduction of CAM components to support μ-EDM despite the usage of vaster expansion of automation of EDM machines and CNC controller's etc. Three-dimensional cavities machining operation is challenging since tools path adopted in existing techniques of CAM is complex [27]. The prevailing practice does not allow electrode wear compensation nor assist slice thickness variation or endorse the distinctive orientation of cut with every piece [28].

3.4 Measurement

The micro attributes pertaining to surface finish is not smooth. There are no legitimate processes of surface roughness assessment which is a fundamental quality of micro tooling [29]. Specialized apparatus is instructed to figure out the (HAZ) heat affected zone and (RL) recast layers which have noticeable effect on work piece surface properties [30]. For feature dimensions, evaluation of electrode is instructed to obtain stronger machining in Micro EDM [31].

3.5 Repeatability and accuracy of machine positioning

It is one of the primary fault causes. ISO 230-2-1997 is employed for measuring the veracity & repeatability of μ-EDM with the cooperation of laser interferometer [32]. For machining a micro hole at a given quantity of arranged electrodes might be adopted therefore the hole position depends on machine position certainty and hence the shape and size of the crater [33].

3.6 Errors in cycle evaluation
The EDM machine processor indicates the signal as brought in through the electrical approach of work piece and electrode through spark [34]. The processor accepts priorities defined to substantiate each signal of machine contact signal and is not chosen continually. Generally, verification of contact signal is drawn every two or five minutes undoubtedly to diminish the machining speed the EDM speed must be lowered as attainable but better enough to bypass stick slip. The voltage exists in between the spindle and table during the stepping cycle. Till contact of electric is made machine drives. The measurement efficiency depends upon the advance of appeal to the work piece surface [35].

3.7 Errors of temperature fluctuation
Owing to variation in room temperature condition, there is variation in work piece and electrode internal grain arrangements and therefore the relative arrange between the whirling head and machine table thereby modifying the covering unit position in reference to machine zero position and electrode. The inequalities can be lowered by emphasizing optimum temperature contained rooms and assuring thermal strength in machine systems [36]. Every EDM machine requires to be justified and setup desires to be maintained for some ambient situations, the assigned fluctuations require to be measured and recognized to diminish electrode and workpiece positional errors [21,26,28].

3.8 Spark gaps
The pulse parameters selection immediately regulates the R_a and MRR desired in traditional EDM. High value of peak current and voltage offers excessive value of MRR in case of tungsten carbide whereas EWR can be lessened by significant value of pulse off time followed by modest values of peak current [37]. In μ-EDM the little spark gap helps in achievements of micro characteristics with less EWR. The specifications of pulse are appointed on the base of surface roughness and on the dressing speed. The rotating electrode must not jolt with the machined surfaces roughness [38].

3.9 Fixtures and Jigs
Ceramic model is the most recognizable material for carrying the WC electrode [39]. Owing to variations in adequate diameter of ceramic guide and electrode, there are chances of electrode tolerance variation. The electrode located within the guide moves to radical positions contributing to peak variation [40].

3.10 Electrode type
The electrodes employed in μ-EDM are extremely basic in shape and during the EDM process are revolving, which accordingly establishes uniform conditions all-round the electrode [41]. These uniform sparking conditions endorse the usage of electrostatic field modelling of the medium between the electrodes. The electric field between the electrodes is interpreted as the electrical force per unit charge [23]. The orientation of the field at a position is described by the orientation of the electric force expended on an effective attempt charge established at that spot [42].

4. Results and Discussions
The review provides an outline of significant complications decreasing usage of μ EDM and affecting the production. The issues cited in this study can use to work out the process within tolerances as per expectations. When allocating tolerances of handling for μ EDM entire process situations such as kind of arranging, kind of electrode grinding, and operation extent must be dealt with, since all such exercises may attend to mistakes giving wrong yields which can't be recognized for investigation purpose. The overall capability of machining relies on a compound affinity between diverse specifications of process and the optimization is built primarily on empirical techniques. The

manufacturing technique handling of μ EDM must utilize stable approaches and innovations with continuous results. The proposed micro milling strategy replaces complicate interpretations of continuing processes with straightforward evaluation of length. This must prepare fresh strategy pleasant to the production & for analysis purpose.

5. Conclusions.

In the existing paper, an effort has been completed to outline the concept of μ EDM in todays industrial environment. The μ EDM has shown considerable involvement in ceramic industries where workpiece needs to be machined in micro sizes. Despite the capacities achieved even though these processes are not utilized extensively because of the accessibility process and due to non-accuracy of machine tools. Other key issues addressed includes development of workpiece and electrode, usage and type of electrode used and different wear ratios. Errors addressed during cycle evaluation, handling issues etc. Without addressing the μ EDM discussed issues, it is not possible for its maximum coverage in micro industries.

References

[1] Ho, K.H., Newman, S.T., 2003. State of the art electrical discharge machining (EDM). International Journal of Machine Tools and Manufacture 43, 1287–1300. DOI:10.1016/S0890-6955(03)00162-7

[2] Schumacher, B.M., 2004. After 60 years of EDM the discharge process remains still disputed. Journal of Materials Processing Technology 149, 376–381. DOI: 10.1016/j.jmatprotec.2003.11.060

[3] Qiu, L., Zhu, N., Feng, Y., Michaelides, E.E., Żyła, G., Jing, D., Zhang, X., Norris, P.M., Markides, C.N., Mahian, O., 2020. A review of recent advances in thermophysical properties at the nanoscale: From solid state to colloids. Physics Reports 843, 1–81. DOI: 10.1016/j.physrep.2019.12.001

[4] Ubaid, A.M., Dweiri, F.T., Aghdeab, S.H., Abdullah Al-Juboori, L., 2018. Optimization of Electro Discharge Machining Process Parameters with Fuzzy Logic for Stainless Steel 304 (ASTM A240). Journal of Manufacturing Science and Engineering 140, 011013. DOI: 10.1115/1.4038139

[5] Abu Qudeiri, J., Saleh, A., Ziout, A., Mourad, A.-H., Abidi, M., Elkaseer, A., 2019. Advanced Electric Discharge Machining of Stainless Steels: Assessment of the State of the Art, Gaps and Future Prospect. Materials 12, 907. DOI: 10.3390/ma12060907

[6] Gohil, V., Puri, Y., 2017. Turning by electrical discharge machining: A review. Proceedings of the Institution of Mechanical Engineers, Part B: Journal of Engineering Manufacture 231, 195–208. DOI: 10.1177/0954405415590560

[7] Dutta, V., Sharma, S., Chopra, K., Singh, B., 2016. Hybrid Electric Discharge Machining Processes for Hard Materials: A Review. Mater focus 5, 202–208. DOI: 10.1166/mat.2016.1361

[8] Reddy, P.V., Kumar, G.S., Krishnudu, D.M., Rao, H.R., 2020. Mechanical and Wear Performances of Aluminium-Based Metal Matrix Composites: A Review. J Bio Tribo Corros 6, 83. DOI:10.1007/s40735-020-00379-2

[9] Bilal, A., Jahan, M., Talamona, D., Perveen, A., 2018. Electro-Discharge Machining of Ceramics: A Review. Micromachines 10, 10. DOI: 10.3390/mi10010010

[10] Mukhopadhyay, P., Biswas, D., Sarkar, B.R., Doloi, B., Bhattacharyya, B., 2019. Influence of Al Powder on Circularity During Micro-Electro-Discharge Machining of Monel K-500: International Journal of Materials Forming and Machining Processes 6, 15–30. DOI: 10.4018/IJMFMP.2019070102

[11] Mallick, B., Biswas, S., Sarkar, B.R., Doloi, B., Bhattacharyya, B., 2020. On performance of electrochemical discharge micro-machining process using different electrolytes and tool shapes. International Journal of Manufacturing, Materials, and Mechanical Engineering 10, 49–63. DOI: 10.4018/IJMMME.2020040103

[12] Tiwary, A.P., Pradhan, B.B., Bhattacharyya, B., 2019. Influence of various metal powder mixed dielectric on micro-EDM characteristics of Ti-6Al-4V. Materials and Manufacturing Processes 34, 1103–1119. DOI: 10.1080/10426914.2019.1628265

[13] Jahan, M.P., Kakavand, P., Alavi, F., 2017. A Comparative Study on Micro-electro-discharge-machined Surface Characteristics of Ni-Ti and Ti-6Al-4V with Respect to Biocompatibility. Procedia Manufacturing 10, 232–242. DOI: 10.1016/j.promfg.2017.07.051

[14] Singh, M.A., Rajbongshi, S.K., Sarma, D.K., Hanzel, O., Sedláček, J., Šajgalík, P., 2019. Surface and

porous recast layer analysis in μ-EDM of MWCNT-Al 2 O 3 composites. Materials and Manufacturing Processes 34, 567–579. DOI: 10.1080/10426914.2019.1566617
[15] Modica, F., Marrocco, V., Fassi, I., 2017. Micro-electro-Discharge Machining (Micro-EDM), in: Fassi, I., Shipley, D. (Eds.), Micro-Manufacturing Technologies and Their Applications, Springer Tracts in Mechanical Engineering. Springer International Publishing, Cham, pp. 149–173. DOI: 10.1007/978-3-319-39651-4_6
[16] Boral, S., Sidhu, S.S., Chatterjee, P., Chakraborty, S., Gugaliya, A., 2019. Multi-response Optimization of Micro-EDM Processes: A State-of-the-Art Review, in: Kibria, G., Jahan, M.P., Bhattacharyya, B. (Eds.), Micro-Electrical Discharge Machining Processes, Materials Forming, Machining and Tribology. Springer Singapore, Singapore, pp. 293–310. DOI: 10.1007/978-981-13-3074-2_13
[17] Ubaid, A.M., Dweiri, F.T., Aghdeab, S.H., Abdullah Al-Juboori, L., 2018. Optimization of Electro Discharge Machining Process Parameters with Fuzzy Logic for Stainless Steel 304 (ASTM A240). Journal of Manufacturing Science and Engineering 140, 011013. DOI: 10.1115/1.4038139
[18] Raju, L., Hiremath, S.S., 2016. A State-of-the-art Review on Micro Electro-discharge Machining. Procedia Technology 25, 1281–1288. DOI: 10.1016/j.protcy.2016.08.222
[19] Liu, K. & Ferraris, Eleonora & Peirs, Jan & Lauwers, Bert & Reynaerts, Dominiek. (2009). Process capabilities of Micro-EDM and its applications. The International Journal of Advanced Manufacturing Technology.
[20] Choudhary, S.K., Jadoun, R.S Jadoun., 2017. Review Study and Importance of Micro Electric Discharge.Chemical and Process Engineering Research 50.
[21] Tofail, S.A.M., Koumoulos, E.P., Bandyopadhyay, A., Bose, S., O'Donoghue, L., Charitidis, C., 2018. Additive manufacturing: scientific and technological challenges, market uptake and opportunities. Materials Today 21, 22–37. DOI: 10.1016/j.mattod.2017.07.001
[22] Chen, J.-P., Gu, L., He, G.-J., 2020. A review on conventional and nonconventional machining of SiC particle-reinforced aluminium matrix composites. Adv. Manuf. 8, 279–315. DOI: 10.1007/s40436-020-00313-2
[23] Kapoor, J., Singh, S., Khamba, J.S., 2012. High-performance wire electrodes for wire electrical-discharge machining – a review. Proceedings of the Institution of Mechanical Engineers, Part B: Journal of Engineering Manufacture 226, 1757–1773. DOI: 10.1177/0954405412460354
[24] Schneider, M.M., Nowak, T.E., Bastian, L., Katthagen, J.C., Isenberg, J., Rommens, P.M., Müller, L.P., Burkhart, K.J., 2014. Tension band wiring in olecranon fractures: the myth of technical simplicity and osteosynthetical perfection. International Orthopaedics (SICOT) 38, 847–855. DOI: 10.1007/s00264-013-2208-7
[25] Slătineanu, L., Dodun, O., Coteaţă, M., Nagîţ, G., Băncescu, I.B., Hriţuc, A., 2020. Wire Electrical Discharge Machining—A Review. Machines 8, 69. DOI: 10.3390/machines8040069
[26] Srivastava, V., Pandey, P.M., 2012. Performance Evaluation of Electrical Discharge Machining (EDM) Process Using Cryogenically Cooled Electrode. Materials and Manufacturing Processes 27, 683–688. DOI: 10.1080/10426914.2011.602790
[27] Rajurkar, K.P., Yu, Z.Y., 2000. 3D Micro-EDM Using CAD/CAM. CIRP Annals 49, 127–130. DOI: 10.1016/S0007-8506(07)62911-4
[28] Sultan, T., Kumar, A., Gupta, R.D., 2014. Material Removal Rate, Electrode Wear Rate, and Surface Roughness Evaluation in Die Sinking EDM with Hollow Tool through Response Surface Methodology. International Journal of Manufacturing Engineering 2014, 1–16. DOI: 10.1155/2014/259129
[29] Tonietto, L., Gonzaga, L., Veronez, M.R., Kazmierczak, C. de S., Arnold, D.C.M., Costa, C.A. da, 2019. New Method for Evaluating Surface Roughness Parameters Acquired by Laser Scanning. Sci Rep 9, 15038. DOI: 10.1038/s41598-019-51545-7
[30] Gholipoor,A., Shabgard, M., Arghavanian, R, 2020. Experimental Investigation of Recast Layer, Heat Affected Zone and Corrosion Resistance in WEDM of Inconel 617. Journal of Scientific & Industrial Research, 79,701-704.
[31] Prakash, V., Kumar, P., Singh, P., Hussain, M., Das, A., Chattopadhyaya, S., 2019. Micro-electrical discharge machining of difficult-to-machine materials: A review. Proceedings of the Institution of Mechanical Engineers, Part B: Journal of Engineering Manufacture 233, 339–370. DOI: 10.1177/0954405417718591
[32] Lee, H.W., Chen, J.R., Pan, S. P., Liou, H.C., Hsu, P., 2016. Relationship between ISO 230-2/-6 Test Results and Positioning Accuracy of Machine Tools Using Laser TRACER. Applied Sciences, 6, 105. DOI:10.3390/app6040105

[33] Pham, D. & Ivanov, Atanas & Bigot, Samuel & Popov, Krastimir & Dimov, Stefan. (2007). An investigation of tube and rod electrode wear in micro EDM drilling. The International Journal of Advanced Manufacturing Technology. 33. 103-109. DOI: 10.1007/s00170-006-0639-7.
[34] Abdullah, M.A.C., Yahya, A., Shukri, W.N.W.M., 2020. Integrated Control of Electrical Discharge Machining (EDM) using PsoC. J. Phys.: Conf. Ser. 1529, 042087. DOI: 10.1088/1742-6596/1529/4/042087
[35] Tiwari, S.K., Sahoo, S., Wang, N., Huczko, A., 2020. Graphene research and their outputs: Status and prospect. Journal of Science: Advanced Materials and Devices 5, 10–29. DOI: 10.1016/j.jsamd.2020.01.006
[36] L. Szabó, G., 2020. Thermo-Chemical Instability and Energy Analysis of Absorption Heat Pumps. Energies 13, 1966. DOI: 10.3390/en13081966
[37] Lim, H.S., Wong, Y.S., Rahman, M., Edwin Lee, M.K., 2003. A study on the machining of high-aspect ratio micro-structures using micro-EDM. Journal of Materials Processing Technology 140, 318–325. DOI: 10.1016/S0924-0136(03)00760-X
[38] Ling, S., Li, M., Liu, Y., Wang, K., Jiang, Y., 2020. Improving Machining Localization and Surface Roughness in Wire Electrochemical Micromachining Using a Rotating Ultrasonic Helix Electrode. Micromachines 11, 698. DOI: 10.3390/mi11070698
[39] D, T., Kumari, S., Masanta, M., 2018. Ceramic-metal Composite Coating on Steel Using a Powder Compact Tool Electrode by the Electro-Discharge Coating Process. Silicon 10, 1625–1637. DOI: 10.1007/s12633-017-9646-6
[40] Sandford, C., Edwards, M.A., Klunder, K.J., Hickey, D.P., Li, M., Barman, K., Sigman, M.S., White, H.S., Minteer, S.D., 2019. A synthetic chemist's guide to electroanalytical tools for studying reaction mechanisms. Chem. Sci. 10, 6404–6422. DOI: 10.1039/C9SC01545K.
[41] Pramanik, A., Basak, A.K., Littlefair, G., Debnath, S., Prakash, C., Singh, M.A., Marla, D., Singh, R.K., 2020. Methods and variables in Electrical discharge machining of titanium alloy – A review. Heliyon 6, e05554. DOI: 10.1016/j.heliyon.2020.e05554
[42] Efendee, A.M., Saifuldin, M., Gebremariam, M., Azhari, A., 2018. Effect of magnetic polarity on surface roughness during magnetic field assisted EDM of tool steel. IOP Conf. Ser.: Mater. Sci. Eng. 342, 012108. DOI: 10.1088/1757-899X/342/1/012108

Low Carbon Production Inventory Model for Imperfect Quality Deteriorating Items with the Screening Process

S R Singh[1] and Rinki Chaudhary[1*]

[1]C.C.S. Campus, Department of Mathematics, Meerut, 250101, India
***crinkichaudhary67@ gmail.com**

Abstract
Carbon emissions play an important role in global warming. Currently, some countries are alert for reducing the carbon emissions. Basically, many countries use carbon cap & tax policies to reduce emission. Observance this issue in mind, we considered a production inventory model for a manageable carbon emissions rate. In this paper, we consider an imperfect production inventory model for decaying product with carbon emission and the defective rate is product dependent. To select the perfect and imperfect items have been inspected the produced items. The defective goods are sold in an only one consignment at the stop of the 100% screening process at a lower cost. The planned process is validated with the help of a numerical example. A solution process is given to establish the most favorable ordering policy. Finally, Sensitivity analysis of different parameters has been approved.
Keywords: Inventory, Deteriorating items, Carbon emission, Imperfect production process, screening process.

1. Introduction

Global warming is a big environmental issue all over the world. Our environment becoming hot day by day by rise in the level of atmospheric carbon emission. Fossil fuels like coal and oil are increase the level of carbon dai oxcide. Many researchers, industries, and other organizations are presenting supply chain management including greening and low carbon emission, to keep the issue of the environmental protection in mind. Basically, carbon tax and cap policy are the key mechanism to reduce carbon emission in many countries. Carbon emissions occur during the three stages of the model: production, holding, and removal of imperfect goods. The first emanation is the production setup that emits constant amounts of CO_2 throughout the production cycle. The second emanation is the production process which may include melting, assembling, chemical reactions, etc. The third emanation is basic machining operations that depend on the time during which the production process is run. Emissions, during the time in which the inventory is stored and disposed, depend on the inventory level and time, and depend on the inventory quantity of disposed product respectively. S.R. Singh [1] established a paper for decaying products under the effect of inflation and considered inventory level depended demand. Hua *et al.* [2] represented a green approach to stock control that the impacts of a carbon cost related to the ordering and holding cost. This model is the extension of the inventory model with carbon emissions under the cap and trade system. Wahab *et al.* [3] established a model to consider the carbon relaited cost from the onward and backward between the retailer and the consumer. Singh and Saxena [4] developed a reverse logistics model for decaying products and for a single item with two different quality standards with remanufacturing. The supplier reproduces only those used materials, whose quality level exceeds the required level required by the market, and it overcame the remaining products. Toptal *et al.* [5] developed an inventory model with three reductions (1. carbon cap; 2. tax; and 3. cap-and-trade policies) of carbon emission speculation policy. Qin *et al.* [6] presented an inventory policy under carbon tax, and carbon cap and trade policy in which demand depend on the permissible delay period. They found from their model that a carbon related variable have a negative effect on the permissible delay period. Hovelaque *et al.* [7] presented a model with carbon emissions considered emissions and price dependent demand. Sarkar *et al.* [8] established a model for defective products with carbon emissions. Singh and Jawla [9] developed a reverse logistic inventory model for the defective product with conservation technology investment under the learning effect and inflation. Bazan *et al.* [10] developed a vendor-managed inventory with carbon emission & energy cost, consignment stock policy. Tiwari *et al.* [11] presented a model for decaying product for two-echelon supply chain including carbon emission. The authors assumed that the buyer performs a complete screening procedure, and the decay items are withdrawn. Kung *et al.* [12] established an EPQ model for decaying product considered machine breakdown, inspection, and partial back-ordering. The authors produced imperfect products with different rates before and after inspection. Bai *et al.* [13] developed the supply chain inventory model considered the effect of carbon emission reduction policy. Carbon cap and trade policy has been used to reduce that carbon emissions are being generated due to production and holding. In this paper, we consider an imperfect production model for decaying items. The defective rate is production dependent and carbon emission are also considered. To select perfect and imperfect items we apply screening process and then inspection. Section 2 &3 describes notations and assumptions, mathematical modeling of this paper respectively. The solution procedure are given in section 5. The model has been illustrated with numerical solution which is represented in section 6. The effect on the model from some influential parameters is represented on section 7 through sensitivity analysis. Observations of this paper are shown in section 8. Finally, conclusions and scope for future research are given in section 9.

2. Assumptions and Notations

2.1 Assumptions

1. The demand rate for buyer and the production rate for supplier are known and constant.
2. The demand rate is lower than the screening process rate.
3. The unit cost of production depends on produced – items, wear –tear and raw material, progress costs and is of the form, $C_P(P)=C_r+\frac{K}{P}+NP$; Where C_r is raw material charge, K is labor charge, and N is tool wear and tear charge respectively.
4. It is supposed that the defective rate is production depended i.e. $\theta=\theta_0-\theta_1/P$, Where θ_0, θ_1>0.

2.2 Notations

D	Demand rate
P	Production rate, P > D
C_S	Set up cost per cycle
C_P	Production cost per unit
h_c	the cost to a hold the inventory ($/unit/time) in production center
h_m	the cost to a hold of perfect quality items ($/unit/time)
$h_d^{'}$	the cost to a hold of imperfect quality items ($/unit/time)
C_d	Deteriorating cost of the inventory ($/unit/time) in production center
$C_d^{'}$	Deteriorating cost of perfect quality product ($/unit/time)
$C_d^{''}$	Deteriorating cost of imperfect quality product ($/unit/time)
C_{Sr}	Screening cost per unit item
S	Selling price for perfect item ($/unit/time)

S'	Selling price for imperfect item (\$/unit/time)

Carbon emission parameter

C_{PE}	Production emission cost per unit $C_{PE} = e_p E_e T_x$
h_{ec}	Inventory holding emissions cost per unit , $h_{ec} = ve_c E_e T_x$
C_{de}	Disposal cost per ton of waste
e_p	Average electricity consumption for production
v	Space occupied by a unit product(m^3/unit)
T_x	carbon emission tax (\$/tonCO_2)
E_e	Standard emissions for electricity generation (ton CO_2/kwh)
e_c	average warehouse energy consumption per unit product (kWh/unit/year);

3. Mathematical modeling:

We consider a production system (Figure 1 and 2) that produces both perfect and defective items up to a time t_1. Here manufacture perform the screening process to select perfect and imperfect items up to time t_2 with a screening rate (x). Per unit time $(1-x)P$ inventory is to be gathering in screening center until a period t_2.

3.1 Mathematically representation of model in production center

At time t = 0, production starts and continues up to t_1. The produced goods incessantly are sending into the screening cell with the rate Px.

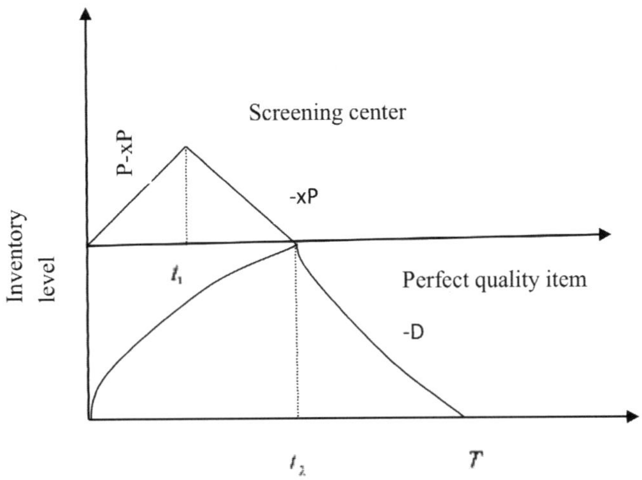

Figure 1: Pictorial representation of production

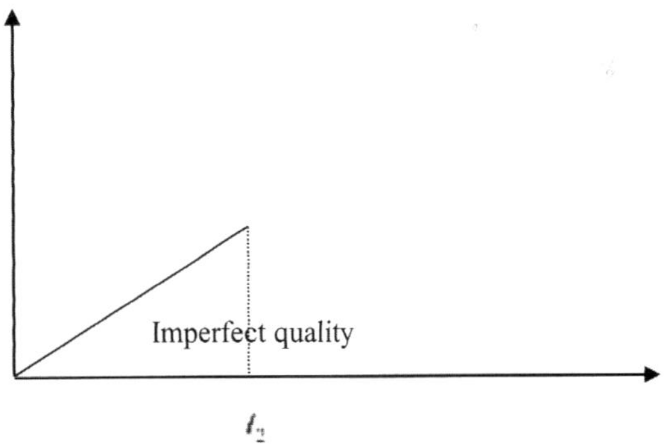

Figure 2: Pictorial representation of Imperfect quality products

The behavior of inventory level w.r.t. to time can be described by the differential equation

$$I'(t) = P - xP \qquad ,0 \le t \le t_1 \tag{3}$$

$$I'(t) = -xP \qquad , \; t_1 \le t \le t_2 \tag{4}$$

With the b. c. $I(0)=0,\; I(t_1)=q$ and $I(t_2)=0$. The solutions of the given equation are as follows

$$I(t) = (1-x)Pt \qquad ,0 \le t \le t_1 \tag{5}$$

$$I(t) = xP(t_2 - t) \qquad ,t_1 \le t \le t_2 \tag{6}$$

Using the condition $I(t_1)=q$ from (5) and (6), We get

$$q = (1-x)Pt_1 = xP(t_2 - t_1) \Rightarrow t_1 = xt_2 \tag{7}$$

3.2 Mathematically representation of model for perfect and imperfect quality item in a screening center

In the screening cell, at time t=0 screening starts and continues up to time t_2. After the screening process, the holding rate of the faultless item is $(1-\theta)Px$ which is larger than the demand of consumer. After satisfying the demand of consumer, the additional inventory $(1-\theta)Px - D(t)$ per unit time is to being stock in best quality item center until period t_2. After the screening process stock becomes zero, due to demand and deterioration at time T. At times t_2 imperfect items are sold in only one consignment with price S'.

So, the variation of perfect quality items $q_1(t)$ w.r.t. time can be described by the following differential equation:

$$I_1'(t) = (1-u)xP - \theta I_1(t) - D, \qquad 0 \le t \le t_2 \tag{8}$$

$$I_2'(t) = -\theta I_2(t) - D, \qquad t_2 \le t \le T \tag{9}$$

Subject to the conditions that $I_1(0)=0,\; I_1(t_2)=I_2(0),\; I_2(T)=0$

$$\frac{dI_d(t)}{dt} = uxP - \theta I_d(t), \qquad 0 \le t \le t_2 \tag{10}$$

s.t. $\quad I_d(0)=0$

$$I_1(t) = \frac{\left[(1-u)xP - D\right]}{\theta}\left(1 - e^{-\theta t}\right) \tag{11}$$

$$I_2(t) = \frac{D}{\theta}\left(e^{\theta(T-t)} - 1\right) \tag{12}$$

$$I_d(t) = \frac{Pux}{\theta}\left(1 - e^{-\theta t}\right) \tag{13}$$

From continuity

$$\frac{\left[(1-u)xP - D\right]}{\theta}\left(1 - e^{-\theta t_2}\right) = \frac{D}{\theta}\left(e^{\theta(T-t_2)} - 1\right) \Rightarrow t_2 = \frac{DT\left(1 + \frac{\theta T}{2}\right)}{\left[(1-u)xP - D\right]} \tag{14}$$

4. **Cost components of this system:**

4.1. Production cost (PC): $= (C_P + C_{PE})\int_0^{t_1} Pdt = (C_P + C_{PE})Pt_1$

4.2. Screening cost (SC) $= C_{Sr}\int_0^{t_2} xPdt = C_{Sr}Pxt_2$

4.3. Set up cost(SUC) : $= C_S$

4.4. Sales Revenue cost (SR): The amount of money which obtained after selling inventory is known as total revenue cost or sales revenue cost. The formula for the cost is given as follows:

$$S\int_0^T D(t)dt + S^{'}\int_0^{t_2} uxPdt = SDT + S^{'}Puxt_2$$

4.5. Holding cost (HC): The amount of money which needs to hold the inventory is known as holding cost. In this model we considered holding cost as time dependent as the expression for finding holdingcost is as follows: $= (h_c + h_{ec})\int_0^{t_2} I(t)dt + (h_m + h_{ec})\int_0^{t_2} I_1(t)dt + (h_m^{'} + h_{ec})\int_{t_2}^{T} I_2(t)dt + (h_d + h_{ec})\int_0^{t_2} I_d(t)dt$

$$= (h_c + h_e)\left((1-x)\frac{Pt_1^2}{2} + xP\frac{(t_2 - t_1)^2}{2}\right) + (h_m + h_e)\frac{\left[(1-u)xP - D\right]}{\theta^2}\left[\theta t_2 + \left(e^{-\theta t_2} - 1\right)\right] +$$

$$(h_m^{'} + h_e)\frac{D}{\theta^2}\left[\left(e^{\theta(T-t_2)} - 1\right) - \theta(T - t_2)\right] + (h_d + h_e)uxP\left[\frac{\theta t_2 + \left(e^{-\theta t_2} - 1\right)}{\theta^2}\right]$$

4.6. Deterioration cost (DC): The amount of money which is lost due to deterioration or damage of inventory is known as deterioration cost. The expression for deteriorating cost is given as follows:

$$(C_d + C_{de})\int_0^{t_2} I_1(t)dt + (C_d^{'} + C_{de})\int_{t_2}^{T} I_2(t)dt + (C_d^{''} + C_{de})\int_0^{t_2} I_d(t)dt$$

$$= (C_d + C_{de})\frac{\left[(1-u)xP - D\right]}{\theta^2}\left[\theta t_2 + \left(e^{-\theta t_2} - 1\right)\right] + (C_d^{'} + C_{de})\frac{D}{\theta^2}\left[\left(e^{\theta(T-t_2)} - 1\right) - \theta(T - t_2)\right] + (C_d^{''} + C_{de})uxP\left[\frac{\theta t_2 + \left(e^{-\theta t_2} - 1\right)}{\theta^2}\right]$$

Total Profit = SR - PC - SC - HC –DC- SUC =

$$\left(SDT + S^{'}Puxt_2\right) - C_{Sr}Pxt_2 - C_S - (C_P + C_{PE})Pxt_2 - \frac{\left(h_d + h_e + C_d^{''} + C_{de}\right)}{2}Pxut_2^2 - (C_d + C_{de})\left[(1-u)xP - D\right]\frac{t_2^2}{2}$$

$$-C_S - \left(h_m^{'} + h_e + C_d^{'} + C_{de}\right)\frac{D}{2}(T - t_2)^2 \quad (15)$$

5. Solution methodology

For the function to be concave, $\frac{\partial AP}{\partial T}=0$, and $\frac{\partial AP}{\partial t_2}=0$ $\exists$ at least one point T & $t_2 \geq 0$ at which the conditions $\left(\frac{\partial^2 AP}{\partial T \partial t_1}\right)^2-\left(\frac{\partial^2 AP}{\partial t_1^2}\right)\left(\frac{\partial^2 AP}{\partial T^2}\right)<0$ must be satisfied. Which shows the optimality condition of total cost function.

6. Numerical examples:

Here, we use MATHEMATICA 9 for numerical examples. We then study the sensitivity analysis of the optimal solution w.r.t. some important parameters.

$C_S = \$300/\text{unit}, T_x = 61.8/\text{tonCO}_2, e_c = 4\text{KWh}/\text{m}^3, E_e = 0.2\text{tonCO}_2/\text{KWh}, \theta_0 = 0.2/\text{units}, \theta_1 = 0.23/\text{units},$

$S = 75/\text{units}, S' = 23/\text{units}, \theta = 0.5/\text{units}, N = 0.02/\text{units}, v = 0.25m^3/\text{unit}, D = \$40 \text{ units}/\text{year}, x = 0.65, K = 400,$

$h_c = 0.5/\text{units}, h_d = 0.13/\text{units}, h_m' = 0.25/\text{units}, C_d = 0.015/\text{units}, C_d' = 0.045/\text{units}, C_d'' = 0.36/\text{units}$

$C_{Sr} = \$0.58/\text{units}, C_r = \$12/\text{units}, C_{de} = \$0.06/\text{units}, e_p = \$20/\text{kwh}$

P= 46units/year , therefore optimal sol is T=1.16754month, Average profit= \$3461.97/month, t_1=0.276452month, t_2=0.368603 month.

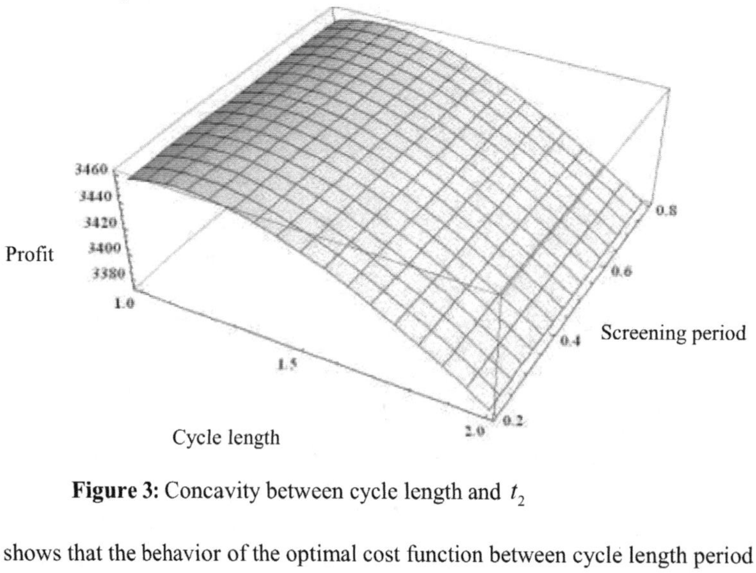

Figure 3: Concavity between cycle length and t_2

Figure 3 shows that the behavior of the optimal cost function between cycle length period and screening period.

7. Sensitivity analysis:

Table 1: Sensitivity analysis w.r.t. different parameter

Variable	% vary of parameter	Value of parameter	t_1 (production run time)	t_2	T(cycle length)	AP (average profit)
	-20	0.52	0.2209380	0.514512	1.13329	3405.73
	-15	0.5525	0.224964	0.469856	1.14114	3417.89
	-10	0.585	0.229358	0.43122	1.14949	3431.21
x	-5	0.6175	0.234179	0.397736	1.15832	3445.84
	5	0.6825	0.245602	0.34312	1.17694	3479.81
	10	0.715	0.252264	0.32078	1.18617	3499.6
	15	0.7475	0.259595	0.300955	1.19454	3521.6
	20	0.78	0.267546	0.283254	1.20086	3546.09
	-20	16	0.221162	0.368603	1.16754	5085.27
	-15	17	0.221162	0.368603	1.16754	4679.44
	-10	18	0.221162	0.368603	1.16754	4273.62
e_p	-5	19	0.221162	0.368603	1.16754	3867.8
	5	21	0.221162	0.368603	1.16754	3056.15
	10	22	0.221162	0.368603	1.16754	2650.32
	15	23	0.221162	0.368603	1.16754	2244.5
	20	24	0.221162	0.368603	1.16754	1838.67
	-20	49.44	0.264705	0.407239	1.28992	3647.69
	-15	51.53	0.259885	0.399823	1.26643	3600.9
	-10	55.62	0.251186	0.38644	1.22404	3542.32
T_x	-5	58.71	0.245177	0.377196	1.19476	3508.02
	5	64.89	0.234378	0.360581	1.14213	3416.2
	10	67.18	0.230733	0.354974	1.12437	3370.74
	15	71.07	0.22492	0.34603	1.09604	3311.07
	20	74.16	0.22061	0.3394	1.07504	3280.83
	-20	3.2	0.264705	0.407239	1.28992	3510.91
	-15	3.4	0.257673	0.39642	1.25565	3498.16
	-10	3.6	0.251186	0.38644	1.22404	3485.78
e_c	-5	3.8	0.245177	0.37719	1.19476	3473.72
	5	4.2	0.234378	0.360581	1.14213	3450.5
	10	4.4	0.229499	0.353076	1.11836	3439.3
	15	4.6	0.22492	0.34603	1.09604	3428.34
	20	4.8	0.22061	0.3394	1.07504	3417.61
	-20	0.16	0.264705	0.407239	1.28992	5134.21
	-15	0.17	0.257673	0.39642	1.25565	4715.64
	-10	0.18	0.251186	0.38644	1.22404	4297.43

E_e	-5	0.19	0.245177	0.37719	1.19476	3879.55
	5	0.21	0.234378	0.360581	1.14213	3044.68
	10	0.22	0.229499	0.353076	1.11836	2627.65
	15	0.23	0.22492	0.34603	1.09604	2210.86
	20	0.24	0.22061	0.3394	1.07504	1794.31
θ	-20	0.4	0.243495	0.374608	1.18656	3466.05
	-15	0.425	0.242514	0.373098	1.18178	3465.04
	-10	0.45	0.241535	0.371593	1.17701	3464.02
	-5	0.475	0.240562	0.370096	1.7227	3463
	5	0.525	0.238625	0.367116	1.16283	3460
	10	0.55	0.237661	0.365632	1.15813	3459.91
	15	0.575	0.236703	0.364158	1.15346	3458.87
	20	0.6	0.235746	0.362686	1.1488	3457.83
v	-20	0.2	0.264705	0.407239	1.28992	3510.91
	-15	0.2125	0.257673	0.39642	1.25565	3498.16
	-10	0.225	0.251186	0.38644	1.22404	3485.78
	-5	0.2375	0.245177	0.377196	1.19476	3473.72
	5	0.2625	0.234378	0.360581	1.14213	3450.5
	10	0.275	0.229499	0.353076	1.11836	3439.3
	15	0.2875	0.22492	0.34603	1.09604	3428.34
	20	0.3	0.22061	0.3394	1.07504	3417.61
P	-20	128	0.228719	0.396881	1.11456	3399.72
	-15	136	0.230977	0.388259	1.12556	3413.55
	-10	144	0.233507	0.380773	1.20609	3428.42
	-5	152	0.236358	0.374273	1.15178	3444.49
	5	160	0.243277	0.363627	1.1855	3481.1
	10	176	0.247502	0.359242	1.20609	3502.18
	15	184	0.252368	0.355349	1.2298	3525.58
	20	192	0.257973	0.351876	1.25711	3551.75

8. Result and discussion:

i. If we increase the screening rate then average profit (AP), cycle length (T), production run time(t_1) also increase because average profit is directly proportional to screening rate from (15) and production run time(t_1) is also directly proportional to screening rate from (7) .

ii. As increase average electricity consumption for production (e_p) , then average profit (AP) decreases but cycle length (T), production run time (t_1) remains the same.

iii. As an increase Production rate (P), then average profit (AP), cycle length (T) and production run time (t_1) are increase because average profit is directly proportional to Production rate.

iv. As an increase Deterioration rate (θ), then average profit (AP), cycle length (T) and production run time (t_1) are decrease.

v. As an increase in carbon tax $\left(T_x\right)$, then average profit (AP), cycle length (T) and production run time (t_1) are decreases.

vi. As increase standard emissions for electricity generation $\left(E_e\right)$, then average profit (AP), cycle length (T) and production run time (t_1) are decrease.

vii. As increase in average electricity consumption per warehouse space unit $\left(e_c\right)$, then average profit (AP), cycle length (T) and production run time (t_1) are decrease.

viii. As increase space occupied by a unit product(v), then average profit (AP), cycle length (T) and production run time (t_1) are decrease.

9. Conclusion:

In this paper, we study an EPQ model with a screening process considering carbon emissions. The objective is to reduce carbon emission costs and maximize profit. This model includes the effect of deterioration, defective products, and carbon emissions. We also examine the effect of energy consumption, and the result confirms the effect of carbon tax regulation in reducing carbon emissions. This paper can be extended by considering technology investment, an adjustable production rate to reduce the possibility of defective and deteriorating goods. Further we can extend included trade credit policy and carbon cap policy with variable parameters.

Reference

[1]. Singh S.R., Kumar N. and Kumari R.(2010). "An inventory model for deteriorating items with shortages and stock dependent demand under inflation for two-shops under one management." *OPSEARCH*, 47(4),311-329.

[2]. Hua G., Cheng, T. C. E., and Wang S. (2011).Managing carbon footprints in inventory management. *International Journal of Production Economics*, 132, 178–185.

[3]. Wahab M.I.M., Mamun S.M.H. and Ongkunarak P., (2011). "EOQ models for a coordinated two-level international supply chain considering imperfect items and environmental impact". *International Journal of Production Economics* 134(1), 151–158.

[4]. Singh S. R., and Saxena N.(2013). "Green Supply Chain Model with Product Remanufacturing under Volume Flexible Environment". *Procedia Technology* 10, 216-226.

[5]. Toptal A., Ozlu H. and Konur D (2014). "Joint decisions on inventory replenishment and emission reduction investment under different emission regulations". *International Journal of Production Research,* 52(1), 243-269.

[6]. Qin J., Bai X., and Xia L. (2015). "Sustainable trade credit and replenishment policies under the cap-and-trade and carbon tax regulations". *Sustainability*, 7, 16340-16361, https://doi.org/10.3390/su71215818.

[7]. Hovelaque V. and Bironneau L. (2015)."The carbon-constrained EOQ model with carbon emission dependent demand". *International Journal of Production Economics,* 164, 285-291.

[8]. Sarkar B., Ganguly B., Sarkar M. and Pareek S., (2016). "Effect of variable transportation and carbon emission in a three-echelon supply chain model". *Transportation Research Part E: Logistics and Transportation,* 91(1), 112–128.

[9]. Singh S. R., and Jawla P. (2016). "A reverse logistics inventory model for imperfect production process with preservation technology investment under learning and inflationary environment". *Uncertain Supply Chain Management,* 4(2), 107-122.

[10]. Bazan E., Jaber M.Y., Zanoni S., 2017. "Carbon emissions and energy effects on a two-level manufacturer-retailer closed-loop supply chain model with remanufacturing subject to different coordination mechanisms". International Journal of Production Economics 183, 394–408.

[11]. Tiwari S., Daryanto Y. and Wee H.M., (2018). "Sustainable inventory management with deteriorating and imperfect quality items considering carbon emissions". *Journal of Cleaner Production* 192, 281–292.

[12]. Kung K.Y., Huang Y.D., Wee H.M. and Daryanto Y.,(2019). "Production-Inventory System for deteriorating Items with machine breakdown, inspection, and partial back ordering". *Mathematics* 7, 616; DOI:10.3390/math7070616.

[13]. Bai Q., Gongb Y., Jin M. and Xue X., (2019). "Effects of carbon emission reduction on supply chain coordination with vendor-managed deteriorating product inventory. *International Journal of Production Economics,* 208, 83-99.

IOP Conf. Series: Materials Science and Engineering 1149 (2021) 012011 doi:10.1088/1757-899X/1149/1/012011

A Comprehensive review on hydroelectric cell: A green energy source for sustainable development

R. K. Singh[1*], A. Manash[1], S.S. Kumar[1], H. K. Satyapal[1], G. Kumar[1] and U. Shankar[1]

[1]Department for NanoScience and NanoTechnology, AKU, Patna, Bihar-800001, India
***rakeshsinghpu@gmail.com**

Abstract A thorough analysis of the hydroelectric cell, in which electric charges are produced by redox reaction, was attempted. As electrodes, zinc & silver are used, with zinc serving as an anode & silver acting as a cathode. Hydroelectric Cell is a capable, environmentally friendly energy generation system that provides several benefits to humans. HEC is a straightforward, low-cost, & novel technique. The recently developed HEC is proving to be superior alternative to environmentally friendly electrical energy products. For the HEC fabrication, various ferrite nanomaterials such as magnesium, nickel, and others were used. The significance of the recent study is that it focuses on green energy generation, which is an alternative source in today's energy demand. In addition, hydroelectric cell output is predictable and consistent, and it has the capability to substitute conventional methods of generating renewable energy in terms of cost & security.

1.Introduction

Energy production is often linked to increased emission levels in the atmosphere. As a result, there is a thrust to develop new eco-friendly sources and devices. Oil (crude) is mostly used for transportation and is substantial source of pollution in the atmosphere. Environmental pollution is wreaking havoc on humanity. As a result, now is the perfect time to innovate and introduce new energy alternatives to minimise emissions. As the global supply of oil (crude) decreases, the usage of EVs (electric vehicles) in place of IC (internal combustion) engines is seen as a feasible future transportation alternative. However, the usage of Li-ion batteries & other EV components creates a tremendous amount of hazardous chemical and electronic waste. As a result, the effect of emerging transportation technologies would exacerbate an already dire situation [1]. The ongoing global energy crisis & the depletion of natural fossil fuels necessitate the invention of alternative fuel systems based on advanced energy materials [2,3]. Solar cells, Lithium batteries, & other electrochemical batteries are currently available energy sources with operating dangers, dumping, and environmental disadvantage concerns [4,5]. Economically, environmentally sustainable, and long-term energy production arrangements focused on renewable, abundant resources such as water, wind, & sun would be highly useful to humanity, but they will have disposal issues [6].Since the last forty years, many research efforts was made on hydrogen-based efficient economical energy advancement using solar/UV irradiation on water in the process of artificial photosynthetic techniques [7-9]. Incredible efforts were made to increase the photoenergy modification performance of photoelectrodes made from cultured oxide in photoelectrochemical cells [10,11]. However, higher costs, the usage of electrolytes, and a lower alteration performance factor when using input solar/UV irradiation were major factors in its lack of industrialization prospects. In this direction, HEC's revolutionary innovation is a capable ecofriendly platform with various benefits for humans. The aim of a recent study is to investigate the various processes that arise in a hydroelectric cell. Secondly, it is alternate low cost, green route as compared to traditional galvanic cell, Daniel, Leclanche & Edison cell. The importance of the recent study is to focus on the green method of energy generation which is an alternative source in today's energy requirement. In addition, hydroelectric cell output is predictable and consistent, and it has the capability to substitute conventional methods of generating renewable energy in terms of cost & security.

2.Literature Review

Before going into detail of HEC, the basis of HEC must be considered, one is water electrolysis & another water photooxidation. According to Fujishima & Honda [12], water electrolysis will occur even without the use of electricity if one of the following three conditions is met. First, In normal conditions, oxygen evolution occurs at potential that is more -ve than hydrogen evolution; Second, hydrogen development happens at a potential higher than oxygen progress under normal conditions; third, the potential for oxygen progress is made more negative, while the potential for hydrogen progression is made more positive, until the former is more negative than the latter.

In 2006, Caser et. al [13] reported on the influence of Si doping on the photo-oxidation efficacy of hematite (Fe_2O_3) electrodes, synthesized via two different techniques. Both techniques yield translucent thin films made of Fe_2O_3 nanoleaflets, showing high degree of photo response to illuminations with global air mass ~ 1.5 solar lights. The morphological as well as photo-response of the thin films got improved with Silicon doping.The revolutionary patent by R. K. Kotnala [14] directed to the basic of HEC & many papers were reported afterwards. This review is an endeavor to understand the HEC phenomenon in detail. In this invention Li replaced Mg ferrite quantifiable based HEC & process of preparation is described. An original galvanic cell technique of producing current (electric) in Di water by Li replaced Mg ferrite HEC was developed. A production method of pellet (ferrite) having Zn anode & Ag inert electrode was also established. The material separates H_2O molecules & conducts ions inside porous ferrite. Separated ions react (electrochemically) with electrodes & forms Zn $(OH)_2$ at anode & H_2 gas at Ag electrode. This HEC generates 5 mA current (short circuit) & 950 mV open cell voltage. Current improved to 20 mA by thermally deposited Zn electrode on ferrite pellet. S. S. Chauhan et. al [15, worked upon HEC made of Li replaced Mg ferrite, as nanoporous circular pellets. The electrodes being used are of elements Zinc and Silver, with Zn working as anode and Ag as cathode. The production of electrical charges being done by associated redox reaction. The physio-absorption of H_2O by the material results into release of H_3O^+ & OH^- ions. The OH^- ions proceed towards Zn electrode to form Zn $(OH)_2$, and release electrons. Whereas H_3O^+ ions move towards silver electrode to release hydrogen gas. Charges form because of the redox reaction, and current begins to flow. RKJ51 generates 920mV exposed cell voltage and 82mA peak current from a 1-inch diameter pellet. Because the voltage provided by HEC is insufficient to power a single LED, it is important to increase the voltage. Hence, for LED lighting current booster circuit are used in Hydroelectric cell. R. K. Kotnala et. al [16] found HEC cell of 17 cm^2 area is capable to produce current (82mA short circuit) & emf (920mV) with max power (74 mW output), which is 3 (three) times more than reported power (1.4 $\mu W/cm^2$ output) generated by water in cement matrix.R. K. Kotnala et. al [17] in another work new atmosphere friendly H_2 manufacturing technique based on newly invented HEC was reported. HEC is the unique, ecofriendly H_2 gas producing method related to photo-biosynthesis, thermo-chemical & photo-electrocatalysis etc. Spontaneous separation of H_2O molecule by HEC help in making of H_2 gas. Because of redox reaction, the pellet (ferrite) combined with Zn & Ag electrodes identified as HEC releases Zn (OH)2 & H2 steam. With improved processing conditions, a nano porous and O2 lacking ferrite pellet was developed. N2 adsorption-desorption isotherm yielded a maximum average nanopore size distribution of 3.8 nm. Unsaturated cation surface hanging bonds and O2 vacancies in ferrite chemidissociate H2O molecule into hydroxide & hydronium ion. By changing the voltage (external) on the HEC, record purity H2 gas is emitted continuously. Chromatography (gas) was used to analyse the collected gas, which was compared to a pure reference H2 gas. This unique method produces large quantity H_2 (1.856 mmol/h) when compared to prevailing photocatalytic ferrite methods. H_2 gas production using HEC is very simple, low cost & unique technique.S. Jain et. al [18] reported O_2-deficient mesoporous magnetite NPs (nanoparticles) was produced using chemical method to manufacture magnetite-based HEC. H_2O molecule chemically separates on Fe cations surface & O_2 vacancies, later the physical absorption of H_2O molecule separation occurs because of charges stuck in mesoporous magnetite. A simple magnetite HEC delivers 50 mA current (peak) when the area is 4.8 cm^2 & max power o/p (38.5 mW). The emf produced by redox reaction at specific electrodes within the cell was 0.77 V. The reduction in cell emf was due to Fe2+ to Fe3+ oxidation, resulting in cells (internal) containing Fe3O4, which obstructs OH ion scattering to the Zn electrode. Zinc hydroxide & H_2 gas are the waste products from HEC which is ecofriendly. Magnetite HEC power generation is a clean, environmentally friendly, and simple technique. Instinctively OH- & H3O+ ions are formed by H_2O molecule dissociation when HEC is partly inserted in Di water. The trapped H3O+ ions in nanopores progress adequate electric field that additionally separates physiosorbed H_2O molecules. Thus, the procedure of H_2O molecule separation is enhanced in a greater way to rise ionic current inside the cell. Inside the cell, electric current and voltage are generated by OH- ion oxidation of the Zn electrode and reduction of H3O+ at the Ag electrode. The HEC cell, with a surface area of 17 cm2, can produce current (82 mA short circuit) and emf (920 mV) with a maximum power output of 74 mW, which is three times higher than the recorded power output of 1.4 W/cm2 (generated by H2O in cement matrix).A. Shukla et. al [19] In this reported work consequence of Li/Mg/Fe doped in TiO_2 has been investigated. Molar ratio of substituent & TiO2 is taken 0.1:0.9. The square pellet (One inch) powder was formed by press (hydraulic) & samples were made by solid state reaction method. HEC was made-up by joining two unlike electrodes zinc & silver on differing end of square pellet. When little quantity of H_2O sprinkled on HEC surface it suddenly dissociates H_2O into OH- and H+ ions without any external source. Dissociated ions transfer to Zn & Ag electrodes and forms Zn $(OH)_2$ & H_2. The potential was established inside the cell & current flows in external circuit because of redox reactions. Mg substituted TiO_2 produces max current (short circuit) 6 mA & 0.925 V (open cell voltage). Mg substitution improved the porosity of TiO_2 which improve uncovered area (surface) in TiO2 for greater H_2O molecule separation, which finally improves current inside the cell.In this study, K. L. Yadav et al. [20] documented an exceptional revolutionary material Ni replaced Li ferrite (LNFO) to produce HEC which generates environmentally friendly electricity. HEC made with LNFO pellets was pre-sintered at 750 °C and 800 °C, respectively, and delivered densities (current) of 3.8 mA/cm2 & 3.6 mA/cm2.According to the PL & XPS spectrum, the decrease in o/p current is due to decrease in defect concentration. Ni replaced Li ferrite based HEC (1.7 mA/cm2) provided densities (current) that are 2 times greater than conveyed in Li replaced Mg ferrite based HEC (1.7 mA/cm2).J. Shah & R. K. Kotnala et. al [21] in this reported study, ZnO nanoparticles was produced using novel, ecofriendly HEC without a current (external) or electrolyte. The HEC had a nano porous Li pellet (ferrite) that interacted with two electrodes: zinc as anode & silver as an inert cathode. When the HEC was injected into Di water, the surface unsaturated cations & O2 vacancies in the nano porous ferrite separated H2O molecules into OH- and H3O+ ions. The OH- ions migrated to Zn electrode and formed Zn (OH)2, while the H3O+ ions advanced to the Ag electrode as H2 steam. Heating the Zn (OH)2 obtained as anode mud to 250 °C transformed the Zn (OH)2 into ZnO nanoparticles. Without using an external current source, this modern, eco-friendly, and low-cost method produced significant yield (0.017 g) of ZnO nanoparticles. S. Jain et al. [22] discovered that using 500 μL of water, hematite - based HEC generates current (30 mA) and emf (0.92 V). The maximum offload power (output) of 27.6 mW generated by a hematite based HEC with an area of 4.84 cm2 is 3.52 times greater than stated power (7.84 mW) produced by a Li-Mg ferrite HEC. For Li 0.2 doped MgFe2O4 sample, Kharbanda et al [23] announced that a distinctive HEC of 2-inch diameter produces peak current (17.1 mA) and voltage (949 mV) with output (maximum) power (15.85 mW).

Table:1 Comparison studies of different authors

Author Name	Material used	Conclusion
Fujishima and Honda	Water electrolysis	Water electrolysis without electrical power.
Caser	Hematite	Highly improved photo response under illumination with global

		air mass equals 1.5 solar light
R. K. Kotnala	Li doped Mg Ferrite	This HEC generates 5 mA current (short circuit) & 950 mV open cell voltage. Current improved to 20 mA by thermally deposited Zn electrode on ferrite pellet
S. S. Chauhan et. al	Li doped Mg Ferrite	RKJ51 whose 1-inch diameter pellet gives 920mV exposed cell voltage and 82mA peak current
A. Shukla	Li/Mg/Fe doped in TiO_2	Mg substituted TiO_2 produces max current (short circuit) 6 mA & open cell voltage 0.925 V.
K. L. Yadav	Ni replaced Li ferrite (LNFO)	Produced densities (current) are 2 times greater than stated in Li replaced Mg ferrite based HEC (1.7 mA/cm2).
R. K. Kotnala	Li doped Mg Ferrite	HEC cell of 17 cm^2 area is capable to produce current (82mA short circuit) & emf (920mV) with max power (74 mW output), which is 3 (three) times more than reported power (1.4 $\mu W/cm^2$ output) generated by water in cement matrix.
R. K. Kotnala	Li doped Mg Ferrite	This unique method produces large quantity H_2 (1.856 mmol/h) when compared to prevailing photocatalytic ferrite methods. H_2 gas production using HEC is very simple, low cost & unique technique.
J. Shah & R. K. Kotnala	Li doped Mg Ferrite	This new, ecofriendly, & cheap method obtained a significant yield of 0.017 g for ZnO nanoparticles without applying an external current source.
S. Jain	Hematite based HEC	Max off-load power (output) 27.6 mW produce by area (4.84 cm^2) hematite based HEC is 3.52 times greater than stated power (7.84 mW) produced by Li-Mg ferrite HEC.
Kharbanda et. al	Li doped Mg Ferrite	HEC of 2-inch diameter

		produces peak current (17.1 mA) & voltage (949 mV) with output (maximum) power (15.85 mW) for Li 0.2 doped MgFe2O4 sample
S. Jain et. al	Li substituted Mg Ferrite	HEC cell of 17 cm^2 area is capable to produce current (82mA short circuit) & emf (920mV) with max power (74 mW output), which is 3 (three) times more than reported power (1.4 $\mu W/cm^2$ output) generated by water in cement matrix.

3. Conclusion

HEC is pathbreaking invented device produces electricity only by water dissociation using specially processed metal oxides. This entirely a new technique from galvanic cell prevailing since 18th century primary cells such as Daniel, Leclanche & Edison primary cell etc. Processing of the cell is very cheap & very sensitive towards electrolysis of H_2O molecules. It is ecofriendly source for producing power & has potential to surpass existing electrochemical batteries. HEC is a capable ecofriendly device with lot of merits for human being. The cell is recyclable, and no harmful element which harm the environment.

4. Acknowledgement

The authors are thankful to Aryabhatta Knowledge University, Patna and Department of Education, Government of Bihar for providing needful facility and establishment of nanotechnology center. The author also expresses sincere gratitude to Dr. R. K. Kotnala, National Physical Laboratory, New Delhi for his innovative research of hydroelectric cell.

5. References

[1] Kotnala R.K, Invention of HEC: A green energy groundbreaking revolution, J Phys Res Appl 2018, 2:1.
[2] Muhumuza R, Zacharopoulos A, Mondol JD, Smyth M, Pugsley A. Energy consumption levels and technical approaches for supporting development of alternative energy technologies for rural sectors of developing countries. *Renew Sustain Energy Rev*. 2018; 97: 90- 102.
[3] Sivula K, Formal FL, Grätzel M. Solar water splitting: progress using hematite (α-Fe2O3) photoelectrodes. *ChemSus Chem*. 2011;4(4):432-449.
[4] Kabir E, Kumar P, Kumar S, Adelodund AA, Kim K. Solar energy: potential and future prospects. *Renew Sustain Energy Rev*. 2018; 82: 894 - 900.
[5] Winslow KM, Laux SJ, Townsend TG. A review on the growing concern and potential management strategies of waste lithiumion batteries. *Resour Conserv Recycl*. 2018; 129:263-277.
[6] Acar C, Dincer I, Naterer GF. Review of photocatalytic water splitting methods for sustainable hydrogen production. *Int J Energy Res*. 2016;40(11):1449-1473.
[7] Li H, Yu Y, Starr MB, Li Z, Wang X. Piezotronic - enhanced photoelectrochemical reactions in Ni (OH)2-decorated ZnO photoanodes. *J Phys Chem Lett*. 2015;6(17):3410-3416.
[8] Wu ZS, Zhou G, Yin LC, Ren W, Li F, Cheng HM. Graphene/metal oxide composite electrode materials for energy storage. *Nano Energy*. 2012;1(1):107-131.
[9] Li H, Li Z, Yu Y, et al. Surface- plasmon- resonance- enhanced photoelectrochemical water splitting from Au- nanoparticledecorated 3D TiO2 nanorod architectures. *J Phys Chem C*. 2017;121(22):12071-12079.

[10] Hou Y, Zuo F, Dagg A, Feng P. Visible light-driven α-Fe2O3 nanorod/graphene/Biv1−xMoxO4 core/shell heterojunction array for efficient photoelectrochemical water splitting. *Nano Lett.* 2012;12(12):6464-6473. JAIN ET AL. 11
[11] Yi SS, Wulan BR, Yan JM, Jiang Q. Highly efficient photoelectrochemical water splitting: surface modification of cobalt-phosphate-loaded Co3O4/Fe2O3 p–n heterojunction nanorod arrays. *Adv Funct Mater*. 2019;29(11):1801902.
[12] A. Fujishima, K. Honda, Electrochemical Photolysis of Water at a Semiconductor Electrode, NATURE *VOL*. 238 JULY 7 1972.
[13] Ilkay Cesar, Andreas Kay, Jose A. Gonzalez Martinez, and Michael Gratzel, Translucent Thin Film Fe2O3 Photoanodes for Efficient Water Splitting by Sunlight: Nanostructure-Directing Effect of Si-Doping, J. AM. CHEM. SOC. 2006, 128, 4582-4583.
[14] R. K. Kotnala, Li substituted Mg ferrite material based Hydroelectric cell and process for preparation thereof, United States Patent publication, Pub No. US 2016/0285121 A1,2016.
[15] S. S. Chauhan, A. Gaur et. al, Application of HEC for LED lamp, 2019 Innovation in power and advanced computing technology (I - PACT).
[16] R. K. Kotnala, J. Shah, Green hydroelectric energy source based on water dissociation by nanoporous ferrite, Int. J. Energy Res. 2016, 40: 1652 -1661.
[17] J. Shah, S. Jain et. al, A facile non-photocatalytic technique for hydrogen gas production for hydroelectric cell, International journal of hydrogen energy 42 (2017) 30584-30590.
[18] S.Jain, J. Shah et. al, Environment – Friendly mesoporous magnetite nanoparticles based HEC, J. Phys. Chem. C, 2018, 122, 5908 -5916.
[19] A. Shukla, R. Das et. al, Substitution of Ti dioxide HEC induces higher green energy, AKJ Manuscript.
[20] S. Jain, J. Shah et. al, Ni substituted oxygen deficient nanoporous Li ferrite based green energy device hydroelectric cell, Journal of alloys and compounds, 2020.
[21] J. Shah, R. K. Kotnala, Rapid green synthesis of ZnO nanoparticles using a HEC without an electrolyte, Journal of physical and chemistry of solids.
[22] S. Jain, J. Shah et .al, Significance of interface barrier at electrode of hematite hydroelectric cell for generating ecopower by water splitting, *Int J Energy Res*. 2019;1–13.
[23] V. Verma, P. Kharbanda, T. Madaan et. al, Ferrites: magnetic materials as an alternate source of green electrical energy, Heliyon 5 (2019) e01151.

A Methodological Study for the Stress Analysis to Evaluate Single Lap Adhesive Joint

Sagarsingh Kushwah[1], Meet Bhatt[1], Chirag Desai[1], Shreyashkumar Parekh[1] and Poojan Joshi[1]

[1] Department of Mechanical Engg, R. N. G. Patel Institute of Technology, Bardoli, Surat, Gujarat, India

sagarsinghkushwah123@gmail.com

Abstract. The use of adhesive bonding is increased in load-bearing applications in place of conventional joining techniques such as bolted or rivets joints. The high-stress concentration occurs at the end of an adhesive-bonded lap joint so all failures begin at the end of the overlap region and get fractured in the end. The single-lap adhesive-bonded joint was used in the current study to examine the peel stress and shear stress results along the overlap region. The comparison of the analytical and experimental result was made to validate the results along with it the load vs displacement curve was obtained from results. Also, stress vs strain curve and peel stress vs overlap length of the one mm thickness of testing specimen was obtained.

1. Introduction

In various applications such as the marine industry, aerospace, cars, trains and civil constructions, adhesive bonding is utilized as a load-carrying joint in structural jointing. Single lap adhesive joint offers many benefits, such as saving time, economizing costs, good damping properties and very high strength in comparison to other conventional joining processes. The load is transmitted smoothly from one plate to another through an adhesive in this joint. An adhesive is a material that binds two or more separate materials together and resists their separations when applied to one or both surfaces [1]. The use of adhesives offers many advantages like less stress concentration, uniform stress distribution, less weight, and ease of manufacture compared to mechanical fastening. It also improves the aesthetic design as well as ergonomic design. The adhesive joint is more widely used over mechanical joints to connect structural components due to more uniform interfacial stress distribution and resistance towards fatigue failure. In various types of joints, structural adhesives are used. The adhesive joint commonly used is illustrated in Figure 1. Among these is the simplicity of the single-lap adhesive joint. The adherent can be a composite with either metal or fibre. A single lap joint is the most common and diverse research is conducted to analyze its strength and characteristics.

Ahmed, E. and Tehami, A. showed that adhesive thickness, curing conditions, adherend thickness, bond length, and joint end geometry are the factors that affect the strength of the adhesive bond [2]. The sizes of the area to which the adhesive is applied and the adhesive used are factors that determine the applications of various adhesives. The adhesive can be applied to either one material or both of the materials being bonded. Shiuh-ChaunHer reveals that in adhesive bonding, the parts are aligned and pressure is applied to assist in adhesion and to eliminate the air bubbles. The applying tools for the adhesive are brushes, rollers, spray guns, applicator guns using pellets. These can be operated either manually or automatically [3].

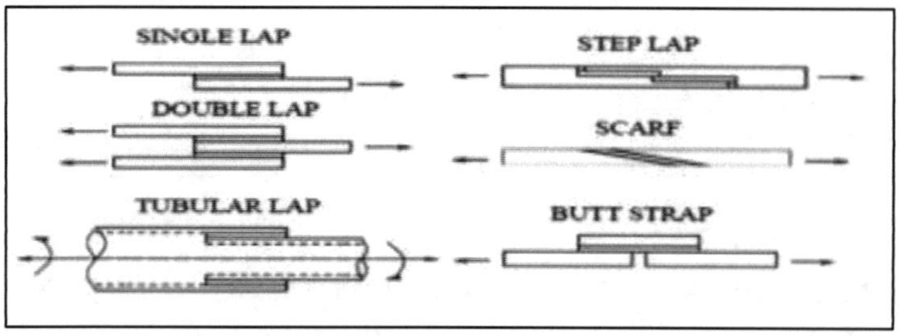

Figure 1 Types of lap joint [14]

The wide applications of composite materials are spacecraft, solar reflectors, satellite structures, radar, rocket engines, etc. A defined degree of anisotropy in the properties of the laminate varying fibre orientation. It is also used in bearing material, pressure vessels, electrical devices, truss members, etc. CRPF is used extensively in Formula one [4].

2. Literature Review

For the determination of the mechanical properties of various adhesives, the single-lap adhesive joint is used as a standard test model. The single-lap adhesive joint is also widely exploited in various applications due to its characteristic's property of efficiency and simplicity.

In the past few years, many researchers investigated the stress analysis of a single lap joint. A variety of studies have been performed on mathematical modelling and stress analysis of adhesive lap joints. Nabil M. Chowdhury has shown that the difficulty of the mathematical model is high, so the solution is performed either numerical or analytical methods. In adhesively bonded joints, the calculation of the stress distribution can be done by various analytical methods [5]. X.Q. Zhang has shown that the analytical approach has been found difficult due to its geometry and complex nonlinear function of material property. Several simplifications and assumptions are made in a wide range of developed solutions [6].

Volkersen method was the first analytical method used to calculate the stress analysis of bonded joints which was introduced in 1938 by Volkersen. This approach describes the principle of differential shear. Volkersen didn't consider the bonding effect because of the eccentric load path [7].

Further research has been carried out by Goland M. and Reissner E. in 1944. The effect of rotation of adhered was initially considered by them. The problem was further divided into two main sections. (1) By applying the principle known as finite deflection principle, applied to bent plates which are cylindrical in geometry, the measurement of joint loads at the edges has been carried out, and (2) Due to the impact of the applied loads, the determination of joint stress has been carried out [8].

In 1973, adhesive plasticity was considered by Hart-Smith, L. J.The problem was further divided into four parts. The determination of bending moment M was the first step. The effects of the bending stress on the strength of the adherents are considered in the second stage. In the third step, an analysis of the adhesive shear stress distribution can be carried out using the elastic-adhesive formulation. The problem of peel stress is addressed in the fourth final stage [9].

Ojalvo I.U and Eidino H.L. gave a new model based on the Reissner and Goland model. The modifications were made by them in the shear stress equations by modifying certain coefficients which considered the new boundary conditions for bonded peel stress calculation, together with the addition of new terms in the differential equation. The prediction of the variations in shear through the thickness of the bond was the main work done [10, 11].

According to the literature review, it reveals that the researchers have carried out most of their work on various lap joints, which are bonded to single adhesives. The overlapping area is the focus of the lap joint analysis. The prime factors for the failure of the lap joint are shear and peel stress. The adhesive is applied in different industries due to its favourable mechanical properties, resistance to chemical and atmospheric conditions, and low weight. The application of adhesive makes it possible to obtain construction materials with high stiffness and strength.

3. Problem Definition

A standard bonded single lap joint is taken into account for the initial study. This lap joint is made up of various analyses. Two identical adherents were taken which were of uniform thickness. They were bonded to the adhesive, which also had a standardized thickness. The thickness and is considered for upper and lower adherents. Figure 2 demonstrates the modelling of the overall geometry of the joint.

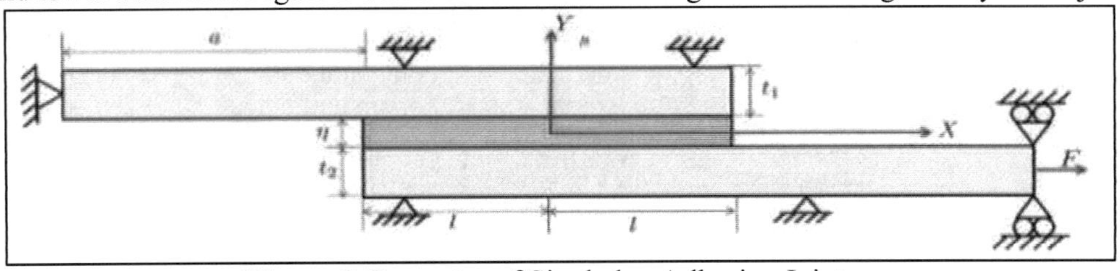

Figure 2 Geometry of Single lap Adhesive Joint

Table 1 Geometry of Single lap Adhesive Joint

Symbol	Single Lap Joint Parameter	Value
t_1	Upper Adherend thickness	1 mm
t_2	Lower Adherend thickness	1 mm
η	The Adhesive thickness of a single adhesive model	1 mm
α	Length of adherend without overlap region	100 mm
21	Length of Overlap	30 mm
F	Applied load	As per the result

The aim of using adhesive is to maximize the strength and minimize the stress concentration of the joint. An attempt is made to provide simple as well as accurate solutions in the adhesive region to determine the peel stress and shear stress. For the particular lap joint, load values have been evaluated to study the behaviour of a single lap adhesive joint. After that, the value of shear and peel stress value can be obtained using numerical solutions through two-dimensional finite element analysis.

4. Material Selection & Methodology

Typically, aluminium, epoxy, FRP and steel, etc. are generally used for a single lap joint. In the present work, epoxy is selected as it has a wide variety of applications in the automobile industry and aerospace engineering. There is a range of thickness is available, but the thickness of 1mm with a length of 130 mm is taken at present work. The Laser cutting operation is carried out to prepare fibre epoxy. Two specimens of fibre epoxy having an identical dimension of thickness 1mm, width 25 mm, and length 130 mm used for making adhesive joints.

Table 2 Material property of composite material

Adherend Material	Young Modulus (E)	Poisson Ratio (v)
Fiber Epoxy	101.97 GPa	0.6

For adhesive work, both adhesives (epoxy) should be selected from each other. Araldite is selected as adhesive because it has wide application in the field of automobile and engineering components. Araldite gives a resilient bond. The key properties of Araldite are high shear strength and peel strength, good temperature resistance, and toughened paste.

Table 3 Materials properties of adhesive

Adhesive Material	Young Modulus (E)	Poisson Ratio (v)
Araldite	4.89 GPa	0.35

The specimen preparation involves the Epoxy sheet which is cut by performing the laser cutting operation. The dimensions should be as per requirement i.e., 130*25 mm as shown in figure 3. The Dogbone is prepared by the same sheet by laser cutting operation as shown in figure 4.

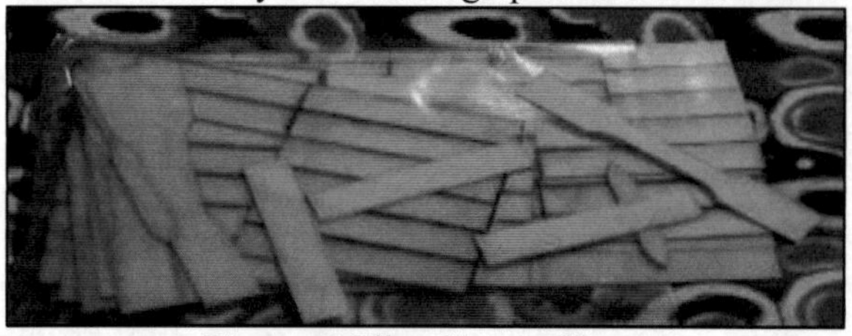

Figure 3 Fibre Epoxy Sheet

Figure 4 Dogbone Specimen

The model typically used for tensile specimens is a dog bone form specimen in accordance with the EN ISO 527-2 standard, which is demonstrated in figure 4 [12, 13].

The Universal Testing Machine based on the present work is having a capacity of 100 tones. For the determination of the young's modulus, the testing speed usually used is 1 mm/min but for failure testing higher-rated can be used. To determine the stress-strain curve, the load and displacements are measured. The procedure of testing is accompanied by grasping the specimen in the longitudinal direction until failure. A specialized fixture is used to grip the specimen in the longitudinal direction as the size of the specimen is too small and it is relatively easy to use for the measurement of adhesive displacement. However, if possible then it is recommended to use a non-contact device to avoid any kind of interference. Setting the appropriate requirement for the machine is necessary for further processes. After that, start the test after feeding all requirements until the specimen breaks. Remove the specimen from the gripper after completing the test. The output data like load-displacement values should be taken into account [14-16].

5. Experimental Result

The result for Dogbone, for the material Fiber epoxy peak load value, reaches up to 460N. This analysis would also use both an analytical and a numerical model to assess the effects of peel and shear stress.

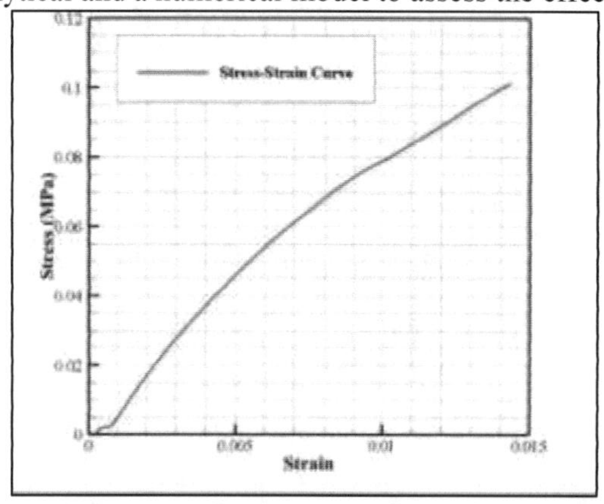

Figure 5 Stress vs Strain curve of a testing specimen

IOP Conf. Series: Materials Science and Engineering 1149 (2021) 012012 doi:10.1088/1757-899X/1149/1/012012

For testing specimen, the Stress vs Strain curve has been shown in figure 5. The specimen's thickness was 1.0 mm. The Result of the specimen, as shown in figure 6, for the adhesive Araldite peak load value, reaches up to 5021.05 N, which is shown in the load vs displacement curve (Figure 7). This analysis would also use both an analytical and a numerical model to assess the effects of peel and shear stress.

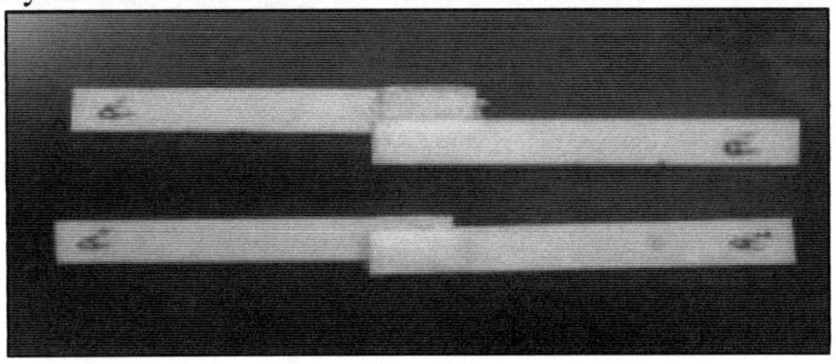

Figure 6 Tested specimen of single lap joint

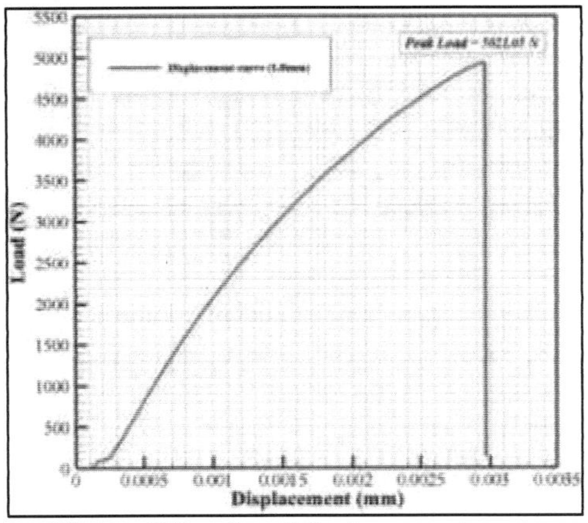

Figure 7 Load vs Displacement curve

From the above graph, the peak load value is 5021.05 N for 1.0 mm of thickness.

6. Validation of Shear Stress & Peel Stress Results :

In a single lap adhesive joint, the distribution of shear stress along the overlap joint is shown in figure 8. It has been found that the occurrence of shear stress is maximum at the free end of the adhesive region that is near to the adherend. The shear stress is maximum at both ends, whereas in the middle of the adhesive there is a decrement of the shear stress. In the case of finite product, the differential shear is more at the edges under loading conditions, such that the stress distribution is not symmetrical in both the right and the left interfaces.

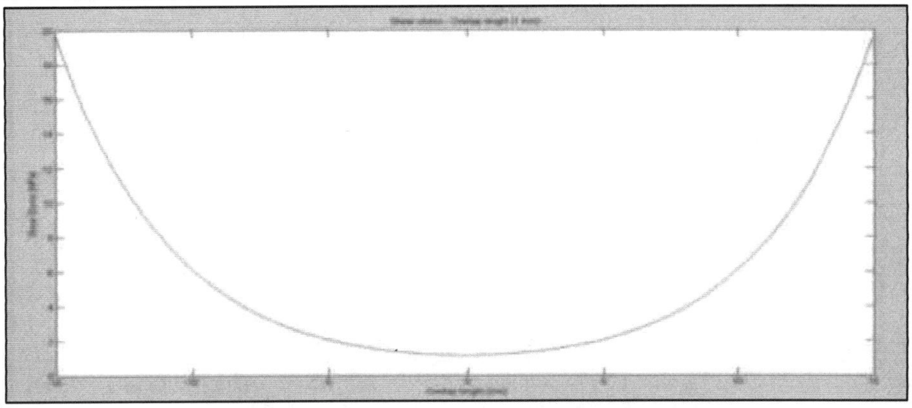

Figure 8 Shear stress vs overlap length (1 mm thick)

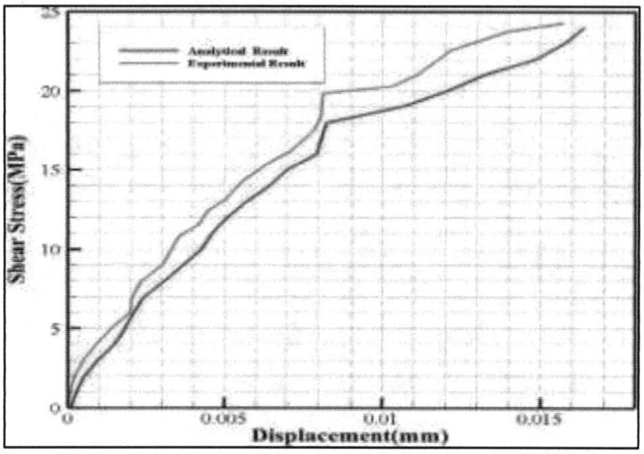

Figure 9 Comparison of Analytical & Experimental Result

The experimental results are compared with the analytical results (Appendix A) in figure 9. The deviation between both curves occurs due to various errors caused by personal bias. The graph of Peel stress vs Overlap Length for 1.0 mm adhesive thickness is shown in figure 10.

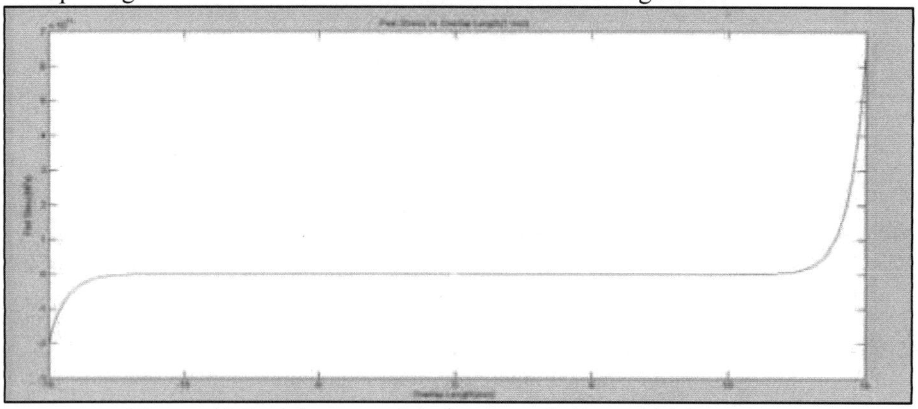

Figure 10 Peel Stress vs Overlap Length (1 mm thickness)

7. Conclusion

It is noteworthy that the critical factor responsible for the design of the lap joint is shear stress.

During the present work, analytical and experimental results indicate the accuracy of work has been compared.

The thickness of the adhesive must be improved or optimized to enhance mechanical stability and increase the resistance to mechanical stresses. The increase in adhesive thickness allows the residual shear stress to decrease. Whenever the thickness of the adhesive increases the residual stress decreases.

References

[1] Kushwah S et al 2021 Optimization of coil spring by finite element analysis method of automobile suspension system using different materials Materials Today: Proceedings **42P2** 827-831. https://doi.org/10.1016/j.matpr.2020.11.415

[2] Ahmed E and Tehami A 2018 Evaluation of Strength of Adhesively Bonded Metallic Single Lap Joint with Different End Geometries *SAE Technical Paper* **01** 1238.

[3] ChaunHer S 1999 On Stress analysis of adhesively-bonded lap joint Composite Structures **47** 673-678.

[4] Jayadeep U B 2020 Lumped parameter models for adhesive contact mechanics Int. J. Adv. Eng. Sci. Appl. Math. **12** 44–50. https://doi.org/10.1007/s12572-020-00265-7

[5] Chowdhury N et al 2016 On Experimental and finite element studies of bolted, bonded and hybrid step lap joints of thick carbon fiber/epoxy panels used in aircraft structures Composites Part B **100** 68-77.
[6] Zhang X Q et al 2015 On Investigation of the fatigue life of pre-and post-drilling hole in dog-bone specimen subjected to laser shot peening Materials and Design **88** 106-114.
[7] Silva D et al 2009 Analytical models of adhesively bonded joints Part I: Literature Survey International Journal of Adhesion and Adhesives **29** 319:330.
[8] Goland M and Reissner E 1994 The Stresses in Cemented Joints Journal of Applied Mechanics **11** A17-A27.
[9] Hart-Smith L J 1973 Adhesive-Bonded Single-Lap Joints NASA 112236.
[10] Ojalvo I U and Eidino H L 1978 Bond Thickness acts upon Stresses in Single-Lap Adhesive Joints American Institute of Aeronautics and Astronautics **16** 204:211.
[11] Tong L and Jain L K 1996 Analysis of adhesive bonded composite lap joints with transverse stitching Appl. Compos. Mater 343-365.
[12] Komorek A Godzimirski J 2020 The Influence of Selected Adhesive Properties and the Manner of Surface Preparation upon Impact Strength of Block Adhesive Joints Strojnícky časopis - Journal of Mechanical Engineering.
[13] Lucas F M et al 2018 Handbook of Adhesion Technology 2nd ed. Springer.
[14] Sampathkumar N Three Dimensional Geometrical and Material Nonlinear Finite Element Analysis of Adhesively Bonded Joints for Marine Structures University Of Southampton.
[15] Agarwal K M et al 2021 Mechanical behaviour of Aluminium Alloy AA6063 processed through ECAP with optimum die design parameters *Advances in Materials and Processing Technologies* DOI: 10.1080/2374068X.2021.1878705
[16] Agarwal K M et al 2020 Deformation analysis of Al Alloy AA2024 through equal channel angular pressing for aircraft structures *Advances in Materials and Processing Technologies* DOI: 10.1080/2374068X.2020.1834756

Appendix A

```
clear all
clc
%-----------------------------------------------------------
% Inpu Data
%-------------------------------------------------------------
ei = 101970 ;
e0 = 101970;
ti = 1;
t0 = 1;
l = input('Enter the value of overlap length: ');
x = -l:0.1:l;
P = 5021.05 /19;
eta = 1;
E = 4890;
nu = 0.35;
G = E/(2*(1+nu));
%---------------------------------------------------------
y = sqrt((G/eta)*((1/(ei*ti))+ (1/(e0*t0))));
a = (P*y)/2;
b = (ei*ti - e0*t0)/(ei*ti + e0*t0);
tau = a*((cosh(y*x)./sinh(y*l))-b*(sinh(y*x)./cosh(y*l)));

figure(1)
plot(x,tau);
xlabel('Distance Along Overlap (mm)')
ylabel('Adhesive Shear Stress (MPA)')
```

Investigating structural, magnetic and multiferroic properties of gadolinium substituted strontium hexaferrite ($SrFe_{12-x}Gd_xO_{19}$)

Singh S Kumar[1], Rakesh K Singh[1,*], Aniket Manash[1], Gaurav Kumar[1], Uday Shankar and Harendra K Satyapal[1]

[1]Aryabhatta Centre for Nanoscience and Nanotechnology, Aryabhatta Knowledge University, Patna, 800001, India
* **rakeshsinghpu@gmail.com**

Abstract. An economical and citrate precursor based sol-gel technique is employed for pure phase synthesis of Gd^{3+} substituted M-type strontium hexaferrite ($SrFe_{12-x}Gd_xO_{19}$) for x = (0.25 - 1.00). The XRD patterns reflected all prominent planes related to hexagonal symmetry ($P6_3$/mmc space group). The crystallite size and induced strains is determined using Williamson-Hall plot. Lattice constants (a) and (c) shows minor increments with Gd^{3+} concentration. Particle distributions in SEM micrograph agree with W-H plot analysis. FTIR analysis is in consistent with X-ray intensity results. The 'Law of Approach' to saturation is applied to estimate parameters such as, magnetization (M_s) of order 40.91 to 57.7 emu/g and retentivity (Mr) ranging from 26.3 to 32.5 emu/g. A remarkable increase in coercivity (H_C) is observed with increasing Gd^{3+} composition, with a value of 2500 to 6310 Gauss. Moreover, incorporation of Gd^{3+} in $SrFe_{12-x}Gd_xO_{19}$ lattice hinders electrical leakage profitably to yield remanent (P_r) and coercive polarization (P_c) of order 0.19 $\mu C/cm^2$ and 17 kV/cm respectively. Thus sol-gel method proves to be useful in synthesizing nanomaterials of enhanced physical properties along with it being cost effective. Therefore, we are reporting tuning of magnetic and multiferroic properties of $SrFe_{12}O_{19}$ due to crystal strain mediated by Gd^{3+} doping at Fe^{3+} site.

1. Introduction

M-type hexaferrites are included in a category of magnetic materials possessing exceptional properties like high coercivity (H_c), magneto anisotropy and high magnetization (M_s) [1]. Hexaferrites are effectively being used as dense recording media, permanent magnets, magneto optical instruments etc. [2, 3, and 4]. The reported H_c and M_s for hexagonal ferrites are 6700 Oe and 72 emu/g respectively. The ferromagnetic to paramagetic transformation point is about 450 °C [5]. The strontium hexaferrite, ($SrFe_{12}O_{19}$) have hexagonal symmetry and included in *$P6_3/mmc$* space group [6]. $SrFe_{12}O_{19}$ is ferromagnetic in nature. The unit cell of $SrFe_{12}O_{19}$ has S and R blocks arranged in a superimposed manner. Its crystal unit have the ordering as RSR*S*, in which S* and R* represents 180 degree rotated S and R blocks [7]. The physical property of $SrFe_{12}O_{19}$ is influenced by domains such as, Fe^{3+} ions occupancy (tetra or octahedral) in the crystal, element being doped and the synthesis method [8]. Researchers have worked up on enhancing physical characteristics of $SrFe_{12}O_{19}$ by substituting with cations such as Sm^{3+}, Mn^{3+}, Ce^{2+}, Al^{3+}, etc. [9-12]. We have preferred citrate based sol-gel method, as this technique is economical and has the benefit of homogeneous particle distribution in synthesized samples [13]. The variant ions replacement develops crystal strains in unit cells, which influences physical properties. Therefore, in this paper we have explored crystallographic changes in $SrFe_{12}O_{19}$ with varying

Gd^{3+} compositions for refining structural, magnetic and multiferroic properties of $SrFe_{12}O_{19}$. The refined physical parameters suggest this material to be effective as highly dense recording media and for dielectric applications. To the best of our beliefs, there is limited literature reported on Gd^{3+} substituted $SrFe_{12}O_{19}$ and discussions on its multiferroic properties along with magnetic property have not been addressed before.

2. Experimental details

The purity grade $SrFe_{12-x}Gd_xO_{19}$ for x = (0.25 - 1.00) was prepared using an economical and chemical based sol-gel technique. The starting nitrates used as reactants were Sr $(NO_3)_2$, Gd $(NO_3)_3$, Fe $(NO_3)_3$ (99.9% purity). A definite amount of these reactants are mixed in DI water in individual beakers. All these solutions are then converted into the citric acid solution. Next, pH of this final solution is shifted to neutral by precisely adding ammonia. Thereafter, we put this mixed solution on hot plate set at 80 °C for (3-4) hours with vigorous stirring, till liquid becomes jelly type. This thick jelly like substance was put in hot air chamber at 120 °C for 1 hour for drying. It is observed that when temperature inside air chamber was increased to 250°C, gel starts swelling like a cake. We wait for 30 minute in this condition so as to allow total volumetric expansion of the jelly product. Then sample is allowed to cool and is collected in small containers after crushing down with mortar-pestle. A sufficient amount of each sample was annealed at 750 °C for 2 hours with a ramp rate of 2°C/min in microprocessor based furnace. The XRD patterns were recorded at a wavelength (λ=1.5406 Å) with the help of the X-ray diffractometer (Bruker). Magnetic hysteresis (M-H) curves were recorded at 300 kelvin using Vibrating Sample Magnetometer, Lakeshore (Model 7500) equipped with creating electromagnetic field of ∓3.2 Tesla. The morphology of samples was checked with Scanning Electron Microscope (Carl Zeiss). The site occupancy of ions within unit cell was identified using FTIR (Perkin Elmer). The ferroelectric polarization patterns were obtained with Radiant Tech Precision 10 kV Multiferroic loop analyzer.

3. Results and discussion

3. 1. X-ray diffraction analysis

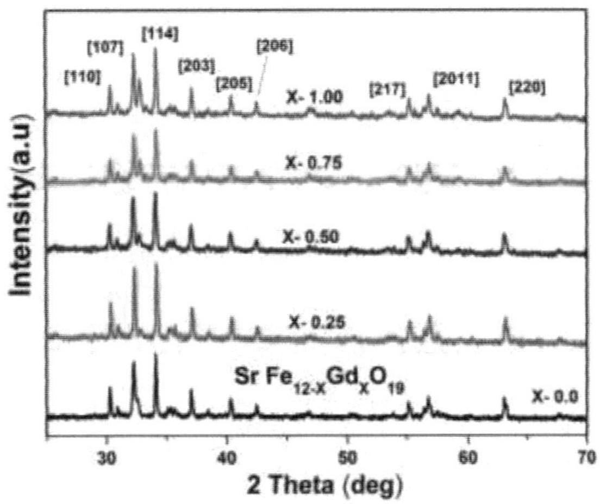

Figure 1. X-ray intensity pattern for $SrFe_{12-x}Gd_xO_{19}$ for (x = 0.25 - 1.00).

As we can see in figure 1, XRD patterns justify pure phase synthesis of sample containing all crystal planes reported for hexagonal crystal structure (*P6₃/mmc* space group). Such intensity patterns have also been reported by research groups [14]. We can see in Fig. 1, that broadening of the prominent XRD peaks decreases, with increment in Gd^{3+} compositions, suggesting increment of particle size with dopants. It can be observed that the intensity of signature planes (107) and (114) increased with Gd^{3+} compositions. Intensity peak near Bragg's angle $2\theta = 32.5$, 35.3, planes (104) and (108) belongs to undesirable phases (α-Fe_2O_3). These intermediate phases were present in samples with higher Gd^{3+} content(x = 0.75, 1.0). Further, heat treatment of samples at 750°C for 2 hrs do affect lattice parameter (a), (c), and so cell volume increases. The concern structural parameters are tabulated in table 1.

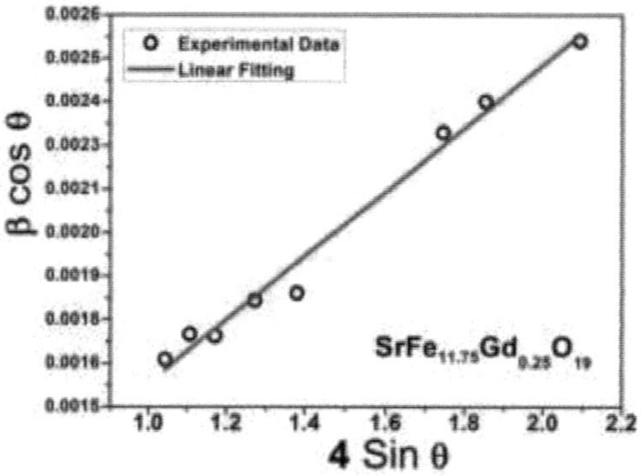

Figure 2. Williamson-Hall Plot for 0.25 mole Gd doped $SrFe_{12-x}Gd_xO_{19}$.

The increased lattice constants (a) and (c) alters the distance between Fe^{3+} ions, causing enhanced ions interactions which further tunes magnetic parameters (sec 3.4). The Williamson-Hall plot is analyzed for finding out lattice strain values. In figure 2, $\beta\cos\theta$ vs. 4 $\sin\theta$ plot is lineally fitted for 0.25 moles doped strontium hexaferrite. Slope of the lineally fitted data (red line) directly quantifies value of lattice stain. For finding crystallite size, y-intercept value is equated to $k\lambda/D$, where D denotes crystal size. These results are included in table 1. We can notice a monotonic increment of strain with the increment of Gd^{3+} ions in unit cell of $SrFe_{12}O_{19}$. The crystal strain of order 3.12×10^{-3} is calculated (W-H plot) for $SrFe_{12}O_{19}$ substituted with 1 mole Gd^{3+}ions. The ionic radius of Gd^{3+} ions (1.05 Å) being almost twice with respect to Fe^{3+} (0.55Å) ions can be the expected reason [15]. Prominent planes in the XRD patterns were analyzed to calculate lattice constants (a), (c) using equation (1). Planes selected for substitution in equation (1) are [107], [114], [203], [205] and [206].

$$\frac{1}{d^2} = \frac{4}{3}\left(\frac{h^2+hk+k^2}{a^2}\right) + \frac{l^2}{c^2} \quad (1)$$

After getting lattice constants values, we calculate the lattice volume as

$$V = \frac{\sqrt{3}}{2}\, a^2 c \quad (2)$$

We can deduce from table 1, that the lattice constants (a), (c) and crystallite size increases with increment of Gd^{3+} content in $SrFe_{12}O_{19}$, due to the increased crystal strains in samples [16].

Table 1. Structural details of $SrFe_{12-x}Gd_xO_{19}$ for (x = 0.25-1.00).

$SrFe_{12-x}Gd_xO_{19}$	*Crystallite size*	Strain*(10^{-3})	*a=b(Å)*	*c(Å)*	*cell volume(Å³)*
X= 0.00	44.1	1.01	5.844	22.83	675.238
X= 0.25	47.8	1.37	5.857	22.98	682.701
X= 0.50	52.2	1.59	5.871	23.08	688.954
X= 0.75	56.4	2.71	5.878	23.13	692.073

X= 1.00	62.1	3.12	5.881	23.21	695.197

3.2 SEM analysis

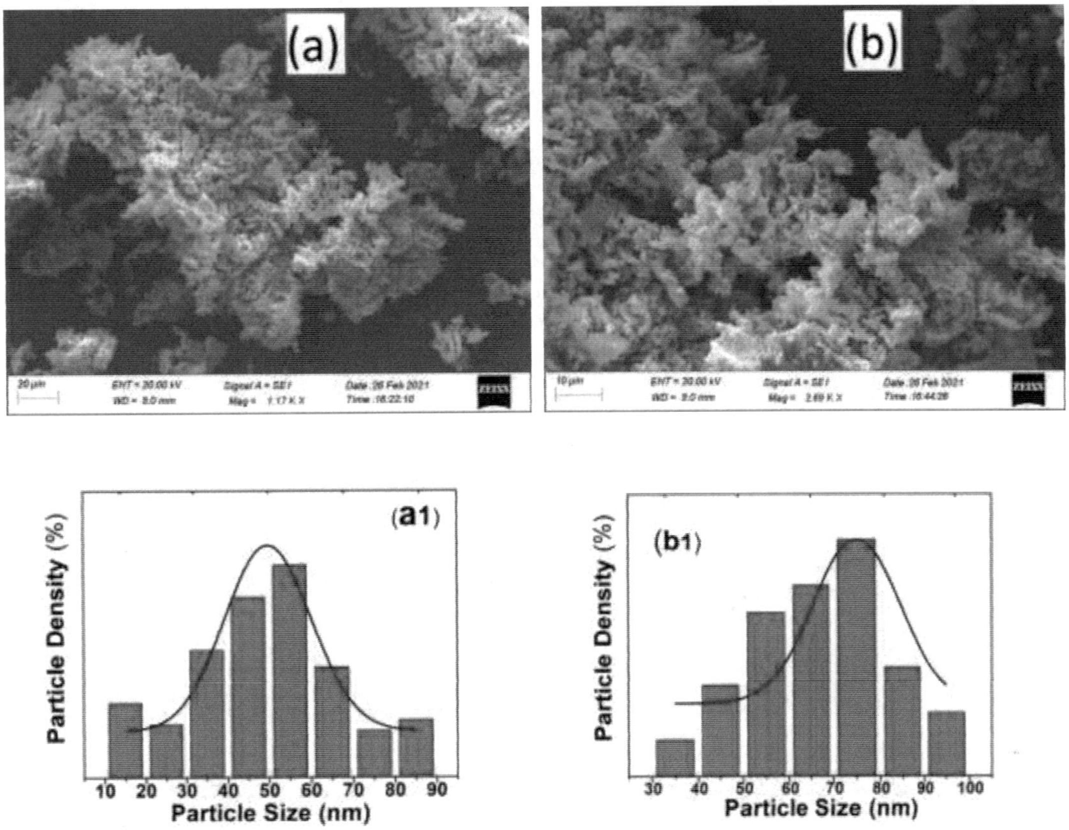

Figure 3. SEM images of $SrFe_{12-x}Gd_xO_{19}$ for composition 3(a) x =0.0 and 3(b) x =1.00 along with histograms.

The surface analysis of samples is done with SEM, which is helpful in supporting crystallite size approximated using W-H plots. Therefore we have shown morphological images of $SrFe_{12}O_{19}$ for two decisive compositions, is shown in figure 3(a) and 3(b). The particles prefer to agglomerate with an objective to minimize its superficial area, with Gd^{3+} content.The approximate particle size for $SrFe_{12-x}Gd_xO_{19}$ at compositions, x= 0.0 and x = 1.00 were 49.52 nm and 87.85 nm respectively. Particle size has been estimated with the help of ImageJ software by selecting numerous scanned areas and analysizing it one by one. As such related histograms are also shown as figure 3(a1) and 3(a2). The elecron microscopy of samples were conducted at applied voltage of 20 EHT and samples were gold coated for 90 seconds by sputtering technique.

3.3 FTIR analysis

The FTIR transmission spectrums of $SrFe_{12-x}Gd_xO_{19}$ for (x = 0.25- 1.00) is shown in figure 4, with wavenumber (400 - 1000 cm^{-1}). For $SrFe_{12}O_{19}$, two absorbance infrared peaks are significant, one at 439 cm^{-1} for occupancy at the octahedral sites in crystal and next one at 602 cm^{-1} for tetrahedral occupied sites [17]. The wave number (418 - 438 cm^{-1}) quantifies octahedral bond dimension [18]. The frequency (585 -

601 cm^{-1}), reveals dimension of tetrahedral bond viz. (Fe/O), and (Sr/O) [19]. As the bond length of octahedral sites > tetrahedral site, we have higher vibrational frequency for tetrahedral sites.

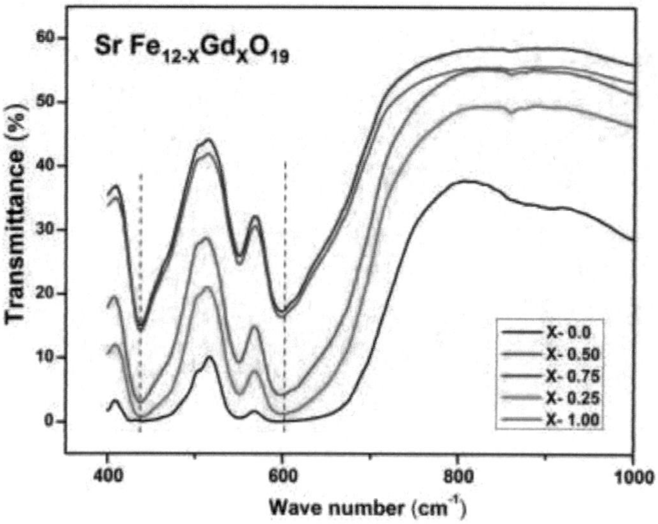

Figure 4. FTIR spectrum of $SrFe_{12-x}Gd_xO_{19}$ for (x = 0.25- 1.00).

The bond length (r) is found using frequency (ν) and force constant (*K*). We firstly determine the effective mass (μ) which is expressed as [20]

$$\mu = \frac{(M_O \times M_{Fe})}{(M_O + M_{Fe})}$$ Where M_o and M_{Fe} = Mass of O and Fe atoms (3)

Now the harmonic frequency equation is

$$\nu = \frac{\sqrt{k/\mu}}{2\pi C}$$ Where C is the speed of light. (4)

Next, (*K)* and (r) are inter-related as

$$K = (17/Ar^3) \quad (5)$$

A minor shift in the transmittance peaks towards higher wave number with respect to Gd^{3+} composition (table 2) is noticed. The stretch in M-O bindings is the outcome of lattice strain created by bigger radius of Gd^{3+} ions (1.05Å). Such behavior is also mentioned by few researchers [21].

Table 2. Site occupancy details for $SrFe_{11}Gd_1O_{19}$.

Site	*Samples*	*Wave number (cm^{-1})*	*Force constant (N/m)*	*Bond dimension(A°)*
Octahedral	X= 1.0	437	174.97	2.21
Tetrahedral	X = 1.0	601	246.88	1.89

3.4 Magnetic Properties

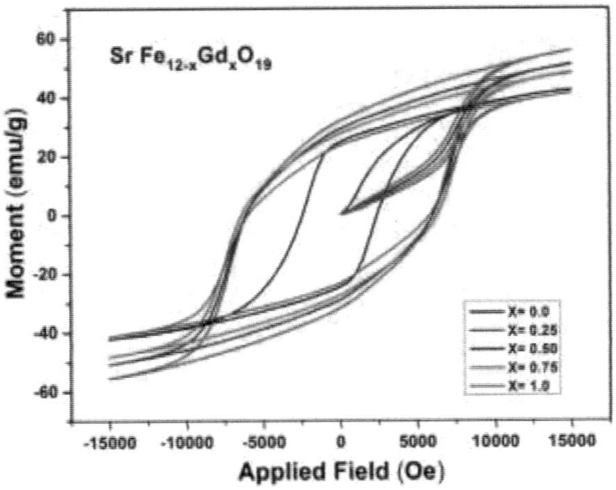

Figure 5. MH loops for $SrFe_{12-x}Gd_xO_{19}$ for x = (0.25 - 1.00).

Moment versus magnetic field (MH) loops of $SrFe_{12-x}Gd_xO_{19}$ for X = (0.25-1.00) is shown in Fig. 5. Magnetization curves confirm ferromagnetic property of all samples and shows high order of coercive field response. Hysteresis loops are observed to get saturated in the magnetic field of ±1.5 Tesla. We can notice from table 3 that M_r and M_s do not follow systematic pattern with varying Gd^{3+} composition. The magnetization is highest for sample $SrFe_{11.75}Gd_{0.25}O_{19}$. The increase in magnetization is the outcome of increased hyperfine field, as we know that M_s varies linearly with hyperfine field [22].

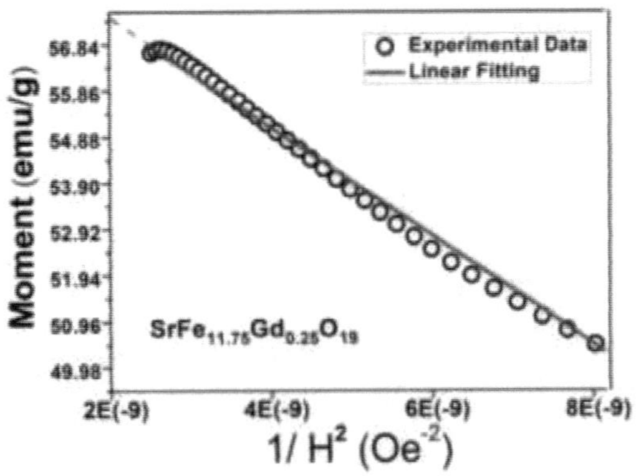

Figure 6. Moment vs. $1/H^2$ plot for $SrFe_{12}O_{19}$ for 0.25 mole Gd^{3+}composition.

Further, M_s is calculated applying the 'The Law of approach'to saturation. This law is effective in finding two important magnetic parameters, magnetization (M_s) and crystal anisotropy constant (K_1). The equation for "Law of approach" is [23].

$$M = M_s\left[1 - \left(\frac{A}{H}\right) - \left(\frac{B}{H^2}\right)\right] + \varkappa H \qquad \text{Where A is const. and B is anisotropy.} \qquad (6)$$

As $\varkappa H$ is applicable in elevated temperature domains and A/H is negligible in current criterion, we get

$$M = M_s [1 - (B/H^2)] \quad (7)$$

The y-axis intercept in moment versus ($1/H^2$) plot (Fig. 6), gives M_s.The coercivity value is maximum for $SrFe_{12}O_{19}$ doped with 0.75 mole Gd^{3+}, owing to the amplified exchange interactions of Fe-O-Fe ions [24]. The higher coercivity of order 6310 Oe signifies prepared sample's effectiveness as dense recording medium. The (M_r / M_s) was greater than 0.5 for pure as well as doped samples (table 3), which justifies that $SrFe_{12-x}Gd_xO_{19}$ will also be effective as permanent magnet [25].

Table 3. Magnetic parameters details of $SrFe_{12-x}Gd_xO_{19}$ for (x = 0.25-1.00).

$SrFe_{12-x}Gd_xO_{19}$	*Magnetization (Ms) (emu/g)*	*Retentivity (Mr) (emu/g)*	*Squareness ratio*	*Coercivity Hc(Oe)*
X=0.0	40.91	29.8	0.728	2500
X= 0.25	57.70	31.9	0.552	6099
X= 0.50	53.34	32.5	0.609	6195
X= 0.75	54.48	28.2	0.517	6310
X =1.00	42.74	26.3	0.615	5912

3.5 Multiferroic Properties

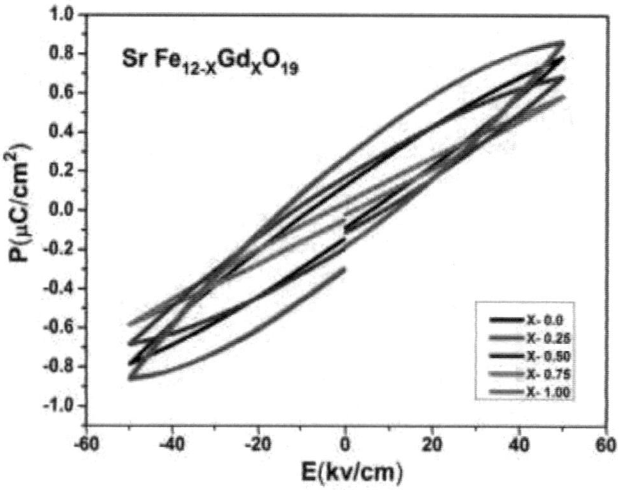

Figure 7. Ferroelectric loops for $SrFe_{12-x}Gd_xO_{19}$ for x = (0.25-1.00).

For multiferroic testing's, few amount of samples (of each batch) has been given capacitor form by compressing them to disc shaped pellets of radius 6 mm and 2.5 mm thickness using hydraulic palletizer. Then these pallets were calcined at 700°C for 4 hrs to remove organic and moisture contents completely. Further, both sides of pallets were coated with silver paste to make it behave like a conducting electrode. Next, P-E loops of $SrFe_{12-x}Gd_xO_{19}$ for (x = 0.25- 1.00) was recorded at 300 Kelvin, figure 7. As we can see, the P-E loop of $SrFe_{12-x}Gd_xO_{19}$ for x = 1.0 is obtained as full convex nature, suggesting the samples have electrical leakage. With increased Gd^{3+} ions in strontium hexaferrite lattice, ferroelectric loop becomes typical, signifying reduction in electrical leakage. It is interesting to observe that Gd^{3+} substitution in $SrFe_{12-x}Gd_xO_{19}$ crystal geometry has considerably enhanced electrical polarization by inhibition of electrical leakages. The incorporation of Gd^{3+} in $SrFe_{12-x}Gd_xO_{19}$ lattice promotes phase purity in samples with remanent (P_r) and coercive polarization (P_c) of order 0.19 μC/cm^2 and 17 kV/cm respectively.

4. Conclusions

Sol-gel technique is employed to synthesize monodispersed samples of strontium hexaferrite $SrFe_{12-x}Gd_xO_{19}$ for x = (0.25 - 1.00). All prepared samples have hexagonal symmetry referred to *P63/mmc* space group. A monotonous increase in the lattice constants (a) and (c) was observed with increasing Gd^{3+} composition in strontium crystals. The particle size and crystal strain in samples amplified with an increasing Gd^{3+} content in lattice. FTIR spectrum confirms tetrahedral and octahedral site occupancy. Magnetization value of about 57.70 emu/g is reflected for $SrFe_{12}O_{19}$ doped with 0.25 mole Gd^{3+} ions. The coercive field of order 6310 Gauss is recorded for $SrFe_{12}O_{19}$ doped with 0.75 mole Gd. Also, occupancy of Gd^{3+} ions in $SrFe_{12-x}Gd_xO_{19}$ crystals successfully enhanced electrical polarization by prohibiting electrical leakages. These tuned physical properties of $SrFe_{12}O_{19}$, due to lattice strain mediated by Gd^{3+} substitution suggests its futuristic technological applications. This study can be further extended to understand rare earth element's doping at strontium site, by manipulating crystalline anisotropy for improving various materials' properties.

Acknowledgements

Authors are grateful to Dept. Science and Technology, Bihar for establishment of Nanotechnology center at Aryabhatta Knowledge University, equipped with important characterization facilities.

References:

[1] M. Siva kumara, A Gedankena, W Zhong, Y.W. Dub, D Bhattacharya, Y Yeshurun and I Felner 2004 *J. Magn. Magn. Materials* **268** 95-104
[2] Y Yang and X Liu 2014 *IEEE Transactions on Magnetics* **50**
[3] M P Sharrock 1989 *IEEE Trans. Magn.* **25** 4374-4389
[4] R Kumar, R K Singh, M K Zope and M Kar 2017 *Material Science Engineering:* B **220** 73-81
[5] C M Fang, F Kools, R Metselaar, G De and R.A. De Groot 2003 *J. Phys. Condens.Matter.* **15** 6229
[6] Ali Ghasemi 2012 *Journal of Magnetism and Magnetic Materials* **324** 1375-1380
[7] L S I Liyanage, S Kim, Y K Hong, J H Park, S C Erwin and S G Kim 2013 *J. Magn. Magn. Mater.* **348** 75
[8] M Zhang, Q Liu, G Zhu and S Xu 2019 *Applied Physics* A **125**,191
[9] M Schmidt, M Hofmann and S J Campbell 2003 *Journal of Phy.Condensed Matter* **15** 8691-8701
[10] R Carey, P A Gago Sandoval, D M Newman and B W J Thomas 1991 *J. Appl. Phys.* **75** 6789
[11] Z Mosleh, P kameli, M Ranjbar and H Salamati 2014 *Ceramics International* **40** 7279-7284
[12] F L Wei, H C Fang, C K Ong, C S Wang and Z.Yang 2000 *J. Applied Physics* **87** 8636-8639
[13] H Yanbing, S Jian, S Lina, T Quan, L Qin, J Hongxiao, J Dingfeng, B hong, G Hongliang and W.Xinqing 2009 *J. Alloys Compounds* **486** 348-351
[14] S Singhal, A N Garg and K Chandra 2005 *J.Magnetism and Magnetic.Materials* **285** 193-198.
[15] H J Kown, J Y Shin and J H Oh 1994 *J. Appl. Phys.* **75** 6109
[16] P Wartewig, M K Krause, P Esquinazi, S Rosler and R Sonntag 1999 *J.Magn.Magn.Mat.***192** 83-99
[17] V V Pankov, M Pernet, P Germi and P Mollard 1993 *J. Magn. Magn.Materials* **120** 69
[18] Yilong J, Lijun J, Yuhang Z and Huaiwu Z 2019 *RSC Advances* **9** 33388–33394
[19] Y Senzaki, J Caruso, Mark J Hampden Smith, Toivo T Kodas and Lu Min wang 1995 *J. American Ceramic Society*, **78**(11) 2973-2976
[20] C Surig, K A Hempel and D Bonnenberg 1994 *J. IEEE Transactions on magnetics* **30** 4092-4094
[21] Neil J Shirtcliffe, Simon Thompson et al. 2007 *Mater. Research Bulletin* **42** 281-287
[22] P.Bahera and S Ravi 2016 *J.Supercond. Nov. Magn.* **30** 1453
[23] Q Mohsen 2010 *American J. Applied Sciences* **7** 914-921
[24] AV Trukhanov, V O Turchenko, I A Bobrikov, S V Trukhanov, I S Kazakevich and A M Balagurov 2015 *J. Magn. Magn.Mater.* **393** 253-259
[25] C H Ri, L Li and Y Qi 2012 *Journal of Magnetism and Magnetic Materials* **324** 1498-1502

Optimization of Fused Deposition Modelling process parameters using Teaching Learning Based Optimization (TLBO) algorithm

Md Nadeem Alam[1], Md Mahfuj Alam[2] and Shafi Ahmad[3]

[1] Department of Mechanical Engineering, Jamia Millia Islamia, New Delhi India.
[2] School of Engineering and Technology, Sharda University, Greater Noida (Uttar Pradesh) India.
[3] Department of Mechanical Engineering, Jamia Millia Islamia, New Delhi India.
mdnadeemalam@gmail.com

Abstract. Fused Deposition Modelling (FDM) is one of the most commonly used Additive Manufacturing (AM) techniques with a wide range of applications in various modern manufacturing industries. It is widely employed to fabricate prototypes where immense surface finish is required. Furthermore, the literature suggests that process parameters such as nozzle temperature (NT), nozzle diameter (ND), and feed rate (FR) have a significant influence on the surface finish achieved in an FDM process. Hence, this work intends to examine the effect of process parameters viz. NT, ND, and FR on the side and top surface roughness of poly-lactic acid (PLA) sample fabricated through FDM process. Experiments are designed as per Taguchi's L18 orthogonal array and a population-based algorithm identified as Teaching Learning Based Optimization (TLBO) algorithm is used to determine the optimal process parameter settings for optimum side and top surface roughness simultaneously. The results of the study reveal that NT of 493 K, ND of 0.4 mm and FR of 60 mm/s results in optimum side and top surface roughness simultaneously.

1. Introduction

Additive Manufacturing (AM) processes are the commonly used modern manufacturing method adopted to fabricate intricate components with increased complexity. Industrial demands of the products with flexible and complex geometries are driving manufacturers to study and improve the specifications of AM process that has potential applications in numerous modern manufacturing industries such as automotive, aerospace, art, jewelry, etc.[1]. One of the most commonly used AM processes is the Fused Deposition Modeling (FDM) technique, which has several driving factors such as availability of a vast variety of materials, low maintenance cost, easy material handling, compact size, low operating temperature, lower tolerance, and rapid production of thin components. However, the surface roughness (Ra) of the product manufactured using FDM is an essential factor that affects the functionality of the product, lower Ra enhances the component accuracy, and reduces the post-processing investment [2,3]. Therefore, investigating the Ra of the products manufactured using FDM process is an area of interest among researchers. Plethora of research work has been done to investigate the influence of various control parameters such as layer thickness, infill percentage, infill pattern, and extrusion temperature on Ra, dimensional accuracy (DA) and mechanical properties of FDM components. But, very few work has been performed to investigate the influence of nozzle temperature (NT), nozzle diameter (ND), and feed rate (FR) on Ra of various faces of FDM components. Deng *et al.* [4] investigated the influence of NT and FR on the emission of particles during FDM of acrylonitrile butadiene styrene (ABS) and polylactic acid (PLA) material and concluded the followings (i) for PLA material low values of NT ranging between 180°C to 200°C were preferable for high particle count, as the decomposition temperature of PLA material is around 220°C, (ii) in case of ABS material the preferable temperature range is found to be 200°C to 240°C, as its decomposition temperature is about 270°C, and (iii) for both the materials, increasing the FR from 30 mm/s to 60 mm/s improve the particle count thereby decreases the Ra of the product but further rise of feed rate to 90 mm/s significantly increase the Ra because the elevated FR increases the material fraction, which facilitates the high printing speed and heat dissipation that margins the thermal decomposition and increases Ra. According to Alsoufi and Elsayed [5] decrease in the layer height and nozzle diameter from 0.3 mm to 0.1 mm, and 0.5 mm to 0.2 mm respectively, significantly reduces the Ra of the component whereas increases the fabrication time. Mahapatra and Sood [6] proposed a Back Propagation Algorithm based Artificial Neural Network (ANN) model to explore the relationship between process parameters including layer thickness, raster angle, orientation, raster width, and air gap with top surface roughness (R_T), back surface roughness (R_B) and side surface roughness (R_S) of the FDM component, and concluded that among the various factors, the raster width and the raster angle has the most significant influence on R_T, whereas the orientation of the part and layer thickness has the most significant influence on R_B and R_S, respectively. From above literatures, it has been witnessed that apart from top surface roughness (R_T), back surface roughness (R_B) and side surface roughness (R_S) have a significant effect on overall product quality.

To achieve the optimum possible Ra through FDM process, the selection of optimum control factors is very important. Literature demonstrates that numerous work has been performed to explore the influence of several control factors on different performance characteristics of FDM process by using single and multi-optimization techniques [7]. However, very less work has been reported to investigate the influence of nozzle temperature (NT), nozzle diameter (ND), and feed rate (FR) on the Ra of different faces of the FDM components. Further, there are very few researches that have used evolutionary algorithms to optimize the multiple output responses simultaneously. With this inspiration, a population based evolutionary algorithm identified as Teaching-Learning Based Optimization (TLBO) algorithm is used in this work to optimize the R_S and R_T simultaneously of the FDM product. TLBO algorithm is a very efficient optimization algorithm for multiple constrained engineering problems that does not require any parameters related to a specific algorithm [8]. Hence, the objective of this work is to analyze the effect of NT, ND, and FR and thereby employing TLBO algorithm for the multi-response optimization of R_S and R_T. The remaining of the paper is structured as follows: section 2 of the paper discusses the materials and methods used in this work. Section 3 of the paper demonstrate the results of the study and finally, section 4 presents the conclusion of the work.

2. Materials and Methods

This study aims to investigate the influence of three process parameters on the surface quality of the products fabricated using FDM process. To achieve the research objectives the research methodology adopted in this work begins from experiment setup and selection of process parameters and their levels. Subsequently, Taguchi's Design of Experiment (DOE) is used to define a well-defined set of experiments based on which experiments are carried out and output performances are measured. Finally, the results were analyzed and TLBO algorithm is used to obtain the optimum combination of process parameters for optimum results of R_S and R_T simultaneously.

2.1. Material Selection

Figure 1 (a) shows a desktop type RxP-2000 FDM 3D printing machine which is used to fabricate the samples. Samples of dimension 16 mm×12 mm×10 mm were fabricated using PLA filament of diameter 3 mm. PLA is one of the most commonly used plastic materials in FDM process due to its lower melting point, low cost, less toxic, good biocompatibility, good biodegradability, and less energy requirement [9]. The experiments have been performed in a fabrication chamber as shown in Figure 1 (b).

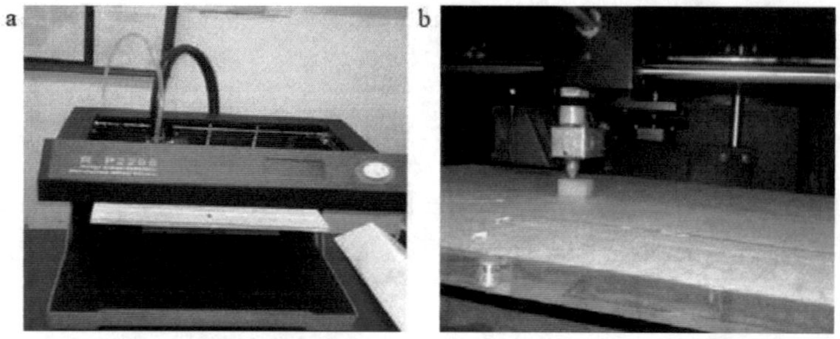

Figure 1. (a) RxP-2000 3D printing machine; (b) Fabrication of component in the chamber

2.2. *Process parameters and their levels*
In light of the literature associated with surface quality improvements of a component in FDM process, the effect of three process parameters viz. nozzle diameter (ND), nozzle temperature (NT), and feed rate (FR) have been investigated in this study. Effect of three levels of ND viz. 0.4 mm, 0.5 mm, and 0.6 mm were examined with the feasible range of temperature from 473 K and 493 K and FR 70 mm/s to 90 mm/s. Three levels of ND and FR and two levels of NT were selected as shown in Table 1.

Table 1. Process parameters and their levels.

Process parameter	Symbol	Units	Level 1	Level 2	Level 3
Nozzle Diameter (ND)	ND	mm	0.4	0.5	0.6
Nozzle Temperature (NT)	NT	K	473	493	-
Feed rate (FR)	FR	mm/s	70	80	90

2.3. *Taguchi Experimental Design*
Design of experimentation is helpful in studies involving process parameters where the analysis of the combined effect of process parameters on an output response is required [10]. Among the different methods of experimental designs, Taguchi experimental design is an efficient and widely accepted method used to investigate the combined effect of process parameters [11,12]. It has the advantage of a cutback in cost, time, and resources as it reduces the number of experiments to be performed. Taguchi proposed standard orthogonal arrays (SOAs) in which several experiments are arranged in rows and process parameters in columns [13]. The selection of orthogonal arrays (OA) for conducting experiments is based on the total degree of freedom (DOF) of the process parameters. It has been suggested that any OA having a DOF equal to or more than the total DOF of the process parameters can be used for performing experiments [14]. The DOF of a specific process parameter is calculated by subtracting one from the number of levels. In this study, three factors are considered with two factors having three levels and one factor having 2 levels. Hence, the DOF of process parameters will be 5 $[(2\times2) + (1\times1)]$. The DOF of SOAs is defined as one less than the total number of experiments in the OA. Therefore, Taguchi's L18 orthogonal array is found suitable for the considered combinations and levels of process parameters.

2.4. *Measurement of output response*
In the present study R_S and R_T are evaluated to analyze the overall quality of the fabricated surface. The measurement of surface roughness of 18 samples was performed on one of the side surface and top surface for each sample using a contact system type Mitutoyo SJ210 Perthometer, model surface roughness tester. It employs the stylus retractable drive unit method of measurement with a stylus tip

radius of 2 μm and tip angle of 60°. A tracing length of 4 mm was used for analysis. The surface roughness of each sample was measured at the center position of the fabricated surface.

2.5. Teaching Learning Based Optimization (TLBO) algorithm

TLBO algorithm was proposed by [8] which is popular among researchers due to its simple concept, no requirement of the algorithm-specific parameter, fast convergence, and easy and effective execution process. It has been widely accepted and positively employed to solve the optimization problem on different knowledge domains [8]. The iterative method of TLBO algorithm is separated into two parts. First part is named as teacher phase which is inspired from learning of students from the teacher. While in the second part regarded as learner's phase, the knowledge is gained by interacting with each other. For an optimization problem with decision variables X = [x1, x2, ..., xn] and an objective function f(X), TLBO algorithm considers desired objective function value as the knowledge level of a learner and the search points in the decision variable space as learners. The computational steps involved in the teacher and learners phase is discussed as follows:

2.5.1. Teaching Phase.

As the name suggests, teaching phase of TLBO algorithm involves knowledge enhancement through teacher. Considering N number of learners (population size = N) with knowledge x1, x2, ..., xn (decision variables) in n subjects. The aim is to maximize the overall knowledge (objective function) of the learners which depends upon the knowledge in each subject. It noteworthy that for a maximising problem highest value and for minimizing problem least value of the objective function shows high overall knowledge. The first step is to identify the teacher which will improve the knowledge in each subjects resulting in overall knowledge enhancement of the learners. In this algorithm, the teacher is identified as the learner with highest overall knowledge. Further the knowledge alteration in each subject by interaction with teacher is computed using Eqn. (1).

$$x_{i,j}^{new} = x_{i,j} + r_i(x_{T,j} - T_F M_j) \tag{1}$$

where, $x_{i,j}^{new}$ and $x_{i,j}$ are the updated and original knowledge of learner i in subject j respectively. $x_{T,j}$ depicts the knowledge of the teacher in subject j. M_j denotes the mean knowledge of the all the learners in subject j.r_i is a random number between 0 and 1. T_F is defined as the teaching factor whose value is determined using Eqn. (2).

$$T_F = round[1 + rand(0,1)] \tag{2}$$

2.5.2. Learning Phase.

In learning phase, each learner enhances his/her knowledge by interacting with each other. This interaction may be partial or complete. In partial interaction, learners interact with each other randomly while in complete interaction each learner interacts with other learner to modify his/her knowledge level. In this work complete interaction has been considered. The modified knowledge in subject "j" of learner "P" when interacts with learner "Q" is computed using Eqn. (3) and Eqn. (4) according to the required objective function value as maximum or minimum respectively.

$$x_{P,j}^{new} = \begin{cases} x_{P,j} + r_i(x_{P,j} - x_{Q,j}) & ;if\ f(X_P) < f(X_Q) \\ x_{P,j} + r_i(x_{Q,j} - x_{P,j}) & ;if\ f(X_Q) < f(X_P) \end{cases} \tag{3}$$

$$x_{P,j}^{new} = \begin{cases} x_{P,j} + r_i(x_{P,j} - x_{Q,j}) & ;if\ f(X_Q) < f(X_P) \\ x_{P,j} + r_i(x_{Q,j} - x_{P,j}) & ;if\ f(X_P) < f(X_Q) \end{cases} \tag{4}$$

where, $x_{P,j}^{new}$ and $x_{P,j}$ are the updated and original knowledge of learner P in subject j after teaching phase respectively. $x_{Q,j}$ is the knowledge of learner Q after teaching phase.

IOP Conf. Series: Materials Science and Engineering 1149 (2021) 012014 doi:10.1088/1757-899X/1149/1/012014

The knowledge level of learners after teacher phase and learner phase together completes the one iteration of the TLBO algorithm. This process is iterated again and again till the knowledge level i.e., objective function value for each learner become equal or reaches a desired level.

3. Result and Discussion

To investigate the influence of various process parameters on the surface quality of PLA test samples fabricated using FDM process experimentations were performed as per design shown in Table 2. While performing the experiments, the experimental sequence was not followed rather they are conducted in randomized manner to reduce the biasness in the results. Further, R_S and R_T of the samples are computed using a contact system type Mitutoyo SJ210 perthometer, model surface roughness tester and the results are shown in Table 2.

Table 2. Experimental Results

Exp No.	ND	NT	FR	R_S	R_T
1	0.4	473	70	15.956	6.956
2	0.4	473	80	18.548	7.545
3	0.4	473	90	20.941	7.842
4	0.5	473	70	25.944	8.245
5	0.5	473	80	26.542	9.445
6	0.5	473	90	28.605	10.509
7	0.6	473	70	31.841	13.245
8	0.6	473	80	33.405	14.455
9	0.6	473	90	35.605	15.509
10	0.4	493	70	12.256	5.85
11	0.4	493	80	16.962	6.21
12	0.4	493	90	18.524	6.695
13	0.5	493	70	23.997	7.192
14	0.5	493	80	24.696	7.506
15	0.5	493	90	26.845	7.905
16	0.6	493	70	30.986	11.592
17	0.6	493	80	31.235	13.141
18	0.6	493	90	33.436	14.305

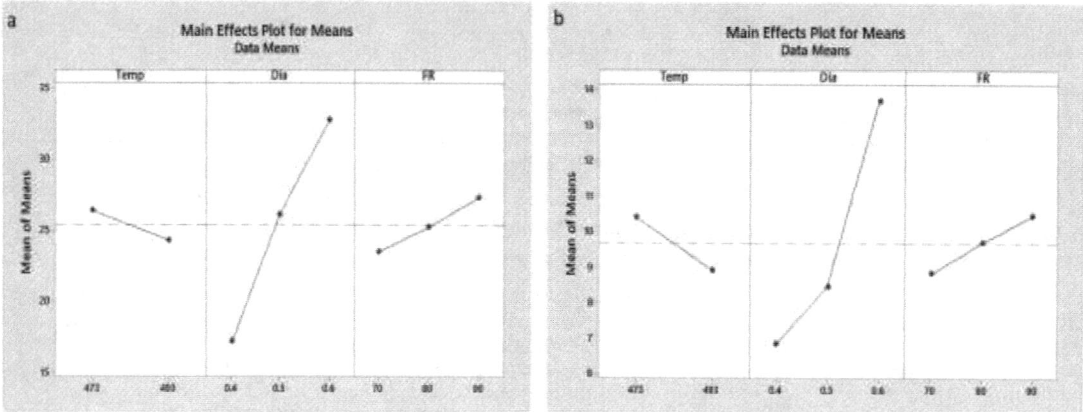

Figure 2 (a) Main effect plot for R_S; (b) Main effect plot for R_T

Figure 2(a) and 2(b) shows that R_S and R_T value rises with increase in ND and FR, however increase in NT reduces the R_S and R_T respectively. Further, to optimize the parameters using TLBO algorithm, it is needed to represent the output response as a function of process parameters. For this purpose, regression analysis was performed. A quadratic regression equation with highest R2 value of 99.22 % and 99.38 % is obtained for of R_S and R_T which is shown in Eqn. (5) and Eqn. (6) respectively.

$$R_S = -32 - 0.1025 * NT + 241.2 * ND + 0.506 * FR - 113 * ND^2 - 0.630 * ND * FR \quad (5)$$

$$R_T = 82.22 - 0.07419 * NT - 179.3 * ND - 0.1222 * FR + 181.2 * ND^2 + 0.406 * ND * FR \quad (6)$$

where, NT = Nozzle Temperature, ND = Nozzle Diameter, FR = Feed rate

Further, the optimization problem for R_S and R_T so formulated is shown in Eqn. (7) and Eqn. (8) respectively.

$$Minimize\ R_S = -32 - 0.1025 * NT + 241.2 * ND + 0.506 * FR - 113 * ND^2 - 0.630 * ND * FR$$
$$\text{Subject to: } 473 \leq NT \leq 493\ ;\ 0.4 \leq ND \leq 0.6\ ;\ 70 \leq FR \leq 90 \quad (7)$$

$$Minimize\ R_T = 82.22 - 0.07419 * NT - 179.3 * ND - 0.1222 * FR + 181.2 * ND^2 + 0.406 * ND * FR$$
$$\text{Subject to: } 473 \leq NT \leq 493\ ;\ 0.4 \leq ND \leq 0.6\ ;\ 70 \leq FR \leq 90 \quad (8)$$

TLBO algorithm was run with a population size of 10 and is iterated 20 times. It was found that using TLBO algorithm the minimum R_S i.e. 13.648 μm is obtained with ND of 0.4 mm, NT of 493K and FR of 70 mm/s. Whereas, minimum R_T i.e. 5.682 μm is obtained with ND of 0.416 mm, NT of 493K and FR of 70 mm/s. Further, to optimize both the parameters simultaneously, the non-dominated sorting TLBO algorithm is employed. According to this approach, the weighted sum of the objective functions are used to formulate a single a single fitness function. It is likely that the different objective functions may be defined over different range of values. Therefore, normalized objective functions are considered to convert them into same range. For this study, the R_S and R_T are normalized and combined according to Eqn. (9)

$$CSR = \frac{R_T}{R_T^*} + \frac{R_S}{R_S^*} \quad (9)$$

where, R_T* and R_S* are the minimum values of top and side surface roughness obtained when they are optimized independently.

Thus, the optimization problem for combined optimization of both R_S and R_T so formulates is shown in Eqn. (10).

$$Minimize\ Z = \frac{R_T}{5.682} + \frac{R_S}{13.648}$$
$$\text{Subject to: } 473 \leq NT \leq 493\ ;\ 0.4 \leq ND \leq 0.6\ ;\ 70 \leq FR \leq 90 \quad (10)$$

Considering a similar population size and number of generation as in individual optimization, the results obtained by TLBO algorithm for combined optimization of R_S and R_T and comparison with the experimental results is shown in Tables 3.

Table 3: Result obtained by TLBO algorithm for top and side surface roughness simultaneously

Parameters	Experimental Results	TLBO results
Diameter	0.4 mm	0.416 mm
Temperature	493 K	493 K
Feed rate	70 mm/s	70 mm/s
Top surface roughness	5.85 μm	5.682 μm
Side surface roughness	12.256 μm	13.648 μm

4. Conclusion

FDM process helps to fabricate products from the elementary design of the components. It allows low cost, lightweight, and rapid prototyping in comparison to other conventional manufacturing process. This work aims to investigate the influence of three process parameters viz. NT, ND, and FR on the surface quality of polylactic acid (PLA) test samples fabricated by FDM process. To achieve the desired objectives of the study, test samples of dimension 16 mm x 12 mm x 10 mm were fabricated using PLA filament of diameter 3 mm on a desktop type RxP-2000 FDM 3D printing machine. Experiments are designed as per L18 Taguchi's orthogonal array which was found suitable for the considered levels of process parameters. Further, output responses viz. side and top surface roughness are measured using a contact system type Mitutoyo SJ210 pertho-meter, model surface roughness tester. Subsequently, TLBO was employed to determine the optimal combination of process parameters. The major conclusions of the conducted study are presented as follows:

- For low side and top surface roughness, lower values of ND and FR, and higher value of NT are desired.
- The optimal combination of process parameters i.e. NT, ND and FR for small top surface roughness are found to be 493 K. 0.4 mm and 70 mm/s respectively.
- A process parameter setting of NT = 493 K, ND = 0.416 mm and FR = 70 mm/s results in lower top surface roughness.
- The multiple optimization of top and side surface roughness using TLBO suggests that optimal combination of process parameter NT, ND and FR is 493 K, 0.4 mm and 70 mm/s respectively.

References

[1] Raju M, Gupta M K, Bhanot N and Sharma V S 2019 A hybrid PSO–BFO evolutionary algorithm for optimization of fused deposition modelling process parameters *J Intell Manuf* **30** 2743–58

[2] Vijay P, Danaiah P and Rajesh K V D 2011 Critical Parameters Effecting the Rapid Prototyping Surface Finish *Journal of Mechanical Engineering and Automation* **1** 17–20

[3] Boschetto A and Bottini L 2015 Roughness prediction in coupled operations of fused deposition modeling and barrel finishing *Journal of Materials Processing Technology* **219** 181–92

[4] Deng Y, Cao S J, Chen A and Guo Y 2016 The impact of manufacturing parameters on submicron particle emissions from a desktop 3D printer in the perspective of emission reduction *Building and Environment*, vol. 104, pp. 311–319, Aug. 2016, doi: 10.1016/j.buildenv.2016.05.021

[5] Alsoufi M S and Elsayed A E 2017 How Surface Roughness Performance of Printed Parts Manufactured by Desktop FDM 3D Printer with PLA+ is Influenced by Measuring Direction *American Journal of Mechanical Engineering*, vol. 5, pp. 211-222.

[6] Mahapatra S S and Sood A K 2012 Bayesian regularization-based Levenberg–Marquardt neural model combined with BFOA for improving surface finish of FDM processed part *Int J Adv Manuf Technol*, vol. 60, pp. 1223–1235, Jun. 2012, doi: 10.1007/s00170-011-3675-x.

[7] Vinodh S and Shinde P 2018 Parametric Optimization of 3D Printing Process Using MCDM Method *Precision Product-Process Design and Optimization* Lecture Notes on Multidisciplinary Industrial Engineering ed S S Pande and U S Dixit (Singapore: Springer Singapore) pp 141–59

[8] Rao R V, Savsani V J and Vakharia D P 2011 Teaching–learning-based optimization: A novel method for constrained mechanical design optimization problems *Computer-Aided Design* **43** 303–15

[9] Alsoufi M S, Alhazmi M W, Suker D K, Alghamdi T A, Sabbagh R A, Felemban M A and Bazuhair F K 2019 Experimental Characterization of the Influence of Nozzle Temperature in FDM 3D Printed Pure PLA and Advanced PLA+ *American Journal of Mechanical Engineering* **7** 45–60

[10] Pomalaza G, Capron M and Dumeignil F 2020 Improving the synthesis of Zn-Ta-TUD-1 for the Lebedev process using the Design of Experiments methodology *Applied Catalysis A: General* **591** 117386

[11] Alafaghani A and Qattawi A 2018 Investigating the effect of fused deposition modeling processing parameters using Taguchi design of experiment method *Journal of Manufacturing Processes* **36** 164–74

[12] Banerjee S, Poria S, Sutradhar G and Sahoo P 2019 Design of Experiments Analysis of Friction Behavior of Mg-WC Nano-composites using Taguchi Methodology *Materials Today: Proceedings* **18** 4026–33

[13] Kacker R N, Lagergren E S and Filliben J J 1991 Taguchi's Orthogonal Arrays Are Classical Designs of Experiments *J Res Natl Inst Stand Technol* **96** 577–91

[14] Ghambarian M, Yamini Y, Saleh A, Shariati S and Yazdanfar N 2009 Taguchi OA16 orthogonal array design for the optimization of cloud point extraction for selenium determination in environmental and biological samples by tungsten-modified tube electrothermal atomic absorption spectrometry *Talanta* **78** 970–6

IOP Conf. Series: Materials Science and Engineering 1149 (2021) 012015 doi:10.1088/1757-899X/1149/1/012015

Effect of varying road profile amplitude on the behavior of a nonlinear quarter car model.

Vipul Paliwal[1], Ritvik Dobriyal[1], Pushpendra Kumar[1] and Avi Raj Manral[1].

[1]Department of Mechanical Engineering, Graphic Era Deemed to be University, Dehradun 248002, Uttarakhand, India
vipulpaliwal@geu.ac.in

Abstract
Vehicle suspensions are one of the most important equipment in an automobile which impart comfortable ride to the passengers. As the passive suspensions are most widely used in the automobiles, therefore this paper deals with the investigations of the suspension behavior at the different values of the road profile amplitude. The effects on the frequency plots are studied by using fast Fourier transform functions. The phase plots are also drawn to understand the periodic and quasi periodic transformations at the different values of road profile amplitude. It is seen that the natural frequency and its harmonics in the spectrum dies out with the rise in the values of the road profile amplitude and also the increase in the strength of forcing is observed. The in harmonics obtained is seen to be withdrawn with the increase in the road profile amplitude.

1. Introduction

Providing comfortable ride to passengers is one of the vital tasks handled by the automobile suspensions by eliminating all types of disturbances issuing from the road perturbations and bumps. Suspensions can be active or passive; generally active suspensions are much expensive due to their complexities but are more effective because of their faster and better response to road disturbances. Most commonly used are the passive suspensions as they are less costlier and more reliable due to their simplicity, therefore, a study has been conducted for a passive suspension and also the behavior of the suspension is investigated at different values of the road profile amplitude with the help of phase plots and fast fourier transform functions.

Mahajan et al. [1] in their study determined the performance of the suspension system for sine input with variable frequencies and had shown that the increase in the frequency of the sine wave will result in the decrease in the settling time of sprung mass acceleration which will directly result in the effect the passenger ride discomfort. Fakhraei et al. [2] examined the sinusoidal road input in the form of speed bump on a non-linear system to study the effect of chaotic vibration on the ride comfort of driver and also, the response of amplitude and frequency on the chaotic behavior are determined specifying the elevated intensity of vehicle vibration in the chaotic regions. Their results are helpful in the successful design of the dynamic vehicle suspensions free from chaotic vibrations thereby increasing the ride comfort of passengers and drivers. Fakhraei et al. [3] in their study had shown that the chaotic vibration nature in the suspensions can be downsized by selecting the pertinent values of the suspension parameters.

Chatterjee. A. [4] discussed the various methods to study and analyze the behavior of non linear dynamic systems. He showed that the feedback of the non linear system involve the frequencies other than the forcing frequency which are the multiples of this forcing frequency. Their study can be used to prove the non linear behavior of the vehicle suspensions.

Li et. al. [5] examined the vibrations of the quarter car model at the different values of the amplitudes concentrating on the synchronized, quasi periodic and chaotic motion. They founded the critical value of the amplitude at which the behavior of the system changes to chaotic which can be lowered by decreasing the value of the damping coefficient. Their study can be used to determine the quasi periodic or chaotic regions for the quarter car vibrations which can be used further to improve the ride quality of the suspension systems.

Zhou et.al. [6] Analyzed the non linear dynamics of the single degree of freedom (SDOF) quarter car model and examined the effect of excitation amplitude on the dynamic response of the system. They examined the behavior at the different values of the frequency and the higher values of the excitation amplitude. Their study shows that the vibration amplitude and the peak resonance increase substantially with the greater values of the excitation amplitude.

Borroweic et.al. [7] Analyzes chaotic vibrations and bifurcations by altering the road profile amplitude and excitation frequency. Their study reveals that with the increase in the excitation amplitude, behavior of the quarter car drifts towards chaotic and quasi periodic regions starts to appear. Their research is used to analyze the regions at which chaos would increase in the quarter car model and the parameters are to be assorted so that lesser chaotic vibrations are obtained which would result in better ride comfort of passengers.

This paper extends this work [5] and [7] and the frequency curves are studied at the lower values of the road profile amplitudes. The behavior of the SDOF quarter car model is analyzed and the values of amplitudes are determined at which the quasi periodic motion transforms to periodic motion resulting in the better design of the vehicle suspension system with the help of the phase plot diagrams.

2. Methodology

The motion of single degree of freedom quarter car model is governed by a second order differential equation as given below.

$$\frac{d^2y}{dt^2} + \omega^2 y + B_1 y^3 + B_2 \frac{dy}{dt} + B_3 \left(\frac{dy}{dt}\right)^3 = -g + A\Omega^2 \sin(\Omega t),$$

where y is the relative displacement of the mass from the road, ω^2, B_1, B_2 and B_3 are k_1/m, k_2/m, c_1/m and c_2/m respectively. k_i and c_i (i=1,2) are the linear and nonlinear coefficients of the spring and damping respectively. This system was proposed by Li et al. [5] and further investigated by Litak et al. [10] and Borowiec et al. [7]. The suspension system was excited harmonically with a road profile containing a single frequency, as shown in figure 1. The system may oscillate at a single frequency (similar to a linear system) or it may undergo quasi periodic or chaotic oscillations depending upon the excitation frequency and the parameter values. Sinusoidal analysis is done so that the effect of non linearity in suspension elements can easily be observed. Since sinusoidal excitation is a signal that contains a single frequency, the analysis done using this can be easily conducted as well as revealing as compared to other non linear analysis.

Dynamics of a non linear system may be qualitatively understood by using the phase plots. Displacement of the sprung mass is plotted against velocity. Phase plot gives an idea of the complicacy of the motion of the sprung mass. Using the phase plot, the range of displacement and velocity and also the periodicity can be judged. Quasi periodic motions occur in case of two incommensurate frequencies. While chaotic oscillations are non repeating and are a periodic. Solutions of equation (1) were investigated numerically by using fourth order Runge Kutta method.

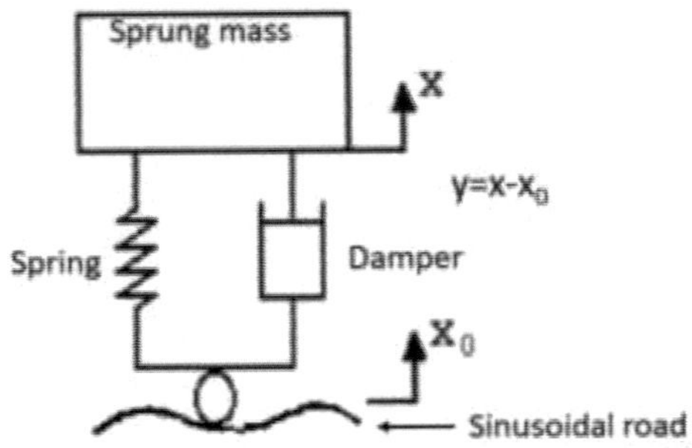

Figure 1. The motion of a single degree of freedom quarter car model over a sinusoidal road

3. Results and discussions

Dynamical behavior of a nonlinear system can be effectively understood from the phase plot diagrams and fft of the time series. Figure 2 (a), (b) (c) shows the time series of relative displacement, phase plot diagrams and the corresponding spectra. The values of spring constant k_1 = 160000 N/m, k_2 = 150000N/m, damping constant c_1 = -250 Ns/m and c_2 = 25 Ns^3/m^3 are used. These values are same as used in [7]. Values of excitation amplitude are varied and the results are plotted as shown below.

In Fig.2 excitation amplitude was selected as 0.07. It is observed that the relative displacement is having various closed profiles. The phase plot in fig. 2(a) is also shows the motion with multiple periodicity. Fig.2(c) is showing a weak sub harmonic and also the sharp peak at the forcing frequency and natural frequency (ω) representing incommensurate frequencies. The complexity of the phase plot is due to these incommensurate frequencies. The inharmonics are also seen at frequency ratio below 1.4 and 2.8 indicating the non linear behavior of the system as in [4] (any departure from the ideal harmonic series is called inharmonicity). In addition second and third harmonic of the natural frequency can also be seen.

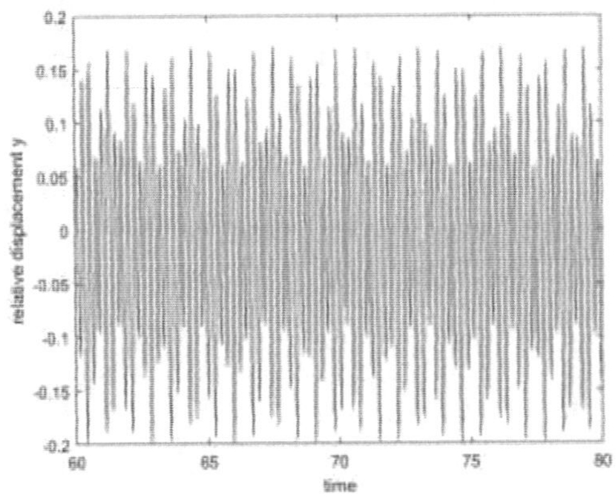

Fig. 2 (a)

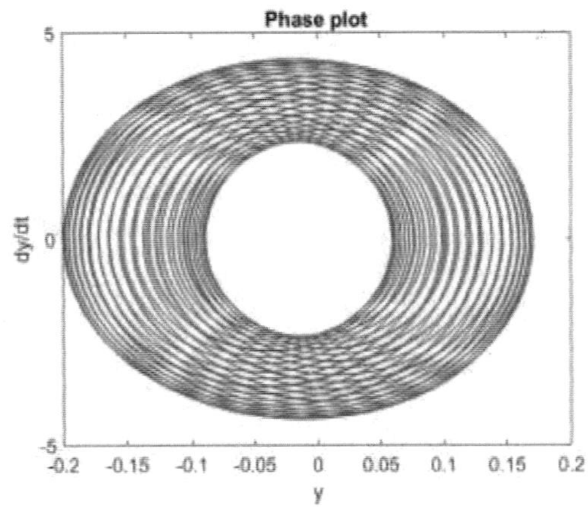

Fig.2 (b)

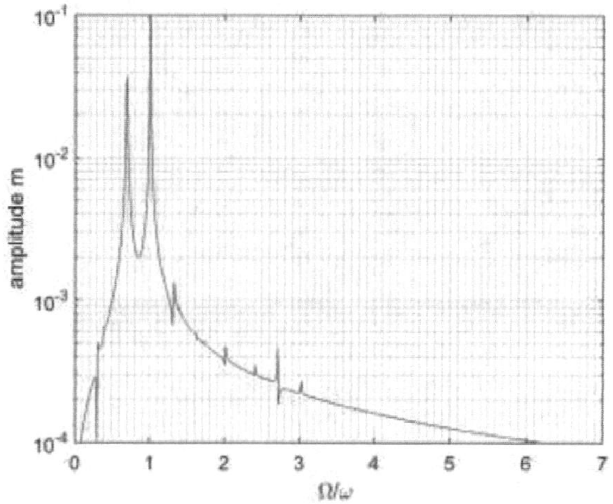

Fig.2 (c)

Figure 2. (a) shows the relative displacement as a function of time while (b) shows the phase plot and (c) shows the spectra corresponding to figure (a) for excitation amplitude of 0.07m.

In Fig.3, excitation amplitude was increased to 0.11m, while keeping the system parameters same as before. In fig.3 (a) less disturbance is observed in the relative displacement profiles. The amplitude at forcing frequency and its first inharmonic is enhanced. A somewhat decrease is seen in the inharmonic at frequency ratio below 2.8. The first harmonic of the natural frequency at frequency ratio of 2 is not active as in the previous case. Multiple peaks as obtained imply the intensified non linear behaviour of the system.

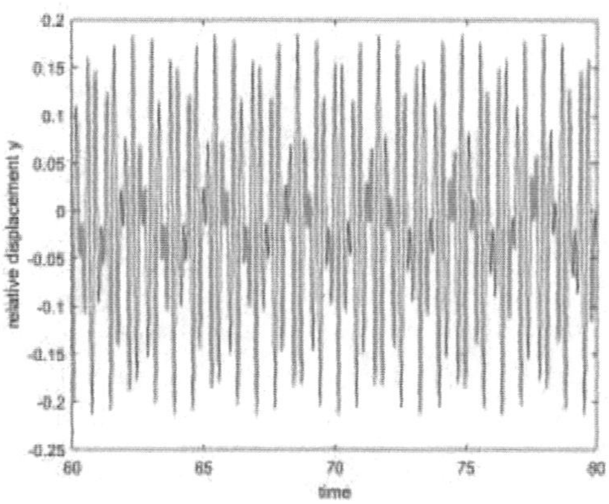

Fig. 3(a)

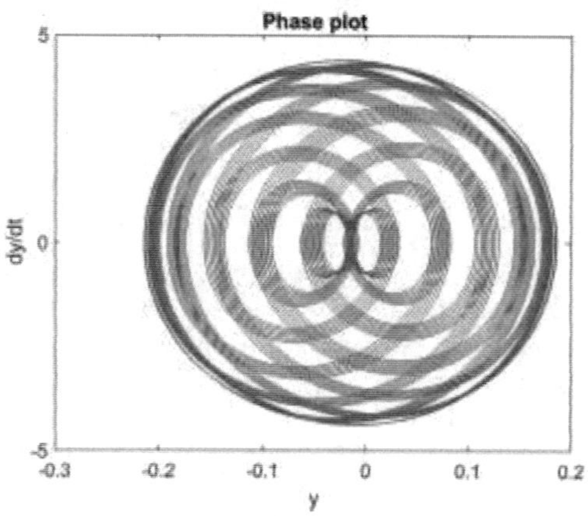

Fig. 3(b)

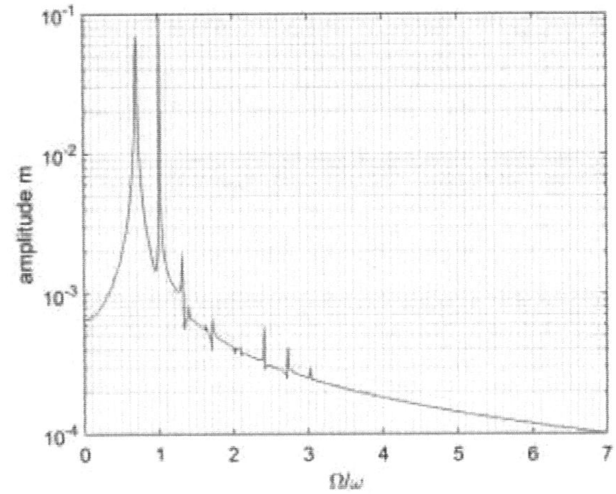

Fig. 3 (c)

Figure 3. (a) shows the relative displacement as a function of time while (b) shows the phase plot and (c) shows the spectra corresponding to figure (a) for excitation amplitude of 0.11m.

In all Fig.4, excitation amplitude was further increased to 0.17m. In fig.4 (a) near sinusoidal profiles of relative displacement are obtained representing a single dominant frequency at higher values of amplitude. In phase plot fig.4 (b), periodic motion can be seen with the same periodicity as that of the forcing frequency. In fig.4(c) it can be seen that the forcing frequency becomes even more dominant as compared to the previous two cases with its second and third harmonic present with the inharmonicity absent in this case along with the natural frequency and its harmonics.

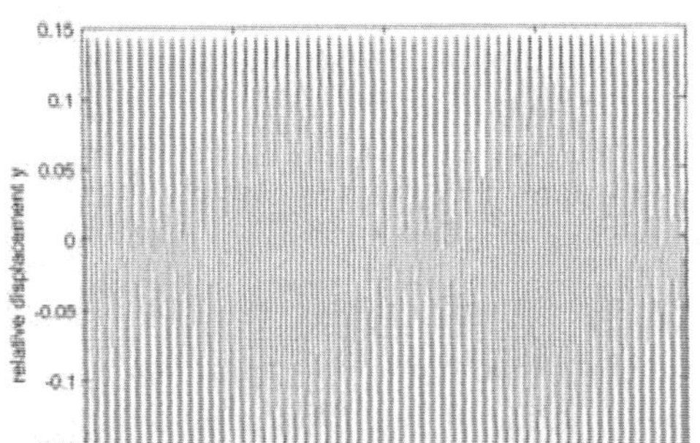

Fig. 4(a)

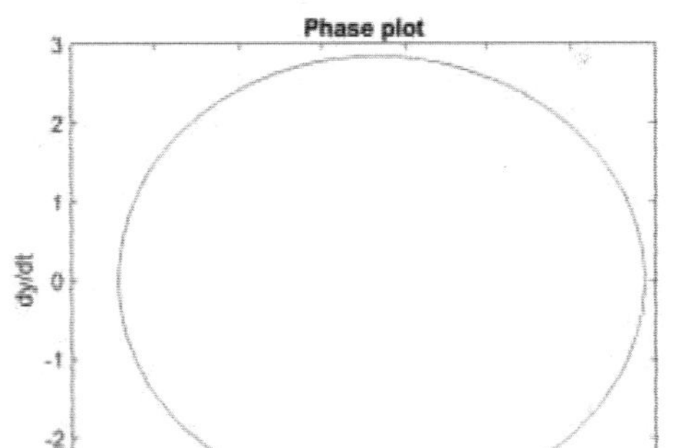

Fig. 4(b)

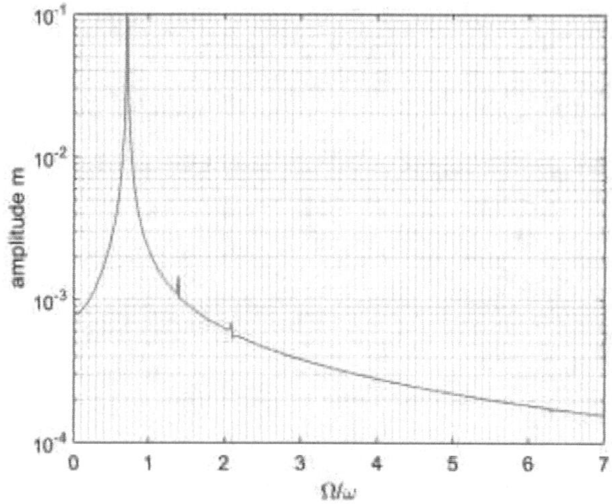

Fig. 4 (c)

Figure 4. (a) shows the relative displacement as a function of time while (b) shows the phase plot and (c) shows the spectra corresponding to figure (a) for excitation amplitude of 0.17m.

4. Conclusion

Behavior of a nonlinear single degree of freedom quarter car model has been investigated by running it over a sinusoidal road profile. The model contains a nonlinear damper with linear and cubic terms. Following conclusions are derived based on the above results.

1. At lower values of the excitation amplitude, the forcing frequency along with the natural frequency mainly contribute to the motion of the sprung mass with the natural frequency dominating. With the increase in amplitude it can be seen that the contribution from the forcing frequency becomes stronger and eventually the natural frequency and its harmonics disappear from the spectrum.
2. This can be understood since at low amplitude of road profile the forcing is weaker and the natural frequency of the oscillator can dominate the output. Upon increasing the road profile amplitude the forcing becomes stronger and finally the output is completely determined by the forcing frequency. Due to low amplitude of road profile the system is not perturbed strongly, therefore the inherent dynamics of the oscillator can come into play and so the natural frequency dominates the output.
3. It should also be noted that the in harmonics are present only in the first two cases when both the forcing frequency and the natural frequency is active. It is suspected that these in harmonics may be due to the interference of the inherent dynamics of the system with the forcing.
4. Increase in the values of road profile amplitude as studied in this paper shows that the system becomes more stable at the amplitude of 0.17, hence resulting in the better ride comfort of the passenger and better vehicle suspension performance.
5. The results can be used in the better understanding of the non linear dynamics, ultimately leading to better designed vehicle suspension systems. Passenger ride comfort can be significantly enhanced if the values of amplitude are kept in the mentioned ranges as shown in the research carried out in this paper.
6. In future work the combined effect of frequency and amplitude variation on the quarter car can be carried out for providing better understanding of the non-linearity in suspension systems.

5. References

[1] Mahajan, B. D., & Divekar, A. A. (2016, May). Modeling and system identi- fication of a quarter car suspension using Simulink. In 2016 IEEE International Conference on Recent Trends in Electronics, Information & Communication Technology (RTEICT) (pp. 180-183). IEEE.

[2] Fakhraei, J., Khanlo, H. M., Ghayour, M., & Faramarzi, K. (2016). The influence of road bumps characteristics on the chaotic vibration of a nonlinear full-vehicle model with driver. International Journal of Bifurcation and Chaos, 26(09), 1650151
[3] Fakhraei, J., Khanlo, H. M., & Ghayour, M. (2017). Chaotic behaviors of a ground vehicle oscillating system with passengers. Scientia Iranica. Transaction B, Mechanical Engineering, 24(3), 1051.
[4] Chatterjee, Anindya. "A brief introduction to nonlinear vibrations." Mechanical Engineering, Indian Institute of Science, Bangalore (2009).
[5] Litak, Grzegorz, et al. "Chaotic vibration of a quarter-car model excited by the road surface profile." Communications in Nonlinear Science and Numerical Simulation 13.7 (2008): 1373-1383.
[6] Zhou, Shihua, et al. "Nonlinear dynamic analysis of a quarter vehicle system with external periodic excitation." International Journal of Non-Linear Mechanics 84 (2016): 82-93.
[7] Borowiec, M., Litak, G., & Friswell, M. I. (2006). Nonlinear response of an oscillator with a magneto-rheological damper subjected to external forcing. In Applied Mechanics and Materials (Vol. 5, pp. 277-284). Trans Tech Publications Ltd.
[8] Zhou, S., Li, Y., Ren, Z., Song, G., & Wen, B. (2019). Nonlinear dynamic analysis of a unilateral vibration vehicle system with structural nonlinearity under harmonic excitation. Mechanical Systems and Signal Processing, 116, 751-771.
[9] Chang, S. C., & Hu, J. F. (2018). Codimension-two bifurcation analysis and chaos synchronization of a quarter-car vehicle model. Advances in Mechanical Engineering, 10(4), 1687814018771764.
[10] Litak, G., Borowiec, M., Friswell, M. I., & Przystupa, W. (2009). Chaotic response of a quarter car model forced by a road profile with a stochastic component. Chaos, Solitons & Fractals, 39(5), 2448-2456.
[11] Chen, C. C., & Chang, S. C. (2018). Stability, Bifurcation, and Quenching Chaos of a Vehicle Suspension System. Journal of Advanced Transportation, 2018.
[12] Chen, Y., Jing, X., & Cheng, L. (2013). Frequency domain analysis and design of nonlinear vehicle suspension systems. Institution of Engineering and Technology.
[13] Dobriyal, R., Paliwal, V., & Kumar, P. (2020, October). Dynamic analysis of a quarter car model using power law damper. In AIP Conference Proceedings (Vol. 2283, No. 1, p. 020009). AIP Publishing LLC.
[14] Paliwal, V., Dobriyal, R., & Kumar, P. (2020). Dynamic Analysis of a Quarter Car Model Moving Over a Gaussian Bump. *Journal of Graphic Era University*, 78-86.
[15] Paliwal V., Dobriyal R., Sengar N., Kumar P. (2021) Optimization of the Suspension Parameters of a Quarter Car Model with Nonlinear Damper. In: Joshi P., Gupta S.S., Shukla A.K., Gautam S.S. (eds) Advances in Engineering Design. Lecture Notes in Mechanical Engineering. Springer,Singapore.

Development and Performance analysis of Hybrid Based Polymer Composite

Kamlesh Kumar Jha[1] , Harsh singh[1] and Manoj Kumar Shukla[1]

[1] Galgotias University, School of Mechanical Engineering, Greater Noida-226001, India.

kjha.bgp@gmail.com

Abstract: Fiber Supported Polymer Network composites are generally utilized as primary materials. These materials trade ordinary materials for severe reasons. What's more, these materials are liked for their lightweight and strength. They additionally find widespread application in fuel pipelines and slurry conveying medium in the mechanical application. This exploration tends to the assembling of glass fiber and standard fiber mix composites. Impact of stacking arrangement on malleable, flexural, and sway estimations of untreated woven jute and Glass texture supported polyester half breed composites have been examined tentatively. Overlays were created by hand lay-up method and followed by restoring at room temperature for 48 h. Separate Jute and Glass fiber overlays were manufactured for correlation reasons. Likewise, these materials are tried for tribological properties like disintegration. Shifting the surface speed, a grouping of slurry and openness time disintegration done. Corrosion-tested examples were oppressed for SEM assessment. SEM micrographs on the FRP showed the misshapen that happened was fundamental because of cutting wear, and the method of disappointment was because of a fragile break. A general examination of the multitude of overlays' properties uncovered that the incorporation of glass in jute covers bring about an equilibrium in ecological effect and cost.

1. Introduction

1.1 General Background

It is an axiom that innovative improvement relies upon propels in the field of materials. Whatever the area might be, the last limit on headway depends upon materials. Composite materials address only a monster step in the always consistent undertaking of streamlining in materials. It is characterized as a multiphase framework that comprises at any rate two unique gatherings of materials that are artificially and particular and isolated by interfaces.

1.2 Fiber Reinforced polymeric (FRP) materials

FRPs comprise either thermosetting or thermoplastic polymers that structure the lattice stage in which filaments are inserted to shape composites. The part of the polymer framework is to move pressure to filaments adequately. It ties the filaments together and shields them from mechanical and substance harm by scraping the area of their surfaces or by the synthetic assault of some unessential ecological matter.

1.3 Classification of FRPs

FRPs are comprehensively named single layer or multi-facet composites. Single-layer composites comprise a few transparent layers, with each layer having a similar direction and properties. The whole cover hence shows up as a single composite. At the point when the constituent materials in each layer are identical, they are called covers. Diverse or mixture composites comprise layers of various constituent materials.

1.4 Hybrid Composite

Natural mindfulness everywhere in the world concentrated on utilizing cellulose fiber as support in the polymer network. The standard fiber-supported composites are sensibly solid, lightweight, and liberated from wellbeing dangers, biodegradable, and subsequently, they can be utilized as building materials and underground pipelines. Despite the benefits, they experience the ill effects of certain constraints, such as lower modulus, helpless dampness protection from assimilation, and quiet strength when contrasted and engineered filaments like glass and carbon.

1.5 Jute-Glass fiber-reinforced composites (Hybrid)

We have utilized jute fiber for such a long time in low-worth items, for example, gunny sacks, twine, and floor covering backing, that we will, in general, think about jute as a lousy quality asset. We have not acknowledged the immense potential for the utilization of jute in fiber-based composites.

2. Literature Review

K. Sabeel Ahmed, 2008 Crossovers have been found to have fantastic potential for more extensive applications. It is a subject of examination, and endeavors have been made to improve regular filaments' properties and make them savvy and arduous. The impact of stacking grouping on tensile, flexural, and interlinear shear properties has yielded significant outcomes. The fuse of glass in jute improves the properties of the coming about the mixture [1]. N. Sgriccia, M. C. Hawley, and M. Mishra, 2008 Examinations were directed to portray the surfaces of treated and untreated strands and to research water assimilation in standard fiber composites. Common fiber composites ingest more water than glass fiber composites. The soluble base treatment eliminates lignin and hemicelluloses from the outside of traditional strands. ESEM showed that the saline treatment covered the strands [2]. K. Sabeel Ahmed and S. Vijayarangan, 2008 Impact of stacking grouping on tensile, flexural, and interlaminar shear properties of woven jute-glass texture built-up isophthalic polyester composites have been tentatively assessed. The joining of glass in jute fiber composites improves the properties of coming about half-breed composites [3].Ha Na Yu, Seong Su Kim, In Uk Hwang and Dai Gil Le,2008 Porousness and primary strength of composite fortifications made from internal jute tangle and external glass fiber was contrasted and glass texture support for re-home of underground lines[4]. J.Stabik, Mackeson, and H.Tomanek, 2007 Disintegration opposition testing of Plastic lines made of PVC and polypropylene loaded up with calcium carbonate were contrasted and consequences of standard scraped spot test and hardness test[5]. G. Srinivasa Rao, Syed Waheedullah Ghori, 2021have studied the fiber tensile, impact and morphological properties of date palm and kenaf reinforced epoxy composite [13]. Z.Leman, E.S.Zainudin ,2021 has performed the experimental investigation and development and performance analysis of hybrid composite side door impact beam [14].

3. Experimental details

3.1 Description of panels fabricated.

The composites manufactured were as a rectangular board of uniform thickness. Cutting of examples for different tests is simple from this shape. Glass and jute utilized for various boards have the measurements demonstrated in Table 1. The thickness of the boards changed from each because of the thickness of the strands used in them. Notwithstanding, thickness above half inches was kept up, taking all things together commissions to investigate the disintegration wear rate adequately.

Table 1: Dimensions of fibers used in laminates

Panels	Length(inches)	Width(inches)	Thickness(mm)	
	L	W	Jute	Glass
Jute	12	10	-------	0.5
Glass	12	10	1.6	0.5
Hybrid 1	12	10	1.6	0.5

Hybrid 2	12	12	0.9	0.5
Hybrid 3	12	12	0.9	0.2

3.2 Material and sample preparation

From our overlays of 10" x 10" size, tests of different sizes were cut for various tests. Long pieces of 30" x 2.5" were cut as rectangular segments. For Flexural and Effect tests, the measures of the examples were 6.4" x 1.25". A hardness test was performed on the 10" x 10" overlay itself. The samples were cut utilizing hacksaw and were additionally machined to accomplish the ideal measurements as in fig 1. The examples for disintegration tests were machined to the element of 6.25" x 2.5". Thickness (x) being equivalent to that of instance yet not surpassing 17.5 mm. Openings were penetrated on it utilizing the boring machine.

Figure 1. Samples for Mechanical & NDT Testing

3.3 Test Exposer

The examination incorporates different sorts of tribological tests, mechanical tests, and portrayal. The other boundaries were thought of, which were fluctuated as needs be to have a comprehensive examination on GFRPs and Half and halves. The accompanying test openings were considered for execution examination. They are as per the following: Radiography Testing (NDT), Tensile Testing, Hardness Testing, Impact Testing, Flexural Testing and Erosion Wear Studies.

3.4 Tensile test and hardness test

A malleable test was done to notice the strength of the delivered material. The example of required measurement was disfigured with an expanding pliable burden applied along with the long hub of material at a consistent rate, as demonstrated in Fig 2. The Universal Testing Machine of KALPAK utilizing 40-ton load was used thinking about the appropriateness for different applications.

The hardness test was performed utilizing Bar Coal Impressed. The gadget's indenter was squeezed against the cover surface, and the hardness esteems got from the dial present on the top in BAI units. Fig 3 shows the device with a testing strategy.

Figure. 2 Universal Testing Machine

Figure 3. Hardness test was done using Bar Coal Indenter

3.5 Radiography Testing (NDT)

This procedure includes the utilization of infiltrating X-beam radiation (electromagnetic radiation of high recurrence and less frequency) to look at parts and items for defects. An X-beam machine appeared in fig 4 is utilized as a wellspring of radiation, and appropriate boundaries like Focus to Film Distance, X-beam tube Voltage, current, and openness time are to be set for the X-beam machine. At that point, X-beam radiation is coordinated along with the hand lay-up onto the X-beam film. After uncovering the film, it must be created and deciphered, as demonstrated in Fig 5.

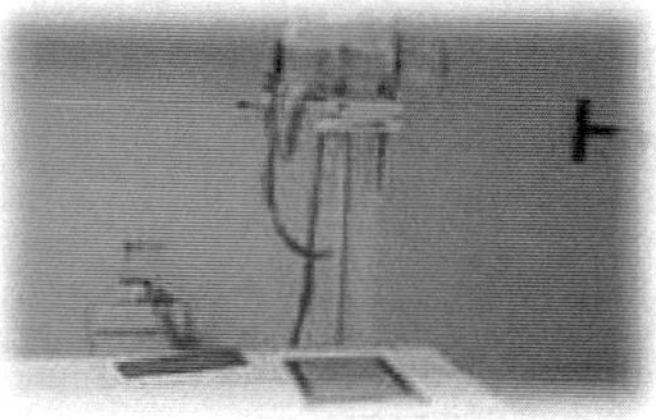

Fig 4. X-Ray Machine

Fig 5. Sample over Photographic Film

3.6 Impact Test and Flexural test

Flexural or curve tests are not difficult to do and can be utilized to determine the composite's fiber/framework interfacial strength. It was performed to decide the energy consumed by the material during the crack. Effect strength (IZOD) was estimated in an effect analyzer (Zed wick, Germany), as demonstrated in Fig 6.

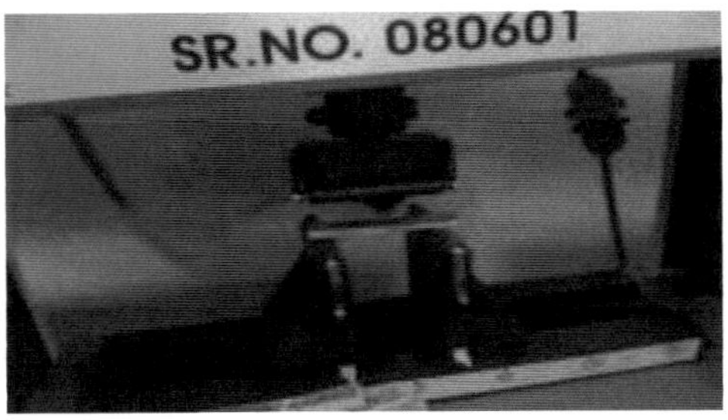

Figure 6. Three-point bending set-up

Figure 7. Specimen fitted in the apparatus.

4. Result and Discussion

4.1 Fiber Volume Fraction and Tensile test

Table 2. Fiber Volume fraction

Panel Type	Fiber vol. fraction (V_f in %)	Type of laminate
Glass	49.63	Thin
Jute	59.71	Thin
Hybrid 1	58.05	Thin
Hybrid 2	51.6	Thin
Hybrid 3	35.3	Thick

Table 3. Result of tensile test

Panel Type	Max Force	Tensile Strength	Young's modulus
	kN	kN/mm^2	kN/mm^2
Glass	80.4	228	18.312
Hybrid 1	20.14	47	6.47
Hybrid 2	20.04	89	5.67
Hybrid 3	9.72	78	7.377

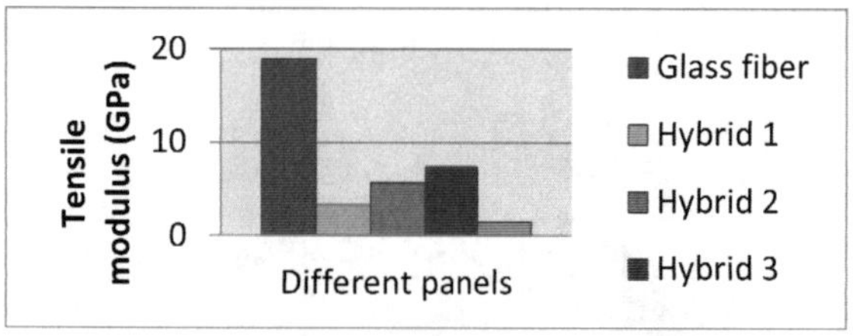

Figure 8. Tensile Modulus comparison chart

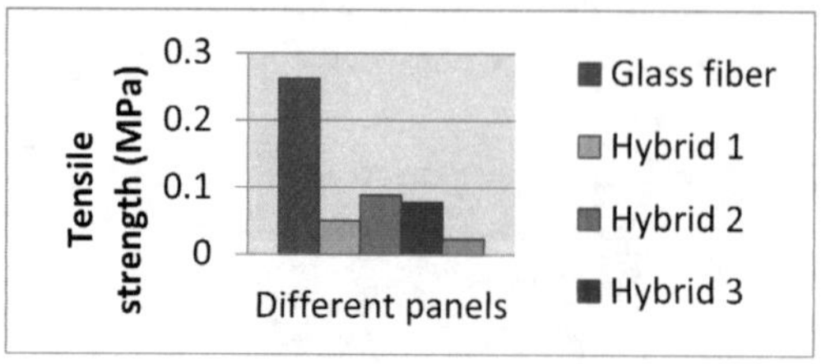

Figure 9. Tensile Strength comparison chart

Tensile test on various examples is impacted by the strength and modulus of the filaments. The rigidity of jute composite shifts between 23.56 – 29.49 MPa and modulus between 1.5 – 6.4 GPa. Sharp expansion in the rigidity of jute fiber happens with the consolidation of glass. They ascribed to the explanation that glass filaments are more grounded than jute strands. Around 42% expansion in elasticity occurred by changing the stacking grouping in half breed 2 from crossover 1. However, in half and half, I 53.8% increment while in crossover II 76% expansion in elasticity is accomplished when contrasted with jute boards. Epoxy mixture showed transitional conduct; thus, the order of tensile strength is: Glass> hybrid-II > Hybrid- I > Hybrid- III > jute

4.2 Hardness Test

Table 4. Result of Hardness test

Panels	Hardness Value (BAIU)	
	Glass top	Jute top
Glass	37.625	-------
Jute	-------	9.25
Hybrid 1	40.124	5.875
Hybrid 2	35.8	9.25
Hybrid 3	37.1	10.5

The results show that the hardness esteem relies on the interfacial holding between the limit filaments and the grid. However, with jute, an increment of 12% in hardness by utilizing epoxy rather than polyester. Stacking succession doesn't acquire a significant change hardness as the limit strands administer it.

4.3 Radiography Test (utilizing X-Beam)

Saw few spots on the uncovered photographic film on every one of the examples. They happened in patches locally and were not generally seen. These are known as Flaws which, for the most part, go with the hand lay-up fabricating measure. Delamination's, assuming any, are viewed as faint lines contrasted with great locales in the film. The accompanying tests were done on acceptable examples to get precise outcomes.

4.4 Impact Test

Table 5. Impact test values

Panel's type	Energy	Impact strength
Glass	24	141.3
Hybrid 1	10.6	47.3
Hybrid 2	7.1	63.43

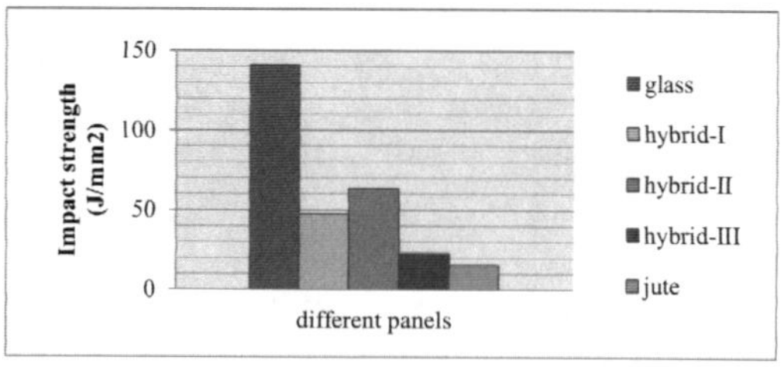

Hybrid 3	1.4	22.76

Figure 10. Impact strength of various panels

Impact test on various examples is reliant on the strength of the fiber. The glass fiber had the option to retain more energy than the half and halves. However, an expansion of the jute board sway test was seen by a fuse of glass filaments. An increment of 39% and 54% is seen in half and a half I and cross breed II individually compared to jute boards.

Table 6. Flexural values

Panels Types	Peak load(N)	σ_{max} (N/mm^2)	E(N/mm^2)
Glass	4616.56	368.624	355.7
Hybrid 1	855.5	42.008	225.8
Hybrid 2	297.15	46.41	513.12
Hybrid 3	446.15	267.695	8100

The fiber could retain more energy in light of solid interfacial holding between the glass fiber and the grid. Hybrid-II is 34% better than Mixture I in supporting effect, subsequently demonstrating that the stacking succession is one of the overseeing elements of effect disappointment. Stacking succession is one of the overseeing elements of effect disappointment.

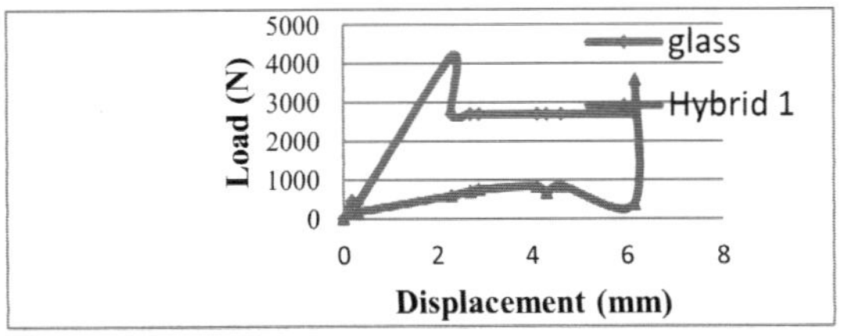

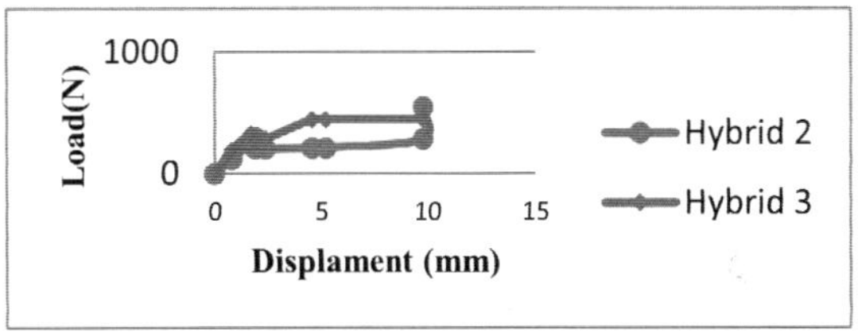

Figure 11. Load vs. dis. of glass and hybrid 1 panels **Figure 12:** Load vs. dis. of hybrid 2 & hybrid 3

The conversations on the flexural test perception are as per the following: Flexural strength and modulus are constrained by the power of the limit layers of support. The mark deviation from the linearity is the sign of disappointment because of the improvement break on the strain side. The heap uprooting chart is no direct under the flexural load. It is in this way hard to find the specific start of fiber disappointment. The breaking strands and fractional draw out of fiber and so forth are associated with disappointment due to flexural. Hybrid-III shows the greatest flexural strength among the distinctive half-breed boards. They show that the kind of pitch impacts flexural strength. It is more for epoxy than polyester.

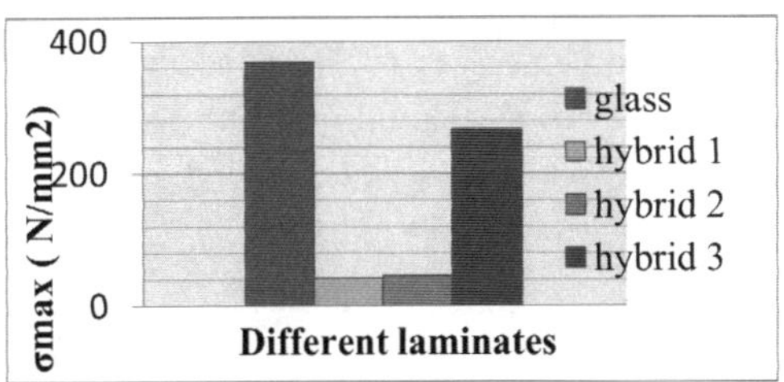

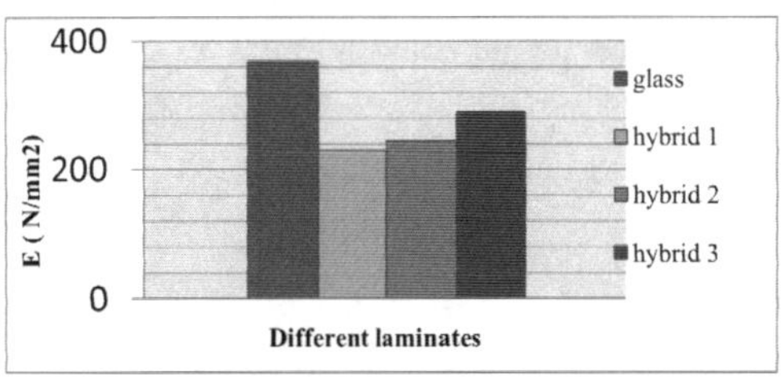

Figure 13. Flexural strength comparison chart **Figure 14.** Flexural Modulus comparison chart

Flexural isn't tremendously influenced by stacking arrangement. Thus, it exclusively relies on fiber and network strength and their interfacial holding.

5. Conclusion

Considering the different examinations on mechanical and tribological properties Sharp expansion in jute fiber's rigidity happens with glass consideration in the jute/glass fiber crossover composites. I 53.8% increment in half breed while in crossover II 76% increment in rigidity is accomplished contrasted with jute boards. Epoxy crossover showed a middle-of-the-road behavior. Around 42% expansion in rigidity occurred by changing the stacking succession in half and half 2 from mixture 1. The hardness esteem relies on the interfacial holding between the limit strands and the lattice. In the above study shows that the jute epoxy exhibited higher tensile and flexural properties. Impact test yields that an increment of 39% and 54% happens in half and a half I and crossover II individually when contrasted with jute boards. Half and half II is 34% better than crossover I in supporting effect in this manner, demonstrating that the stacking arrangement is one of the overseeing components of effect disappointment. Flexural strength and modulus are constrained by the strength of the limit layers of support and aren't tremendously influenced by stacking succession. They rely upon the interfacial holding between the fiber and network. Factors like time length, sway speed, speed, and slurry fixation are critical control factors influencing the mass misfortune and disintegration wear rate. The morphologies of dissolved surfaces saw by SEM recommend that general disintegration harm of composites comprises of framework evacuation and openness of fiber, fiber breaking, and expulsion of broken filaments. It is shown that hybridization with engineered fiber is a reasonable methodology for improving mechanical properties and characteristic strands' solidness. As an outcome, equilibrium in climate effect, execution, and cost could be accomplished.The above results of this research indicate better mechanical properties for jute-epoxy, which makes it better suited for the automotive applications rather than jute-polyester composites. Though the composites have some merits and demerits, the combination of the useful properties of two different materials, quicker processing time, lower manufacturing cost, etc., make them as a versatile material in the field of engineering and technology. Hence with this conclusion, it is sure that the technology shows composite is the most wanted material in the recent trend.

5.1 Scope of future work

In future this study can be extended to new hybrid composites using filler fiber combinations like Al2O3 and Sic, etc. The fiber, matrix interface in natural fiber composites can be enhanced by using advanced treatment methods of natural fibers. New technologies for decreasing the absorption property of jute can greatly expand its possibilities in new exciting and creative designs possible never before. In order to expand the use of jute fiber-based composites in adverse environments, it is necessary to interfere with natures recycling chemistry. High performance adhesives along with fiber modification can be used to manufacture structural jute-fiber based composite materials with uniform densities, durable in adverse environments, and high strength.

6. References

IOP Conf. Series: Materials Science and Engineering 1149 (2021) 012016 doi:10.1088/1757-899X/1149/1/012016

[1] T. Munikenchi Gowda, A.C.B. Naidu, Rajput Chhaya, Some Mechanical Properties of untreated Jute fabric reinforced polyester composites, Composites: Part A 30,1999, pp. 277-284

[2] N. Sgriccia, M.C. Hauley (2008), Characterization of Natural Fiber Surfaces and natural fiber composites, Composites Part A: Applied Science & Manufacturing Volume 39,2008, pp.1632- 1637.

[3] Ha Na Yu, Seong Su Kim, In Uk huang, Dai Gil Leel, Application of fiber composites to trenchless rehabilitation of underground pipes, Composites Structures, Vol 86, Issue 1-3, 2008, pp. 285-290.

[4] K. Sabeel Ahmed, S. Vijayarangan, A.C.B Naidu, Elastic Properties, notched strength, and fracture criterion in untreated woven jute-glass fabric reinforced polyester hybrid composite,2007, pp.2287-94.

[5] O.A.Khandoker, U.S.Ishiaku, A. Nakai, H. Hamada , A Novel processing technique for thermoplastic manufacturing of unidirectional composites reinforced with jute yarns, Composites Part A: Applied Science & Manufacturing, Volume 37, Issue 12,2006, pp. 2274-2284.

[6] K. Sabeel Ahmed, S. Vijayarangan, Tensile, Flexural and Interlaminar shear properties of woven jute and jute-glass fabric reinforced polyester composites, Journal of Materials Processing Technology, Volume 207, Issue 1-3,2008, pp. 330-335

[7] Ariel Stocchi, Bernd Lauke, Analia Vazquez, Celina Bernal, A novel fiber treatment applied to woven jute fabric/vinyl ester laminates, Composites Part A: Applied Science & Manufacturing, Volume 38, Issue5, 2007, pp.1337-1343.

[8] P.N.B Reis, J.A.M. Ferriera, F.V. Antunes, J.D.M. Costa, Flexural behavior of hybrid laminated composites, Composites Part A: Applied Science and Manufacturing, Issue 6, 2007, pp. 1612-1620.

[9] A. Bessadok, S. Roudesli, S. Marais, N. Follain, L. Lebrun, Alfa fibers for unsaturated polyester composites reinforcement, Composites Part A: Applied Science & Manufacturing, Vol 40, Issue 2,2009, pp. 184-195

[10] A.P. Harsha, Sanjeev Kr. Jha, Erosive Wear Studies of Epoxy based composites at normal incidence", Wear, volume 265, 2008, pp.1129-1125.

[11] O.P.Khanna,M.Lal,A Textbook of Material Science & Metallurgy, Dhanat Rai Publications,2006.

[12] U.S. Tewari. A.P. Harsha, A.M. Hager, Fredrich, Solid Particle Erosion of Unidirectional Carbon Reinforced Polyether ketone composites, 2002, pp. 992-1000

[13] G. Srinivasa Rao, Syed Waheedullah Ghori , Fiber Loading of Date Palm and Kenaf Reinforced Epoxy Composites: Tensile, Impact and Morphological Properties,2021.

[14] Z.Leman, E.S.Zainudin, Development and performance analysis of hybrid composite side door impact beam: An experimental investigation, 2021, pp. 173-197.

IOP Conf. Series: Materials Science and Engineering 1149 (2021) 012017 doi:10.1088/1757-899X/1149/1/012017

Critical Factors of Post Implementation of ERP in Higher Education Systems Survey Review

Manjeet Kumar[1], Adarsh Garg[1] and Avadhesh Kumar[1]

[1]SOB, Galgotias University, Uttar Pradesh, India
manjeet.kumar@galgotiasuniversity.edu.in

Abstract. ERP system is an automated system which helps in streamlining the process and/or operations and organise data of a business. Keeping this in mind the educational institutions have been rapidly inclined towards adopting the system for their organizations. Initially the ERP systems were designed for business sector, their adaptation in educational institutions require modifying the basic architecture of the system design which has to engulf the three major nexus of academic institution, v viz., finance, students and human resource. Along with this, the ERP implementation calls in a huge investment from the adopting institution. Each additional domain cost more. In addition to this the training of thousands of end-users is also a tedious task. Also, in any academic institution the security of the data related to student's examination and employee files is a major concern. This paper is a result of a survey conducted amongst the end-users of six major colleges and universities of Delhi/NCR who has already adopted the ERP system and is currently using the same. The paper covers the aforesaid issues and the opinion from the sight of end-users.

1. Introduction

ERP or enterprise-resource-planning, is a unified system via which a business can accomplish the efficient functioning the working units, logistics, manufacture, economic, management, and human reserve. ERP is being widely adapted by different business houses due the advantage of single-application-multiple operations [1]. ERP is a module-based software system whose architecture unifies the major operational divisions of any business operation. ERP architecture majorly incapacitate a central software component which are known as modules whose main aim is to interconnect the business organization functioning areas, such as finances, accounts, human-resource, manufacturing and raw-material management, client-customer-relationship management and supply-chain-management. Any business house if free to choose the type and number of modules of its needs from the pool of modules offered by the out-sourcing company, or if they want to design their own set-up that can also be done by the parent organization from scratch [2]. ERP architecture is one of the major reasons which determine the accomplishment and sustainability for a business house. Technically, ERP architecture is of 4 major types, viz., three-tier-architecture, web-based-architecture, service-oriented-architecture and cloud-computing-architecture [3].

In three-tire-architecture, there is no direct communication between the client and database. In this design, presentation layer is used for data browsing and also provides a user interface. The next layer, the application layer wherein the business rules and logic are implemented and all the data can be transferred or retrieved from the database servers [4]. The web-based-architecture, allows its users to have remote access to their ERP. Here the presentation is divided into two layers one is responsible for web-services and other is responsible for web-browser. Generally, the business houses do not support this type of ERP as they do not want to provide access to their system from outside their premises, but this model is very apt for usage in the educational institutions [5]. The third architecture model, service-oriented-architecture model, is entirely designed basing on the services rendered by the ERP and is not concerned with the network method used. In this model, for each service there is different designated system so that the functioning of the services is not affected [6].

IOP Conf. Series: Materials Science and Engineering 1149 (2021) 012017 doi:10.1088/1757-899X/1149/1/012017

The last architecture model type, cloud-computing-architecture is the result of the increased usage of cloud computing technology. Here the entire system and storage is done on cloud to ease the access and make the system process faster. Due to the success achieved by the ERP in the business sector, the educational institutions have now adapting the ERP rapidly to reorganize the functioning of their organization. But the reality is completely different as the ERP model for academic sector is entirely different from business sector [7]. From literature review it has been observed that approximately 40 percent of academic school have started using ERP. Due to the fact that the ERP was initially designed for the business sector, for implementation in the academic sector various different hosting services are available, the leading amongst these is SAP, Oracle, Microsoft. The implementation process for academic sector involves the basic steps such as: Preparation & Planning; Designing; Development; Testing; Deployment; Operation [8].

It has been also observed that since the majority of academic institutions re-engineer the ERP from the pre-existing business models which is entirely different from their requirement leads to the unsuccessful implementation of ERP in academic sector. Education ERP software provides a technical provision for the educational organizations to automate their database. Due to the increasing demand of educating the next generation, there is huge flow of data which has to be managed in a systematic, easy to access form. Along with this the data related to campus administration, student and/or staff management, their attendance, managing inventory, Finance and payroll procedures [9]. There are several advantages of using ERP in academic institutions, such as:

1.) As all the records are saved into the digital form it helps in saving the paper.
2.) Its implementation reduces the long lines for deposition of fees, which can be done using e-portals.
3.) Connects different departments of the campus with the additional support to the administration for handling the multiple locations, if any.
4.) Helps in improving employees working efficiency.
5.) When ERP has enabled the mobile application in the system, the clients can access the database at any time from any location.
6.) The human resource division can regulate the employee's attendance and in the end the payroll system can be automated [10].

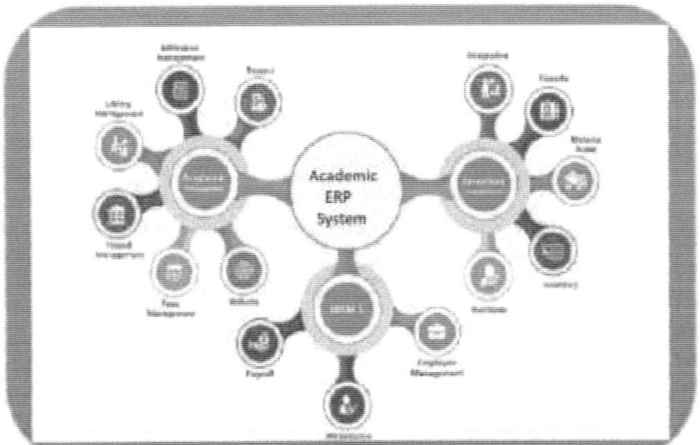

Figure 1: ERP Modules in Educational Institution

Figure 1 explains the common modules and sub-modules which are applicable for an academic ERP. These are explained as follows:

- *Student-Information-Module:* It is responsible for creating and managing the database related to students personal & academic details at the time of admission which includes - Student's photographs, earlier educational marksheets and certificates, other necessary needed proofs at the time of admission. It also sustains all-inclusive record of all students enrolled in the institution. It includes sub-modules such as: a) Student-Registration, b) Generation of Identity Cards, c) Student-database-enrolment [11].

- *Attendance-Management-Module:* It helps the staff and faculty to maintain the attendance with precision and with real-time access.
- *Fee-Management-Module:* It enables the access to the parents and/or guardians for making fee payments and along-with they can also access their wards performance in the organization. It also reorganizes the fee procedure by collecting different types of fees applicable in each student case individually. It includes following sub modules: a) Online-fee-payment, b) Fee's reminder and dues notification and details, c) Total Fees Collected Report [12]
- *Hostel-Management-Module:* It is an all-inclusive module that includes number of hostels available in the organization, students housed in each hostel, room allocation, fee handling and other hostel amenities. It helps the administration to keep a track of daily expenditure of hostel in real-time. It encompasses the following sub-modules: a) Admission, b) Room allotment, c) Hostel Mess management [12].
- *Library-Management-Module:* Helps in maintaining the records of library catalogue along-with books availability and issue-return details. Also permits the search of library-assets. Also produces automatically created real-time reports for total number of books, available books at that particular time, issued-books [12].
- *Course-Management-Module:* It permits the faculty to create time-table, faculty allocation, generation of quiz, test, assignments, enrolling of students in courses.
- *Staff-Management-Module:* It offers the end-users to access details of their payroll. This module can also be integrated with a bio-metric system to maintain the real—time attendance records of the employees. It enables the employees to apply for their leaves and keep a record of their vacation days. There are different classes of sub-modules in this domain: a) Employee-Information-Management b) Time-and-Attendance, c) Payroll, d) Employee-Self-Service [13].
- *Inventory-&-Purchase-Management-Module:* It keeps a record of all the purchases made, stock available and/or consumption records, movement of assets within and/or outside the organization [14].

2. Related work

ERP as generally considered is not a relatively new concept. It begins during 1960s as inventory control system which later on evolved into MRP I (1970s) to MRP II (1980s), to streamline and increase the efficiency of production operations of business as it required arranging and managing huge databases. Slowly and steadily ERP system was developed which founded its roots in the foundation of almost all business practices nowadays. [15]. The journey of ERP adaptation in the business sector began when a German software company, Siemens collaborated with SAP in 1987 and implemented ERP. In 1988, Dow Chemical designed their own ERP and implemented it to enhance globalization of their business. It has been estimated that the ERP industry reached 70 billion dollars by 2004 [16]. Since then, the ERP industry is growing continuously and is now not just limited to business sector but is now expanded to education sector also. Furthermore, due to its effective usage the significance of post-implementation success of ERP is a critical pointer for success of an organization. It has also been efficiently pointed out that nowadays the organizations are more focusing on establishing an effective ERP rather than thinking about whether they need it or not. In other words, nowadays ERP has become an integral part of a business and how effectively the business can establish the ERP is a major concern [17].

ERP distinctly impact the functioning of the whole organization with usage of above 1000 modules & 10,000 program applications which normally cost from 3 million to 4 million for smaller units to as much as above 1 billion for larger units, this cost covers the maintenance, development of ERP for 1 to

4 years depending on how complex the ERP is. This cost investment affects the ERP adaption, and is one of the major critical success factors [18]. The effectiveness and its measurement have been studied by various researchers and their results have rendered the ERP system unsuccessful.

From the review of literature, the critical success factors can be classified into:

a) Effect on data processing, labor decrease, time saving i.e., the unit work perspective.
b) End-user standpoint – includes reliability, ease-of-use, security.
c) The cost of ERP – which includes cost-benefit analysis, return of amount invested by organization [19].

In a survey by Tsichritzis, 1999, it was found that 40 percent of instigated ERP are fully effective and 20 percent are sparred as full failures [20]. Whereas Ptak and Schragenheim, 1999, reported that 60 percent to 90 percent of executed ERP are less operative than the anticipated [21]. It is a general conception that the education is an important factor which affects the training of end-users. But one should understand that both training and education are completely different from each other [22]. On one hand is education, which teaches the concept, how ERP affects functioning of an organization, end-user's familiarization with ERP amongst others. On the other hand, is training which is in actual is a hands-on operating experience of working with the ERP system including data input, understanding the functioning of each module of the implemented system in the organization, amongst others. Therefore, both education and training are two separate entities while analyzing the critical success factors [23].

The Educational ERP is an automated system devised specifically for the academic institutions for the smooth functioning of the organization. There are various advantages of using ERP in educational institutions, such as cost savings, data can be accessed with ease, organized database for each member of the institution, easy synchronization of updated data, etc. Along with this the fraternity members of the institution including administration staff, faculties, students, non-teaching staff, each can have access to the required database for their easy working [24]. During the literature survey, it was distinctly observed that ERP are generally employed in the technical-educational-institutions (TEIs). ERP is adopted for integrating the functioning of different divisions of any technical-educational-institution (TEI) and to use a centralized database for conduction of the workings of divisions with precision and suitability [25]. Generally, ERP comprises of two parts viz., hardware & software. Hardware includes the infrastructure of the ERP, which consists of computers, servers, network – intranet & internet, databases. Software consists of the information and data of all the modules functioning in the organization. Any failure in either of these two components of ERP leads to the failure of the system. Hence, both the hardware and the software are also the critical success factors of ERP adaptation. Another critical success factor is the security system of the ERP which includes the security of the database, information stored, access to the different modules of the system and admin control. Any breach into these domains directly affect the success rate of the adaptation process [26].

In spite of the challenges faced during and after adaption of ERP, business sector units are continuously using the ERP and elucidating their benefits continuously for last two decades. This have resulted in the adaption of ERP in the higher-education-institutions (HEIs) with an aim to increase the working efficiency of the organization. In 2004, Pollock & Cornford, discussed that the adaptation of ERP results in generating a tension amongst the employees of the organization which affects the working of the HEIs [27]. Same was pointed out by Feemster earlier in 2000, wherein it was described that when the ERP was adopted in college in U.S.A. as a merging manner with old and new database resulted in retraining of employees and instigating massive cost & pain [28]. In a study conducted by Frantz in 2002 it was pointed out that the HEIs prefer using the merge method of adoption as have achieved success in doing so [29]. Noaman, enlisted the major advantages of using ERP system in HEIs, viz.,

(a) Accessing different divisions data-base in real-time by the end-users.
(b) Usage of latest web and mobile-phone technology to allow the end-users and stakeholders to interact with the organization.
(c) Ease of access to the information related to the organization planning and management

(d) Reduced security related risks
(e) Improved working efficiency of the end-users [30].

Sabau, et.al, in 2009 also added the technical usage and improved business as the benefits of adoption of ERP in the educational sector [31].

3. Methods and methodology

For the purpose of the research work the data collection was using the random snowball sampling technique. Firstly, the educational institutions wherein the ERP system has been implemented were identified in the region of Delhi-NCR. A total of six such higher educational institutions were selected, all these HEIs are autonomous-UGC approved universities, with different vendors for ERP system. Furthermore, the respondents from these universities were contacted to fill-in an online-based questionnaire survey which was specifically designed to gain information for three types of responders, viz., faculty, students and administration employees. The purpose of collecting data from faculty and students is to survey the end-users for determining post-implementation success and from administration employees to survey the implementation procedure. The total sample size including all responders is 427 from six universities.

4. Results and Discussion

Using three different sets of questionnaires responses were collected from three separate sets of end-users – administration employees, faculties and students respectively. The questionnaires were specifically designed for the purpose of data collection in relation with to determining the adequate critical success factors – cost, user-friendly, end-user training, security, work efficiency, ease-of-access, database, information availability. On reviewing the data collected it was observed that although the administration vows for the secure database, the other end-users are doubtful towards the security of the data. The end-users who have received technical education consider the ERP user-friendly. Those users who have been trained in the ERP holds the same opinion but for others the results vary. It is extremely clear from the data analyzed that the work efficiency of the employees is increased after the adoption of ERP in their organization. In terms of feedback and registering problems in relation with the ERP., the majority of universities are using digital mode of communication for registering the complaints, this also helps in keeping the track of the types of complaints received and the solutions provided which can be centrally monitored by the university administration. Despite of the user-friendliness of the ERP a major number of responders still wants to save the data with themselves as they are concerned about the security and sustainability of the ERP.

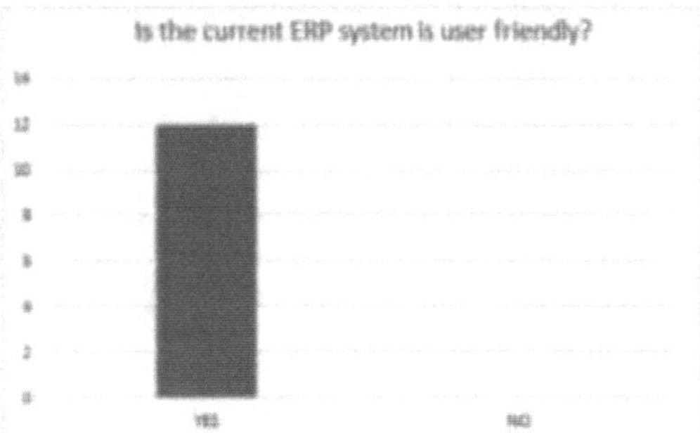

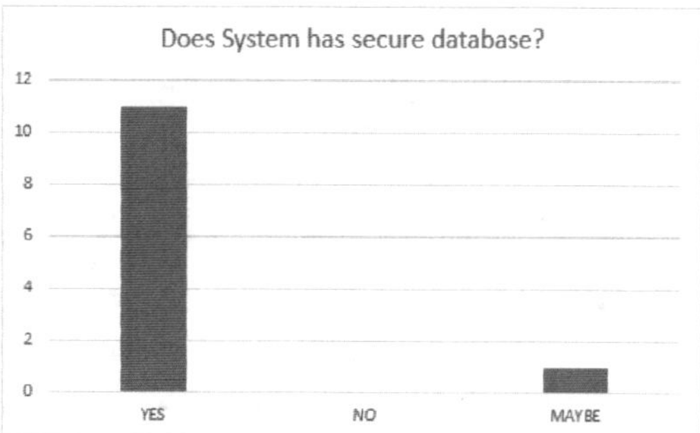

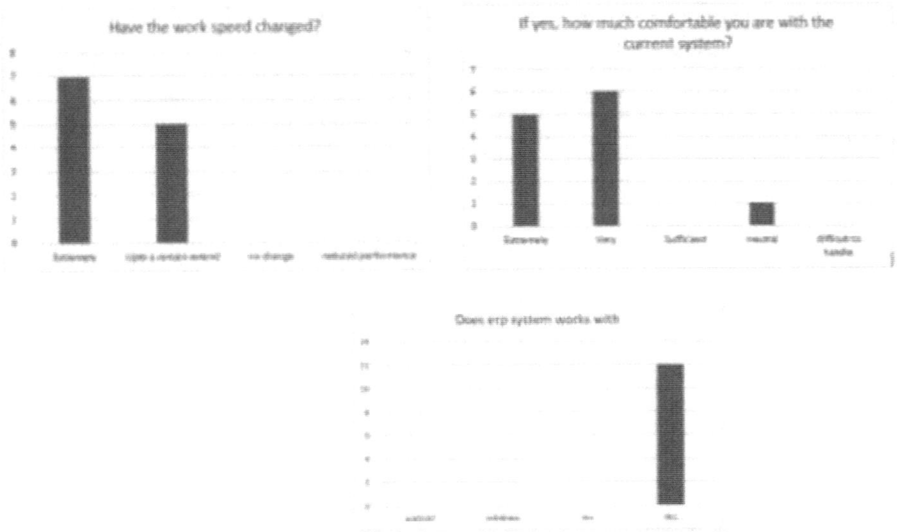

Figure 2: Responses of Administration Employees

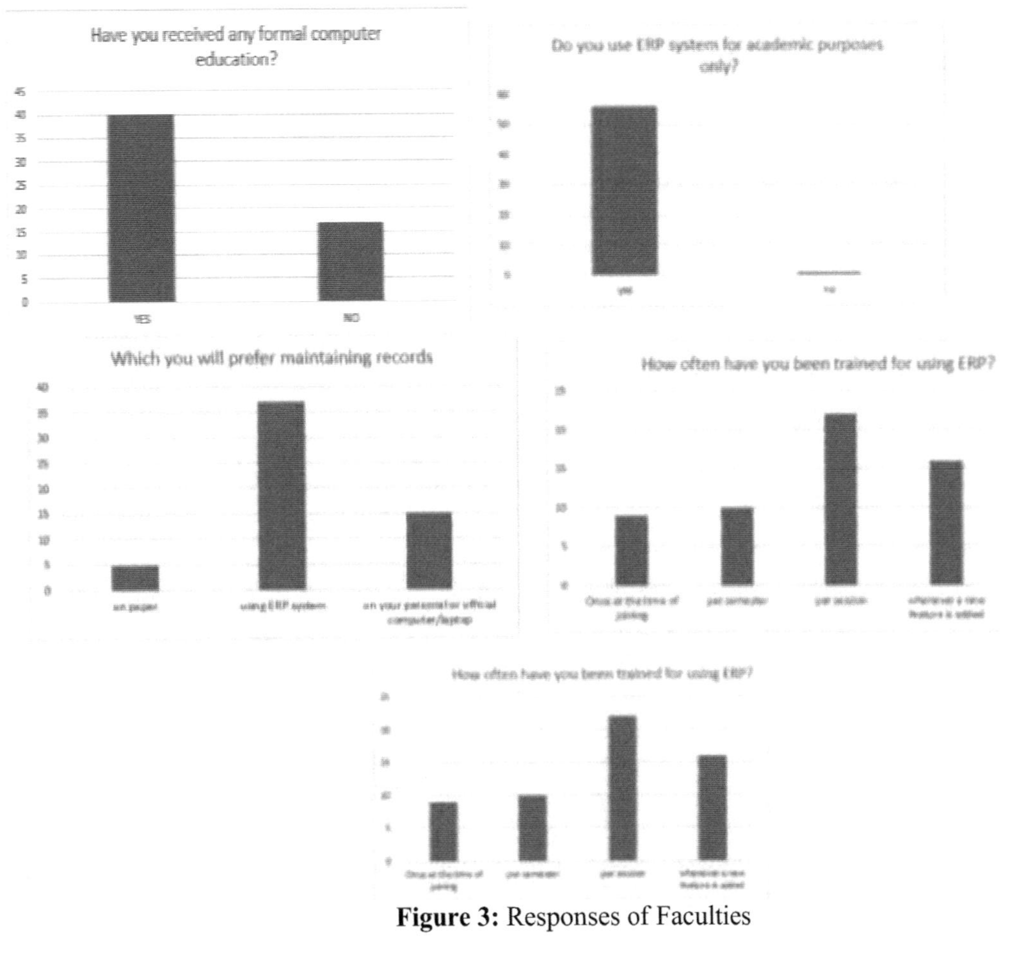

Figure 3: Responses of Faculties

IOP Conf. Series: Materials Science and Engineering 1149 (2021) 012017 doi:10.1088/1757-899X/1149/1/012017

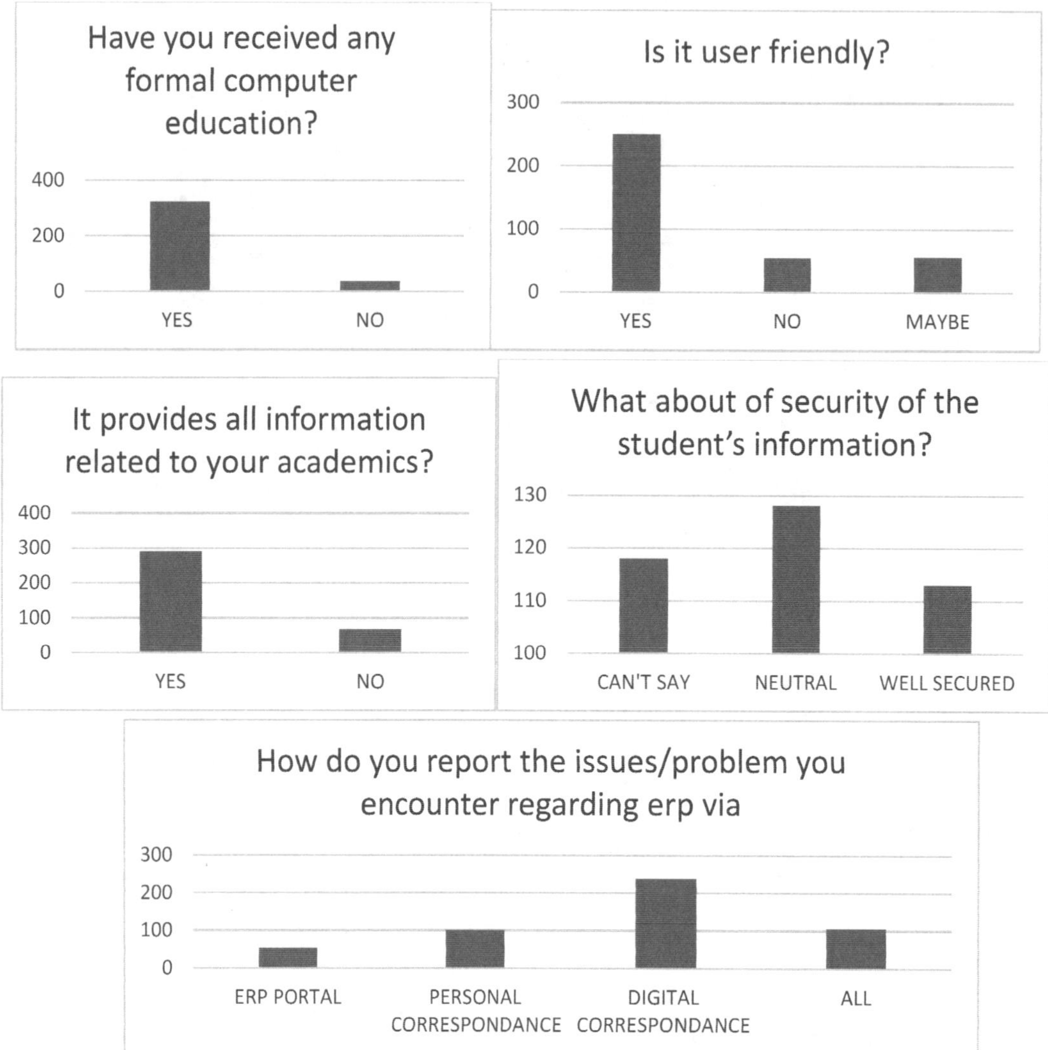

Figure 4: Responses of Students

5. Conclusion

ERP implementation is not only a tedious task as well as very expensive investments the risk factor is also very high, and having failure rate of ERP exceed the costs of system significantly. Ensuring successful Enterprise Resource Planning ERP system implementation in any domain of business need to required considerable research work in the present time and feature also. Although, Post Implementation success/failure critical factors distinctly contrarily defined by different sectors, organizations, units, which have adopted ERP but satisfaction of the operational system with usefulness. In addition to this, usual ERP employment study can be measured as a statistic-factor-research, which is unable to elucidate the dynamics of the execution procedure. This study has revealed that the critical success factors for determining the effectiveness of ERP adaptation is greatly dependent on the end-user training, cost input, step-wise implementation rather than the big-bang approach and lastly the technical training of the users.

References

[1] Allen, D., Kern, T., & Havenhand, M. [2002]. ERP critical success factors: An exploration of the contextual factors in public sector institutions. System Sciences, 2002. HICSS. Proceedings of the 35th Annual Hawaii International Conference on, 3062-3071.

[2] Birnbaum, R., & Edelson, P. J. [1989]. How colleges work: The cybernetics of academic organization and leadership. The Journal of Continuing Higher Education, 37[3], 27-29.

[3] Chapter 7 An Overview of Enterprise Resource Planning [ERP], PRIME VISION / C.A. FINAL / ISCA / AN OVERVIEW OF ENTERPRISE RESOURCE PLANNING [ERP]

[4] Heiskanen, A., Newman, M., & Similä, J. [2000]. The social dynamics of software development. Accounting, Management and Information Technologies, 10[1], 1-32. doi:10.1016/S0959-8022[99]00013-2

[5] Minou Parhizkar, Marco Comuzzi, [2017]. Impact analysis of ERP post-implementation modifications: design, tool support and evaluation. J. Computers in industry [elsevier], 2014, 84 [pp 25-38]

[6] Murphy, C. [2004]. ERP: The once and future king of campus computing. Syllabus-Sunnyvale then Chatsworth-, 17[7], 29-30.

[7] Ali Tarhini, Hussain Ammar, Takwa Tarhini, Ra'ed Masa'deh, [2015]. Analysis of the critical success factors for enterprise resource planning implementation from stakeholders' perspective: a systematic review. J. International business research. 2015, vol. 8 No. 4.

[8] Mohmed Y. Mohmed Al-Sabaawi. [2015] Critical success factors for enterprise resource planning implemantation success. International Journal Of Advances In Engineering And Technology . 2015 August, vol 8 issue 4 pp 496-506.

[9] Abugabah, A., & Sanzogni, L. [2010]. Enterprise resource planning [ERP] system in higher education: A literature review and implications. World Academy of Science, Engineering and Technology, 71

[10] King, P. [2002]. The promise and performance of enterprise systems in higher education. EDUCAUSE Quarterly,

[11] Divya Tuteja. [2014]. Implementation and updation of ERP systems in India: A survey. International Journal For Advance Research In Engineering And Technology. 2014, vol 2 issue III

[12] Mehlinger, L. [2006]. Indicators of Successful Enterprise Technology Implementations in Higher Education Business Morgan State Morgan State University.PhD Thesis

[13] Fisher, M. D. [2006]. Staff Perceptions of an Enterprise Resource Planning System Implementation: A Case Study of Three Australian Universities

[14] Habadi, A., et.al. [2017]. An Introduction to ERP Systems: Architecture, Implementation and Impacts. International Journal of Computer Applications [0975 – 8887] Volume 167 – No.9, June 2017

[15] Pollock, N., & Cornford, J. [2005]. Implications of enterprise resource planning systems for universities: An analysis of benefits and risks.

[16] Amgad Badewi, Essam Shehab. [2013]. Cost, benefit and financial risk [COBEFR] of ERP implementation. Proceedings of the 11th international conference on manufacturing research [ICMR 2013], 19th-20th sept 2013 pp 207-212

[17] Ashwaq AlQashami, Heba Mohammad, [2015]. Critical success factors [CSFs] of enterprise resource planning [ERP] system implementation in higher education institutions [HEIs]: concepts and literature review. J. Computer science & information technology [CS&IT]. 2015, 10.5121/csit.2015.51508.

[18] Hamzah Altamony, Dr Ali Tarhini, Dr Zahran Al-Salti, Ala'a Hamdi Gharaibeh, Dr Tariq Elyas. [2016]. The relationship between change management strategy and successful enterprise resource planning [ERP] implementations: a theoretical perspective. International Journal of Business Management And Economic Research. 2016. vol 7[4] pp 690-703.

[19] Atif Ali Gill, Arfan Shahzad, Subramanium Sri Ramalu [Jun-19]. An examination of post implementation success determinants of enterprise resource planning: insights from industrial sector of Pakistan. International Journal of Supply Chain Management, 2019 vol 8 no. 3.
[20] Tsichritzis, D. [1999]. Reengineering the university. Communications of the ACM, 42[6], 93-100.
[21] Ptak, C.A. and Schragenheim, E. [1999], "ERP: tools, techniques, and applications for integrating the supply chain", CRC Press-St Lucie Press
[22] Raafat Saade Harshjot Nijher. [2016]. Critical success factors in enterprise resource planning implementation: a review of case studies. Journal of Enterprise Information Management. 2016, vol 29 issue 1.
[23] Emad Abu-Shanab, Rasha Abu-Shehab, Mousa Khairallah. [2015]. Critical success factors for ERP implementation: the case of Jordan. International Arab Journal of e-technology. 2015, vol 4 no 1 January.
[24] Goel, et.al. [2012]. "Vulnerability Management for an Enterprise Resource Planning System." arXiv preprint arXiv: 1209.6484 [2012].
[25] Rabaa'i, A. A., Bandara, W., & Gable, G. [2009]. ERP systems in the higher education sector: A descriptive study. Proceedings of the 20th Australasian Conference on Information Systems, 456-470.
[26] Mustafa Agaoglu, E. Serra Yurtkoru, Ash Kucukaslan Ekmekci. [2015]. The effect of ERP implementation CSFs on business performance: an empirical study on users' perception. 4th international conference on leadership, technology, innovation and business management [science direct] [procedia - socila and behavioral sciences]. 2015, 210 [pp 35-42]
[27] Pollock, N., & Cornford, J. [2004]. ERP systems and the university as a "unique" organisation. Information Technology & People, 17[1], 31-52.
[28] Feemster, R. [2000]. Taming the software monster. University Business, 2[10]
[29] Frantz, R. [2002]. John stuart mill as an anti-intuitionist social reformer. The Journal of Socio-Economics, 31[2], 125-136.
[30] Noaman, A.Y., Ahmed, F.F., [2015]. ERP Systems Functionalities in Higher Education. International Conference on Communication, Management and Information Technology [ICCMIT 2015]. 1877-0509 © 2015 The Authors. Published by Elsevier B.V doi: 10.1016/j.procs.2015.09.100
[31] Sabau, G., Munten, M., Bologa, A., Bologa, R., & Surcel, T. [2009]. An evaluation framework for higher education ERP systems. WSEAS Transactions on Computers, 8[11], 1790-1799.

IOP Conf. Series: Materials Science and Engineering 1149 (2021) 012018 doi:10.1088/1757-899X/1149/1/012018

Development of advanced composite materials: Defining the fabrication process for hybridization of graphene and natural silk reinforced epoxy composites

K N Sanjeev Kumar[1*], S Sharma[2], P B Sharma[3] and A-H I Mourad[4]

[1]PhD Student, ASET, Amity University, Gurgaon, Haryana, 122051, India
[2]Associate Professor, Head of Mechanical Engineering, ASET, Amity University, Gurgaon, Haryana, 122051, India
[3]Vice Chancellor, Amity University, Gurgaon, Haryana, 122051, India
[4]Professor, Mechanical Engineering Department, College of Engineering, United Arab Emirates University, Al-Ain, PO Box 15551, UAE
knsanjeev@gmail.com

Abstract: In structural applications, Epoxy resin finds extensive application for its low cost, high strength to low weight ratio, higher amenability with superior dimensional stability in elevated temperatures, and chemical inertness. Furthermore, graphene has been extensively used as reinforcement materials to enhance strength and tensile/flexural stiffness. In contrast, natural silk within the other natural fibres showed higher impact strength and ductility, good biodegradability, and excellent biocompatibility for a reinforcement filler material. Several ongoing field research on graphene and natural silk as an individual reinforcement in the epoxy composites have shown significant positive results. This paper concentrates on the research gap of hybridization of graphene and natural silk fibres (SF) as a combined reinforcement material to develop lightweight and high-performance epoxy composite material for structural applications. This paper discusses the suitable material selection and fabrication process for hybrid GO/SF epoxy composite. Graphene Oxide (GO) choice for the reinforcement on epoxy resin by (0,0.5,1,1.5,2,3 and 5) % volume fractions via wet transfer ultrasonication process and plain-woven Bombyx mori silk for 60% SF reinforcement discussed in the paper. The optimum volume fraction of GO and SF promoting significant improvements of the epoxy resin composites' mechanical properties needs further evaluation during material characterization.

Keywords: Graphene, Natural Silk, Epoxy composites, Advanced Composite Materials

1. Introduction:

Epoxy resins find extensive application for the production of lightweight structural composite materials due to their characteristic properties – High amenability with dimensional stability, good adhesion, chemical resistance, higher temperature extreme tolerance, high strength to weight ratio, low moisture absorption, along with fair mechanical properties at low cost [1-2]. Nevertheless, when get cured, the epoxies become brittle from the high degree of cross-linking leading to lower impact strength. Therefore, there is a necessity to modify monomers in the epoxy to upgrade the anticipated mechanical properties like a low creep, high modulus, strength, and outstanding thermal and chemical stability for structural application usage [3-4]. Epoxy resin is compatible with all standard reinforcements to enhance its impact strength and toughness [5-8].

Graphene is a single layer of carbon atoms into 2D honeycomb lattice packing with sp2 bonds. Graphene is very light (0.77 mg per m2), very flexible, and still has an excellent tensile strength of 130 GPa. This makes it has an excellent reinforcement material for polymer composites. The graphene has a flat structure with an intuitive higher surface area (2630 square meters per gram). This characteristic property enables graphene to get dispersed into the matrix without any surface modifications and makes it a good reinforcement material for the composites [9]. The mechanical and chemical properties of the matrix material become superior upon the addition of graphene due to the enhancement in strength and bonding between the graphene and matrix material [10]. Graphene fillers in the epoxy resin matrix increase the impact strength and toughness [11]. Further, graphene enhances the fire retardant for the epoxy resin matrix, which is essential for various industrial applications [12].

Silk is a natural protein fibre and is one of the strongest natural fibres. Silk has a lower density (1.3 x 103 Kgm-3), higher tensile strength (650-750 MPa), higher breaking energy (60 to 80 MJm-3), good elasticity, and excellent resilience. Natural silks are applied as a reinforcement filler medium, either as a single medium or combined with other filler materials in epoxy resin composites, to advance the mechanical properties, especially flexural and impact strength [13]. Inherently, silk fibres show excellent mechanical toughness as they can absorb and dissipate energy in parallel during deformation [14]. Silk as reinforcement filler improves the tensile and bending strength of epoxy composites [15]. Further, the usage of natural fibres in composites provides an option for biodegradable and bio-derived products.

Several ongoing field research on graphene and natural silk as an individual reinforcement in the epoxy composites have shown significant positive results. The present study focuses on the research gap of hybridization of graphene and natural silk fibres (SF) as a combined reinforcement material to develop lightweight and high-performance epoxy composite material for structural applications. The present study explores the novel concept for successfully integrating resilient and tough natural silk and strong and stiff graphene as reinforcement filler for epoxy resin composites for custom-built mechanical properties [16-17]. Moreover, based on the detailed review, it is expected that integration of tough natural silk and strong graphene could harmonize to fabricate advanced composite materials with enhanced properties fit for a broader scope of structural applications [16-17]. This paper defines the suitable material selection and fabrication process for hybrid GO/SF epoxy composite. However, the agglomeration effect and the optimum volume fraction of GO and SF promoting significant improvements of the epoxy resin composites' mechanical properties need to be studied during material characterization to understand the reinforcement's interfacial interaction with the epoxy resin matrix.

2. Material selection for the hybridization of graphene and natural silk reinforced epoxy composite

2.1 Graphene

Graphene and its derivatives have been applied openly in high-performance material research for property enhancements. In graphene, the stability of the sp2 bonds forming the hexagonal lattice counters the range of in-plane deformations, accounting for outstanding mechanical properties [18]. However, pure graphene is inherently unstable owing to the presence of the delocalized π electrons and agglomerates, apparently due to van der Waals interactions and high specific surface area [18]. Graphene oxide (GO) is stable with oxygen functional groups on the basal plane that take away the π-π stacking interactions and aggregation noted with pure graphene. GO is an extremely flexible material with the attributes of pure graphene and relatively simpler to process. GO is preferred over graphene nanoplatelets regularly due to the functional groups permitting filler and matrix bonding interactions, although the modulus of GO is only about 25% of that of monolayer graphene[18]. The reinforcement efficiency of GO does not get impacted by the number of GO layers [19]. GO has a similar structure to graphene with active hydroxyl, carboxyl, and epoxy groups on the surface and edge, allowing for strong hydrophilicity and provide an active structural basis for surface modification [20]. These characteristics give GO satisfactory composite capability and improve the dispersion of GO in the matrix resin. Moreover, GO is preferred for integrating with silk fibres due to its good uniform distribution solution characteristics [21]. The study shows that Graphene with a higher surface area and under 2% of volume fraction would provide higher results for augmentation of the mechanical properties [22]. During the hybridization, homogenous graphene dispersion in the matrix material needs to be ensured to avoid the formation of aggregates and failure points. The high surface area of reinforcement fillers enables load distribution over a large surface resulting in better performance.[16-17]. Hence GO reinforcement on epoxy resin by (0, 0.5, 1, 1.5, 2, 3, and 5) % volume fractions shall be considered for the study.

2.2 Natural Silk

Bombyx mori (BM) silk materials are often integrated with Graphene, GO, and Carbon Nanotubes to construct mechanically enhanced materials. BM silk fibre (SF) is a prominent natural fibre that exhibits excellent mechanical properties comparable to those of other plant fibres [23]. BM silk fibre has prominent features such as high ductility and strength due to its high crystallinity in molecular structure. The adding of silk fibre volume fractions in the epoxy composite enhanced the impact properties [24]. The silk fibre volume fraction of 30% by weight provided peak enhancements in young modulus, tensile strength, and elongation at break than the neat epoxy. The interfacial adhesions were improved, primarily resulting in improvement of the impact strength [25].SF reinforced epoxy composites display considerable higher tensile strength and flexural strains at break than plant fibre and glass fibre reinforced composites [26]. Several noncovalent interactions in the SF/GO system enhance the composites' strength and toughness than their reinforcements [27]. Silk is also a suitable reinforcement medium as it permits a higher volume fraction (30 to 70%). A higher volume fraction of silk fibres expand the impact and flexural strength, tensile strength, and toughness of the matrix material. SF reinforcement to the extent of 70 vol% has been achieved in epoxy composites using compression moulding, and six times higher impact strength than neat epoxy @ 60 vol% of SF reinforcement was achieved [23]. A balanced weave pattern and the silk surface treatment are critical for stronger SF/epoxy interfacial adhesion and the advancement of mechanical properties. The moisture in silk fabric needs to be controlled before laminate fabrication. In another study, woven silk preform was a better choice than random silk preforms due to higher thickness variations and lower compaction, leading to lower fibre content [28]. Pressurized vacuum-assisted resin transfer moulding (VARTM) @ 30 psi allowed for uniform impregnation over silk preform resulting in the lesser void formation and strong adhesion. The specific flexural strength and modulus were increased by 30% and 65% over plain epoxy [24]. Hence, for the study, a plain-woven BM silk fabric around 90 GSM shall be used.

3. Fabrication of GO/SF epoxy composite

The GO/SF epoxy composite fabrication process has been briefed below [29-39] from the above literature study. The fabrication process for the hybrid composite is schematically shown in Figure 1.

1. The diglycidyl ether of Bisphenol-A (DGEBA) based epoxy resin with low viscosity shall be used for the matrix medium. The low viscosity enables easy flow of resin mixture in the VARTM. The curing ratio and the curing conditions shall be as per the resin manufacture recommendations.
2. A plain-woven BM silk fabric around 90 GSM shall be used for SF reinforcement.
3. The volume concentration of GO considered for the study is 0 %, 0.5%, 1% 1.5 %, 2%, 3%, and 5% by weight in the epoxy matrix.
4. GO volume fraction shall be introduced into epoxy resin via wet transfer by dispersing GO in ethanol through ultrasonication process for 2 hours before mixing with epoxy resin. The resultant mixture shall be placed in the oven at 60° C to remove excess ethanol, and stirring shall be continued for 24 hours to get a homogeneous mixture by maintaining at 50° C.
5. Hybrid GO/SF epoxy composite laminate comprising Silk fibre layers with a volume concentration of 60% shall be fabricated for various GO/epoxy volume concentrations using a VARTM set- up @ 30 psi to obtain a higher SF volume fraction with improved interfacial adhesion.

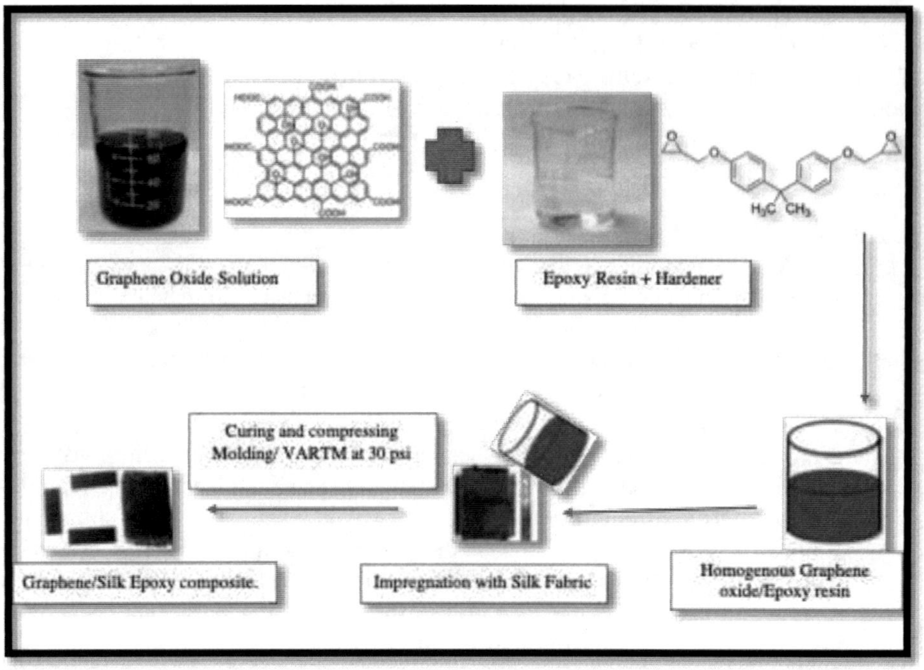

Figure 1 Schematic diagram of processing of hybrid GO/SF epoxy composite.

4.Conclusion

The study's work reveals a definite scope for positive hybridization effect, resulting in enhanced mechanical properties like modulus, ultimate stress, and toughness of Graphene/Silk epoxy composite compared with their reinforcement with epoxy composites. However, the characterization and agglomeration effect after combined reinforcement of Graphene and Silk to be studied in detail post the fabrications of the test samples to understand the interfacial interaction between the reinforcement fillers and epoxy resin matrix.

In this paper, the following conclusions on the material selection and fabrication process are arrived at for the hybridization of the Graphene/Silk epoxy composite, based on the research studies:

1. GO and BM silk selected for reinforcement with the low viscosity DGEBA epoxy resin. The concentration of GO shall be 0 %, 0.5%, 1% 1.5 %, 2%, 3%, and 5% by weight in the epoxy matrix. A plain-woven BM silk fabric around 90 GSM was selected for the SF reinforcement.
2. GO will be introduced into epoxy resin via wet transfer and ultrasonication process.
3. Hybrid GO/SF epoxy composite laminate comprising SF layers with a volume concentration of 60% shall be fabricated for various GO/epoxy volume concentrations using a pressurized vacuum-assisted resin transfer moulding (VARTM) @ 30 psi and not through the wet-lay process, and
4. The schematic diagram (Figure 1) was established to process the hybrid GO/SF epoxy composite.

Acknowledgement
This paper and the related research are the outcomes of the exceptional support from the supervisors.

References

[1] Harke 2006 *Epoxy resin systems for composites* Harke Chemical GmbH (Thailand) pp 1-20.

[2] A.B. Maureen, J M Cary and D N John Epoxy Resin Hexcel Corporation.
[3] Njuguna J 2016 *Lightweight composite structures in transport: design, manufacturing, analysis and performance* (Woodhead Publishing)
[4] Pielichowski K and Njuguna J 2005 *Thermal degradation of polymeric materials* (Shawbury: Smithers Rapra Publishing)
[5] Mourad A-H I, Bekheet N, El-Butch, A A-Latif, Nafee D and Barton D C 2005 *The effects of process parameters on the mechanical properties of die drawn polypropylene* Polymer Testing 24(2) pp169-180
[6] Babaghayou M I, Mourad A-H I, Lorenzo V, Chabira S F and Sebaa M 2018 *Anisotropy evolution of low-density polyethylene greenhouse covering films during their service life* Polymer Testing 66 pp146-154.
[7] Babaghayou M I, Mourad A-H I, Ochoa A, Beltrán F and Cherupurakal N 2020 *Study on the thermal stability of stabilized and unstabilized low-density polyethylene films* Polymer Bulletin pp 1-17
[8] Babaghayou M I, Mourad A-H I, Lorenzo V, de la Orden M U, Urreaga J M, Chabira S F and Sebaa M, 2016 *Photodegradation characterization and heterogeneity evaluation of the exposed and unexposed faces of stabilized and unstabilized LDPE films*, Materials & Design, 111 pp279-290.
[9] Yuchi F, Lianjun W and Jianlin 2010 *Preparation and electrical properties of graphene nano sheet/Al2O3composites* Carbon Science 48(10) pp 1743-49.
[10] Kuilla T, Bhadra S, Yao D, Kim N H, Bose S and Lee J H 2010 *Recent advances in graphene-based polymer composites* Progress in Polymer Science vol35 (11) pp1350–75.
[11] Reis J.M.L 2012 *Effect of temperature on the mechanical properties of polymer mortars* Materials Research 15(4)
[12] Liu S, Yan H, Fang Z, et al. 2014 *Effect of graphene nanosheets on morphology, thermal stability, and flame retardancy of epoxy resin*. Compos Sci Technol 90 pp 40–47.
[13] Shah DU, Porter D, Vollrath F 2014 *Opportunities for silk textiles in reinforced bio composites: studying through-thickness compaction behaviour.* Compos A Appl Sci Manuf 62:pp 1–10.
[14] Guan J, Porter D, and Vollrath F 2012 *Silks cope with stress by tuning their mechanical properties under load* Polymer 53(13) pp 2717–26.
[15] Shah DU, Porter D and Vollrath F 2014 *Can silk become an effective reinforcing fibre? A property comparison with flax and glass reinforced composites*. Compos.Sci.Technol. 101 pp 173–183
[16] Sanjeev K K N, Sanjeev S, Mourad A-H I, PB Sharma (press, 2021) *A literature review for development of advanced composites materials by reinforcement of epoxy composites with graphene and natural silk*, Lecture notes in Mechanical Engineering.
[17] Sanjeev K K N, Sanjeev S, Mourad A-H. I., PB Sharma 2020 A*dvanced composite materials: methodology for hybridization of graphene and natural silk reinforced epoxy composites* Proceedings of the national conference on advancements & modern innovations in engineering and technology (Gurgaon- AMIET 2020) pp 62-65
[18] Dimitrios G P, I A Kinloch, and R J Young 2017 *Mechanical properties of graphene and graphene-based nanocomposites* Progress in Materials Science 90 pp 75–127
[19] Z Li, I A Kinloch, and R J Young 2016 *The role of interlayer adhesion in graphene oxide upon its reinforcement of nanocomposites* Philos Trans R Soc A 374
[20] Y H Fan, S W Yu, H M Wang, Y H Yao, Y Wang and C H Wang 2019 *Study on preparation and properties of graphene reinforced epoxy resin composites* IOP Conf. Series: Materials Science and Engineering **634** 012044
[21] Y Wang, R Ma, K Hu, S Kim, G Fang, Z Shao and V V Tsukruk *2016 Dramatic enhancement of graphene oxide/silk nanocomposite membranes: increasing toughness, strength, and young's modulus via annealing of interfacial structures* ACS Appl Mater Interfaces 8(37):24962-73
[22] Annamahesh A, Sunitha K, and Rangarajan S 2019 *Study on mechanical behavior of graphene*

based polymer composites IJITEE V8 I4.

[23] Yang K, Robert O R, Yizhuo G, Su J W and Juan G 2016 *High volume-fraction silk fabric reinforcements can improve the key mechanical properties of epoxy resin composites* Materials & Design 108 pp 470-478.

[24] Yang K, Sujun W, Juan G, Zhengzhong S and Robert O R 2017 *Enhancing the mechanical toughness of epoxy-resin composites using natural silk reinforcements* Scientific reports 7, no. 1 pp 1-9.

[25] Zuraidah Z, Noor I S, Mohd Y and Mat U W 2020, *Characteristics of continuous unidirectional silk fibre reinforced epoxy composites*, Journal of Advanced Research in Materials Science 69, Issue 1 pp 16-28

[26] Shah DU, Porter D and Vollrath F 2014 *Can silk become an effective reinforcing fibre? A property comparison with flax and glass reinforced composites*. Compos.Sci.Technol. 101, pp 173–183

[27] Ren J, Liu Y, Kaplan D.L. *et al.* 2019 *Interplay of structure and mechanics in silk/carbon nanocomposites* MRS Bulletin **44,** pp 53–58

[28] Youssef K H, M. A Yalcinkaya, Gorkem E G ,Maya P M and Amirkhosravi, M C A 2018 34th International Conference of the Polymer Processing Society, (Taipei, Taiwan)

[29] Mourad A-H I, Cherupurakal N, Hafeez F B, I A Genena F, S Al Mansoori M and A Al Marzooqi L 2020. *Impact Strengthening of Laminated Kevlar/Epoxy Composites by Nanoparticle Reinforcement* Polymers 12(12) pp.2814.

[30] Mourad A-H I, Zaaroura N. and Cherupurakal N 2019 *Wet lay-up technique for manufacturing of advanced laminated composites* Karbala International Journal of Modern Science 5(1) p.5

[31] Mourad A-H I, Al Mansoori M S, Al Marzooqi LA, Genena F A and Cherupurakal N 2018 *Optimization of curing conditions and nanofiller incorporation for production of high performance laminated kevlar/epoxy nanocomposites* ASME Pressure Vessels and Piping Conference American Society of Mechanical Engineers Digital Collection.

[32] Fouad H, Mourad A-H I, ALshammari B A, Hassan M K, Abdallah M Y and Hashem M 2020 *Fracture toughness, vibration modal analysis and viscoelastic behavior of kevlar, glass, and carbon fibre/epoxy composites for dental-post applications* Journal of the mechanical behavior of biomedical materials 101 p.103456.

[33] Mourad A-H I, Idrisi A H, Zaaroura N, Sherif M M and Fouad H 2020 *Damage assessment of nanofiller-reinforced woven kevlar KM2plus/epoxy resin laminated composites* Polymer Testing, 86, p.106501.

[34] Mourad A-H I, Ayad O G, Rahman A, Hilal-Alnaqbi A and Abu-Jdayil B I 2016 *Experimental investigation of kevlar KM2Plus nano-reinforced laminated composite thermo-mechanical properties* ASME 2016 pressure vessels and piping conference American Society of Mechanical Engineers Digital Collection.

[35] Pathak A K, Borah M, Gupta A, Yokozeki T and Dhakate S R 2016 *Improved mechanical properties of carbon fibre/graphene oxide-epoxy hybrid composites* Compos Sci Technol 135 pp. 28-38.

[36] Change W, Kang Y,Yizhuo G,Jun X, Robert O.R and Juan G 2019 *Mechanical properties and impact performance of silk-epoxy resin composites modulated by flax fibres,* Composites Part A Applied science and manuf., V117,pp 357-368.

[37] Jamali J, Mourad A-H I, Fan Y, and Wood J T 2016 *Through-thickness fracture behavior of unidirectional glass fibres/epoxy composites under various in-plane loading using the CTS test* Engineering Fracture Mechanics 156 pp83-95.

[38] Ahmed W K, Mourad A-H I 2012. *Using fibre reinforced polymer to restore deteriorated structural members*. Int. J. Mat. Mech. Eng 1 1-7.

[39] Idrisi A H, Mourad A-H I, Abdel-Magid B, Mozumder M, and Afifi Y 2019 *Impact of the harsh environment on E-glass epoxy composite* ASME Pressure Vessels & Piping Conference American Society of Mechanical Engineers Digital Collection.

IOP Conf. Series: Materials Science and Engineering 1149 (2021) 012019 doi:10.1088/1757-899X/1149/1/012019

Computational Fluid Dynamic Study of Blockage in Artery With and Without Bypass

Affan Ahmed Ansari [1] and Faisal Shameem[1]

[1] School of Mechanical Engineering, Galgotias University, U.P, India
affanahmedansari500@gmail.com

Abstract: In complex vessel geometry different temporal, spatial variation takes place which occur under pulsatile conditions. The high Wall Shear Stress (WSS) at the wall affected by the disturbed flow may ultimately lead to endothelial cell dysfunction, which leads to atherogenesis and thrombosis. The present study observes WSS and streamlines due to stenosis with and without By-pass Artery. It was found that by adding bypass Artery there is tremendous decrease in wall shear stress in the stenosis. Reverse flow of blood through stenosis was observed in 30^0 Bypass Models. Hence the blood flow stabilizes and can be reached at optimum level in 50% stenosed artery. Based on this paper findings, it can be deduced that there would be a low risk of further atherosclerosis when the bypass artery is added.

1. Introduction

Heart diseases are a fatal factor when weighed against an average human's life expectancy. Sudden deaths, mostly, happen due to these heart diseases. For the incident of heart failure, sudden heart death is one of the major causes, which could occur for a number of reasons. "Aging" is statistically one of the highest contributors for a sudden heart death, categorized under the reasons of "natural death". In an experiment performed on an age group ranging from 45 to 65, men had a 7 times higher chance of dying due to sudden heart death as compared to women. Engineering sciences has intertwined with the "medicine" branch which has created a catacomb of opportunities in medical problem solving and diagnosis. There are many cardiovascular diseases which affects people and atherosclerosis is the most common form of cardiovascular disease. Atherosclerosis causes several deaths per year. Blood movement decides the progression and initiation of such cardiovascular disease and decide the behavior of the Wall of the vessel. According to major Medical Services, about 80% of the mortality is caused due to cardiovascular system it is mostly caused due to any increment in the blood resistance which causes the loss of blood flow in the concerned vascular bed.

Bypass Method in coronary artery goes by another name coronary artery bypass graft. Fundamentally, an artery is abstracted from a different part of the patient body as a graft. This created graft is known to divert blood from around the narrowed components of the specific artery. The techniques used in the process are considered consequential [1]. The categorization is done twofold: firstly, individual technique and secondly, sequential technique. The individual technique thoroughly depends on the number of arteries that were blocked and henceforth utilizes the same number of graphs. Here, in this technique the surgeon is supposed to connect each and every graft in an end to side manner of the arteries' vessel. On the contrary sequential technique requires only one graft which is connected to a whole narrowed down artery by a side-to-side cessation and/or a culminate to side cessation. This graft and technique is to be operated from experience. The hemodynamic as well as the flow pattern of the graft is linked to its patency rate and therefore, it is consequential. The reason behind it is that after the bypass surgery, the newly created bypass graft gets directly or indirectly sabotaged at the distal anastomosis [2]. Therefore, the key points that this project is set to illuminate is to utilize CFD or computational fluid dynamics techniques to curate a simulation of the blood flow in a constructed coronary artery and henceforth, analyzing the actual flow in the CABG (Coronary Artery Bypass Graft).

2. Literature Review

Several research papers were studied, it's been found that crucial blockage in blood vessels and showed that when 70 % blockage within the vessel section, there'll be respectable decrease in blood flow of vessels [3]. Few studies also centered their models on varying graft angles 45^0 ,75^0 ,60^0 with 70% stenosis [4]. Jae-Sung et al. have concluded that the best model of Bypass artery will be when diameter of artery is equivalent to diameter of Bypass [5]. P. Halder et. al carried out comparative analysis of non- Newtonian and Newtonian blood flow in 50 % stenosis by using Carreau-Yasuda Model and focused on velocity fields, Wall Shear Stress and found high WSS near stenosis, comparatively more deflation of velocity profile near stenosis in Non-Newtonian that Newtonian fluid [6]. Some researchers also focused on calculation Wall Shear Stress (WSS) and flow separation zone of 3 dimensional stenosed tube with different severity and found that the oscillation of WSS is depends upon severity of stenosis and axial position of pipe [7]. Some researchers also attempted on studying the CABG by using Computational Fluid Dynamics (CFD), attempted to explain flow behaviour in eccentric and concentric stenosed model for various graft angles [8]. Blood flow was resolved in the coronary arteries with successive blockages in which three consecutive sinusoidal blockages were taken in two different concentrations of 84% and 70% and solid vessels [9]. It was found that an increase in shear pressure and a decrease in normal pressure became severe by increasing the blocking rate. These changes are most severe when the flow of blood is at its highest level. Investigations show that there is no direct vertical flow after the blockage due to Reynolds' low number [9]. By taking non-Newtonian blood structures, Tu and Deville mimic blood stenosis using Herschel-Bulkley, Bingham and the law of force in a solid circular tube with a partial closure [10]. They have admitted that the disturbance is strong with vorticity and continuous behind the geometric barrier especially for severe stenosis. In an experiment Tang et al measured off-line stress relationships and expressed that the distribution of pressure has a very high pattern and both the intensity and intensity of compression occur within the stenotic phase [11]. T. Shaw et al investigated the flow of Casson fluid through a frozen bifurcated artery and found that in both the female and coronary arteries the differences in axial velocity and the rate of flow of yield pressure were the same [12].

Very few works are reported in literature for a stenosed artery with bypass using sinusoidal inlet profile. In this study we have simulated the blood flow from 50% stenosis to study the effects of Bypass at different Graft Angles and compare the results of 50% stenosis without Bypass.

3. Methodology

Symmetrically stenosed artery was taken for numerical simulation. In (r, z) coordinate system the profile of stenosis can be represented mathematically as given in [13]:

$$\frac{r}{r_0} = 1 - \frac{\delta}{2r_0}\left(1 + cos\frac{\pi z}{z_0}\right) \quad (1)$$

$-z_0 \leq z \leq z_0$

r = artery radius with the function of z in the stenosis region, L = length of the artery and r_0 = radius of the artery. In this case, $\delta = 0.5$ represents the 50% severity of stenosis & $z_0 = 1.0$ as given in [13].

Fig. 2(a) and 2(b) represents Schematic diagram and mesh structure of 50% stenosis and Grid distribution. The 2D CAD model was setup in Solidworks and exported in igs format to ANSYS for meshing.

The Outlet's Static Pressure was kept at 13332 Pa. The governing equations of continuity and momentum for the model can be represented as:

$$\frac{\partial \rho}{\partial t} + \nabla.(\rho U) = 0 \quad (2)$$

$$\frac{\partial(\rho U)}{\partial t} + \nabla.(\rho U \nabla U) = -\nabla p + \nabla \tau \quad (3)$$

Where stress tensor, τ and strain rate are related by:

$$\tau = \mu(\nabla U + (\nabla U)^T) - \frac{2}{3}\delta \nabla U \quad (4)$$

Above equation were solved for a Non-Newtonian fluid, for unsteady laminar flow yields the primitive variables *p:* pressure, *ρ: density, U: velocity.* t denotes time here.

For the artery wall, the condition of no slip boundary was applied and at the inlet boundary, time dependent velocity was imposed. The outlet, whereas, had an assigned condition of an opening type of boundary with pressure outlet.

For Inlet, a sinusoidal velocity profile is being used. The velocity variation is given in Fig. 1. The simple pulsatile can be given as:

$$U(t) = 1 + sin(\frac{2\pi t}{T}) \quad (5)$$

U: velocity, T: Time Period. t denotes time here

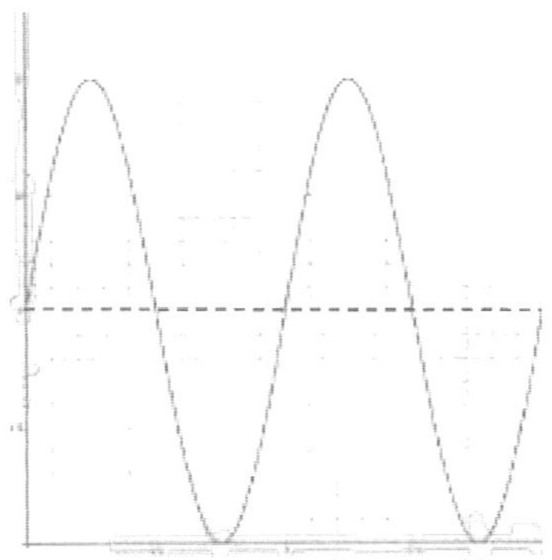

Fig. 1. Sinusoidal unsteady inlet velocity used in Model computation.

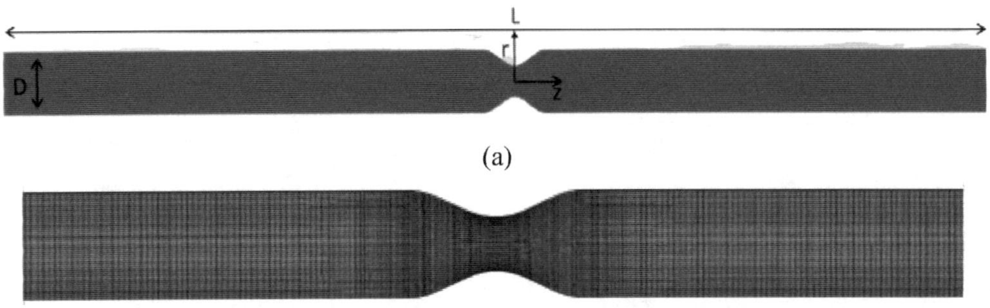

(a)

(b)

Fig 2. (a) Schematic diagram artery with 50% stenosis, here r: radial direction, D: Diameter, L: length of artery; and **(b)** Grid distribution

Bypass was added in this same artery model [modal without bypass as shown in Fig 2(a) and its grid in 2(b)] at different Graft Angle (A) and graft diameter d as shown in Fig. 3(a). Grid distribution for bypass is shown in Fig 3 (b). All boundary conditions were same as model without bypass model. Table 1 gives the details of different modal of artery with bypass.

(a) (b)

Fig. 3 (a) Schematic diagram of Artery with By-pass (b) Grid distribution

Table 1: Model Dimension of the bypass Graft

No. of Models	Graft Angle (A)	D (mm)	d (mm)
1	30	2	2
2	45	2	2
3	60	2	2
4	75	2	2

The ANSYS Mesh used to build the mesh for the proper analysis. No. of nodes and elements are increased gradually from 71325 to 446481 and 91217 to 446481 respectively for independence of Grid (Fig. 4). For the alteration of the mesh nodes from 157450 to 446481, the result displayed a not so exceptional difference in the velocity profiles. A grid with anywhere about 200000 elements is ought to be the optimal grid for such usage (Fig. 4).

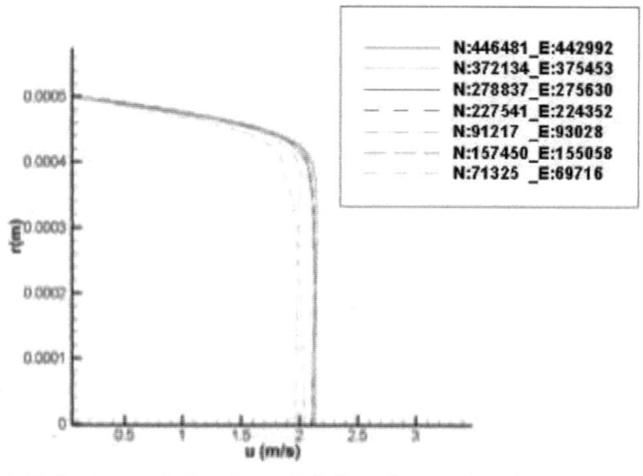

Fig 4. Velocity variation in radial direction at location z = 0

Dynamic viscosity for the Non-Newtonian fluids was implemented using the Carreau-Yasuda model as

$$\mu = \mu_\infty + \frac{(\mu_0 - \mu_\infty)}{(1+(\lambda\dot{\gamma})^a)^{\frac{1-n}{a}}} \quad (6)$$

Table 2: Values used in Carreau-Yasuda Model for mimicking blood

Variables	Value
low shear viscosity, μ_o	0.056 Pa.s
high shear viscosity, μ_∞	0.00345 Pa.s
Time Constant, λ	3.3313 s
Power Law Index, n	0.3568
Yasuda Exponent, a	2
Flow Shear Rate, $\dot{\gamma}$	≥ 0

Different variables and constant values of Carreau-Yasuda model is given in Table 2. This fluid viscosity caused by a low shear thinning mimics well blood flow model [14-15].

4. Results and Discussion

The validation is of Numerical model without Bypass has been done by comparing the velocity profile of once reported in literature by P. Halder et al [6]. The present model velocity profile is in good agreement with literature.

The fully developed steady state of flow through proper simulations and the results were compared with non-Newtonian fluid with previous founding in literature Fig. 5 shows velocity profile at Re = 200 and Fig. 6 streamline distribution of the artery after one cycle of pulsatile flow at Re=350. The Non -Newtonian blood flow gives a deflated velocity profile with negative velocity near the wall.

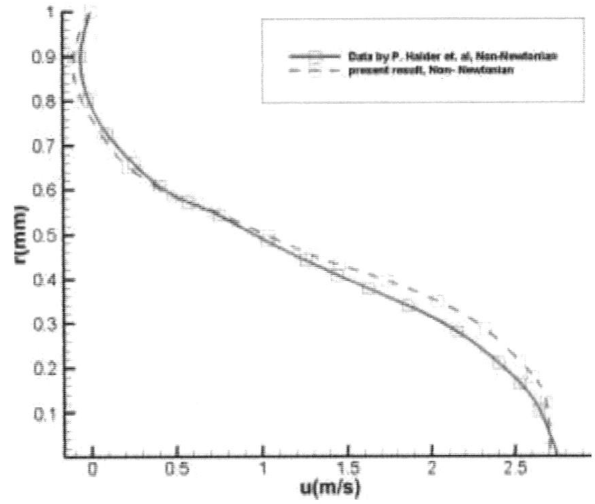

Fig. 5. For validation, velocity profile at Z = 0.001. at Re = 200

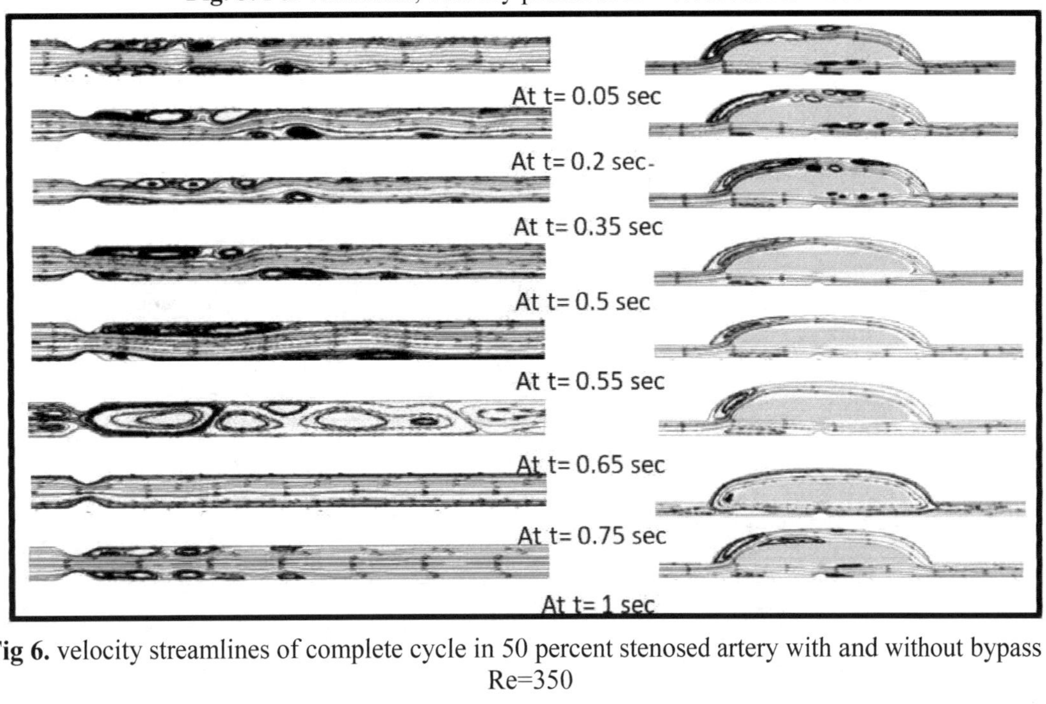

Fig 6. velocity streamlines of complete cycle in 50 percent stenosed artery with and without bypass at Re=350

Fig. 6 shows some of the instances of blood flow for complete one cycle in an artery with and without bypass. At t = 0.05 s 7 vortexes are observed near the walls, by observing fig 6 from t= 0.5 to t=7 sec there is an increase in whirling motion of blood and all vortex have merged to form bigger vortex. At t=7 sec it is observed that vortex was formed even before stenosis and blood flow is highly unstable. Finally, at t=0.75 s there is a uniform blood flow with no whirling motion but then again at t=8 it is observed that vortexes are generated near walls and it keeps on increasing till t=1sec and this cycle keeps on happening. After flow of one cycle, 9 vortices generated near the walls, whirling motion of blood is witnessed which is not a stable flow in an artery.

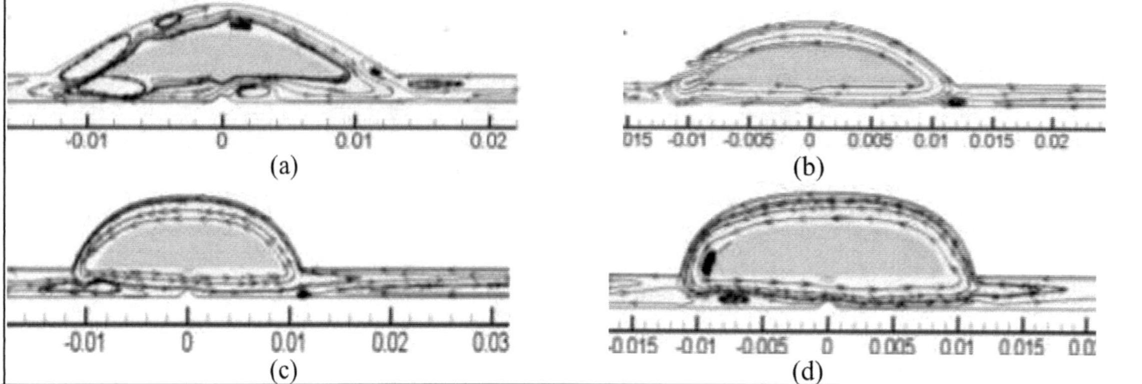

Fig. 7. (a)30^0 Graft Angle of by-pass artery 50% stenosis (b)45^0 Graft Angle of by-pass artery 50% stenosis (c)60^0 Graft Angle of by-pass artery 50% stenosis (d)75^0 Graft Angle of by-pass artery 50% stenosis, at Re= 350, at t=0.75 sec

Fig. 7 shows streamline distribution of all models of bypass artery at t = 0.75 sec. At this time step it has been observed that at 30^0 Graft Angle there is a clockwise flow of blood through stenosed artery and by-pass, but at Graft Angle 45^0, 60^0, 75^0 blood flows in anti-clockwise direction in the same time step (t=0.75).

Table 3: Wall Shear Stress (WSS) at all models at stenosis location (0,0)

Graft Angle	WSS at Stenosis	
	Upper Wall	Lower Wall
1: without By-pass	237.4959825	229.4830533
2: 30^0	82.31438772	76.60927164
3: 45^0	90.73629973	85.08632808
4: 60^0	99.42359236	94.00288312
5: 75^0	102.523754	97.2067871

Table 4: Maximum WSS of bypass Artery

	WSS
Graft Angle	Upper wall
1: 30^0	145.6877
2: 45^0	225.7709
3: 60^0	141.5199
4: 75^0	108.5532

As it can be observed at table 3 that by applying bypass WSS is reduced at stenosis (0,0) tremendously. But it has also been found that as Graft angle increases WSS also increases by some value. As shown in table 4, Model 3 (Graft angle = 45^0) has maximum Wall Shear Stress at Upper wall as compared to other bypass.

5. Conclusion and future work

The present work of numerical modelling of sinusoidal blood flows through constricted stenosed artery with and without Bypass to understand hemodynamics. The models were studied for non - Newtonian fluid flows with Carreau-Yasuda Model. Conclusions that are drawn based on our research:

- Clockwise blood flow in Artery with bypass (30^0 Graft Angle) was observed, hence there is a reverse flow of blood in stenosed Artery at t=0.75 sec
- Formation of vortexes before stenosis was observed at t= 0.65 sec in Artery model 1 which is without Bypass.
- WSS can be reduced tremendously by adding bypass.
- Maximum Wall Shear Stress was observed at upper wall with Bypass Artery of Graft Angle 45^0. So, it is not feasible to use Bypass with Graft Angle 45^0.
- In future it is proposed to perform simulation with different degree of stenosis and to see their effect with bypass graft.

References

[1] CC. Canver 1995 Conduit options in coronary artery bypass surgery, *Chest* **108** 1150-1155.
[2] PD. Ballyk, C. Walsh, J. Butany, and M. Ojha 1997 Compliance mismatch may promote graft-artery intimal hyperplasia by altering suture-line stresses, *J Biomech* **31** 229-237.
[3] Mann R, Zhuang L, Lu X, Wang W 2002 Breaking Symmetry in Non-Planar Bifurcation: Distribution of Flow and Wall shear Stress. *Biorheology*, **39** 431-436.
[4] Ko T, Ting K, Yeh H, 2008. A Numerical Study on the Effects of Anastomotic Angle on the Flow Fields in a Stenosed Artery with a Complete Bypass Graft. *International Communications in Heat and Mass Transfer*. **35** 1360–1367.
[5] Jae-Sung Choi, Sung Chul Hong, Hyuck Moon Kwon, Sang-Ho Suh and Jeong Sang Lee 2011 Influences of Geometric Configurations of Bypass Grafts on Hemodynamics in End-to-Side Anastomosis. *Korean J Thorac Cardiovasc Surg*; **44** 89-98.
[6] P. Halder, A. Husain, M. Zunaid and A. Samad 2017 Newtonian and Non-Newtonian Pulsatile Flows through an Artery with Stenosis. *The Journal of Engineering Research (TJER),* **14** 191-205
[7] Long, Q., Xu, X.Y., Ramnarine, K.V., Hoskins, P., 2001. Numerical investigation of physiologically realistic pulsatile flow through arterial stenosis. *Journal of Biomechanics* **34**, 1229–1242.
[8] Shah Mohammed Abdul Khader, Goutam Thakur, Raghuvir Pai, Mahima Kapoor, Soujata Borbaruah, Keshava Ragavendra, 2018, Numerical Study on Eccentric and Concentric Vascular Grafts Prototypes. *Journal of Advanced Research in Fluid Mechanics and Thermal Sciences* **51** 99-107.
[9] Ebadi A, Qalichi F, 2005. Numeric Simulation of Blood Current in Coronary Vessels with Successive Blockages. *10th Iran National Congress of Chemistry Engineering, Iran, Sistan-Baluchistan University* **1(2)** 99–104.
[10] Tu, C., Deville, M., 1996. Pulsatile flow of non-Newtonian fluids through arterial stenoses. *Journal of Biomechanics* **29**, 899–908.

[11] Tang, D., Yang, C., Kobayashi, S., Ku, D.N., 2001. Steady flow and wall compression in stenotic arteries: a three-dimensional thick-wall model with fluid–wall interactions. *Journal of Biomechanical Engineering* **123(6)** 548-57

[12] Shaw, S., Gorla, R.S.R., Murthy, P.V.S.N., Ng, C.O., 2009. Pulsatile Casson fluid flow through a stenosed bifurcated artery. *International Journal of Fluid Mechanics Research* **36 (1)**, 43–63.

[13] Liao W, Lee TS, Low HT 2004, Numerical studies of physiological pulsatile flow through constricted tube. *International Journal of Numerical Methods for Heat and Fluid Flow* **14** 689–713.

[14] Sahar Ahsaas, Shaligram Tiwari S 2016, Numerical Simulation of Blood Flow through Asymmetric and Symmetric Occlusion in Carotid Artery. *3rd International Conference on Fluid Flow, Heat and Mass Transfer, Ottawa, Canada* 170 – 8.

[15] Razavi A, Shirani E, Sadeghi MR 2011, Numerical simulation of blood pulsatile flow in a stenosed carotid artery using different rheological models. *Journal of Biomechanics* **44** 2021–2030.

Implication of Renewable Energy in Sustainable Development in India: Future Strategy

Sunita Pachar[1], Rubee Singh[1*] and Mohd Atif Wahid[2]

[1]Institute of Business Management, GLA University, Mathura, India
[2]Department of Mechanical and Automation Engineering, Delhi Technical Campus, India
***dr.rubeerajput@gmail.com**

Abstract: Renewable energy (RE) can play an important role in Sustainable Development in India. India is a growing economy, and it would need an assured supply of 3–4 times than the whole energy consumed today. This research paper discussed about RE sources, drivers, challenges and policies. It can be concluded from review of literature and other information provided in paper that RE have a huge potential in India and with latest technologies use it can fulfill India's energy demand in future. Quality of life can be improved with the help of RE. It is predictable that large number of domestic jobs can be created in future in green energy and RE. The study found that RE can reduce environment pollution issues, carbon emission and scarcity of nonrenewable energy sources .RE has a positive relation with economic growth, job creation and welfare. This paper recommends some suggestions for strategic policy in sustainable energy sources.

1. Introduction

The power requirements will always be on the rise and the power generation capacity will also be increasing year on year to meet the demands of power. It is wise to build capacities in a sustainable way so that the unfavorable climatic changes can be mitigated drastically. Energy is the key factor in any country for industrialization, Urbanization and economic growth [1]. It affects many development issues like economic, livelihoods & agricultural productivity. Energy is required for many functions like healthcare, lighting, manufacturing and running many electronic or digital items. It is expected that green energy requirement by 2022 will be 175GW [2]. In green energy government is promoting FDI and allowed in RE recreation or circulation almost 100 % FDI. Solar energy as an alternative source would help us to leave a clean, safe and healthier world for our future generations. In next four years RE sector growth is expected up to US $ 80 billion. In country non renewable sources like coal, oil etc is used for its energy demand fulfillment or consumption [3, 4]. In India energy issue can be solved by sustainable energy sources like solar energy, biomass or geothermal. It is expected that nowadays RE is contributing more than half energy execution in different sources [5]. Sustainable development is a big challenge for any country nowadays. Electronic vehicles are a buzz word nowadays because of its ecofriendly nature approach.

These days' consumers are also very much aware about green products and day by day green products or RE products demand is increasing. In automobile industry trends like e-vehicle is prevalent now days and future growth in this category is expected [6]. Government is promoting nowadays emission free products and they are providing essential guidelines and standards to manufacturers to minimize environment pollution. Nowadays corporate sector is working sustainable resources model and they are trying to use in their processes more and more RE or green energy techniques or equipment's. Well-structured strategy is required for country like India to gain sustainability in business processes and household consumption of energy [7]. Green energy is solution for many challenges like global warming, environmental issues or increasing demand of fossil fuels. Government need to invest more and more in RE initiatives so that in future we can improve better uplifting in economic and social development [8].

2. Review of Literature

2.1 RE sustainability issues

Sustainable energy as, "a full of life synchronization between the even handed accessibility of energy-intensive goods and services to people and protection of the earth for next generations" [2]. Pachar S [3] mentioned in their study that in sustainability source of energy solar and wind energy are contributing more and more now days. Hydro energy India need to focus because this is not contributing significantly. In green energy India is growing fast. Government is taking many initiatives to improve energy supply. In energy supply initiatives Bio mass energy can play a important role. For any country sustainable development energy supply is important aspect. India is having geographical advantage which can be good in sustainable development model. Green energy is now only solution for carbon emission. Traditional energy sources cannot help to meet future demand of industry and household so India needs more initiatives and strategic planning to fulfill future demand of energy. Only solution to improve environment and industry energy need is RE or green energy [3]. Hák et al [9] focused mainly on The Sustainable Development Goals (SDGs) of India, which mainly focused on energy supply and green energy issues. Sustainable energy is dynamic harmony between people need and supply which can help in future generations need fulfilment. Nowadays Government is more concerned for country future need fulfilment. Recently launched SDGs main goals for India is to help in climate change. To provide sustainable resources to future generations [9].

2.2 Economic Development & RE connection

For any country development and sustainability, green energy can be the best strategy. This can help in development and need fulfilment. Millennium Development Goals also includes issues like climate change, renewable energy and modelling [5]. Edenhofer et al. [10] discussed that RE sources are accessible everywhere rather than tradition sources of energy. Economy can get benefit by improving supply of RE and reducing import of petrol and coal. Human development, economic productivity and country economic performance majorly dependent on energy supply. Green energy has constructive impact on development as well. RE can provide many new job opportunities it is expected that RE contributed 2.3 million jobs Worldwide. Financially viable development and energy use a have positive association and it can contribute globally in per capita income increase also.

2.3 RE role in climate change or Energy security

Twidell & Weir [11] discussed that industry pollution or emission of green house gases can be solved by using RE. It is natural resource and available all time and best solution for environment protection. Biodiversity of any country should not affected by energy supply. Every country facing issues related with environment pollution or global warming because use of traditional energy sources use like coal and petroleum. Considered RE technologies as clean sources of energy [12]. Future economic and social needs can be fulfilled by it and it can also make our environment pollution free. Renewable energy technologies are better solution for reducing global warming and greenhouse gas emission. So adopting renewable or green energy policy is best strategy to overcome the challenges related with environmental factors. Renewable energy technologies will improve energy supplies [13]. Fossil fuel energy should be replaced with green energy sources which consist solar energy, Hydropower, ocean energy (tide and wave), geothermal energy and wind [7].

3. Methodology of the Study

This research paper looks for study the Significance of RE in Sustainable Development in India and its potential in future strategic sustainable development. To achieve the objective peer reviewed journals have been analysed and reviewed and data have been taken from many authenticated government sources. This study is result of review of secondary sources like research paper or government report on RE or green energy or web policy documents etc. The present paper is an effort, to identify and discuss renewable energy status and potential in India. To be specific research paper has following objectives: (a) To know the RE future Scope in India (b) To discuss the benefits of RE in India (c) The paper also analyses regulatory framework, issues concerned with renewable sources (d) To recommend suggestions to improve RE contribution in county growth & environment pollution control.

4. Renewable Energy Sources and Technology

In India (MNRE) Ministry of New and Renewable Energy is authorised body for RE in different areas like electricity, manufacturing and transportation etc. This department is involved in energy activities like R & D, certification or equipments standardization. The MNRE contains the all activities of green energy production. The government is working on special plans to set up an indigenous manufacturing capacity grid till 2022. In RE 75 % contribution is from wind program which is fastest contributing [9]. Government is working on many incubation schemes to set up enterprise in green energy

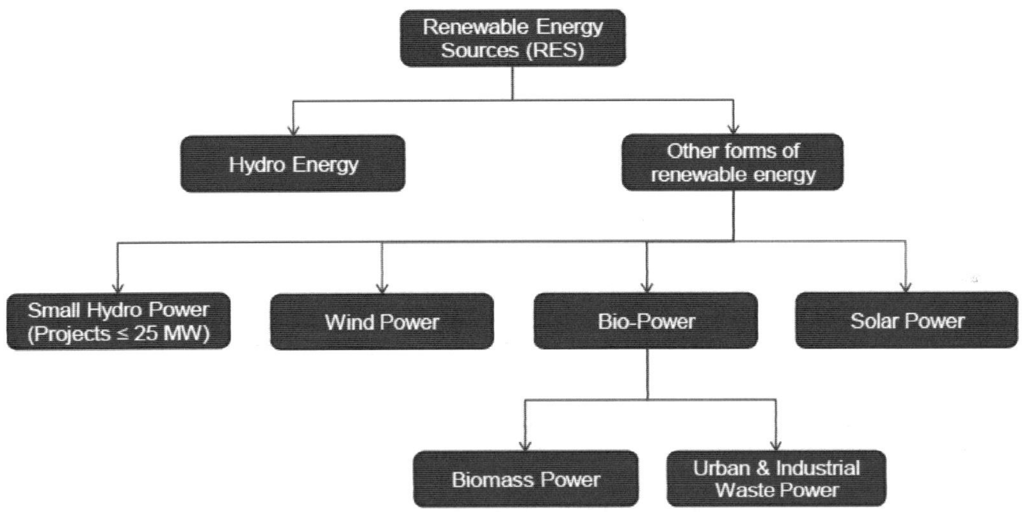

Figure: 1 *Type of RES*

5. Projections

Figure 2 shows the status of renewable energy solar power capacity in India. Solar energy can contribute a major part in energy need fulfillment and it is available almost in every state of India. Thus there is a strong need of better policies related with energy development which can help India to fulfilling energy need. Indian Ministry has started many training programmes to fulfill human resource requirements in renewable area.

A country demand for energy is directly proportionate to its population and India is ranked second, of the most populous country. By the end of 2021, the capacity addition in the utility scale is expected to be at 47,000 MW and the rooftop capacity is expected to reach 12,500 MW. This in turn translates to reduction of carbon dioxide which would have otherwise resulted in the emission of 87,000 KT of carbon dioxide from the coal plants [14].

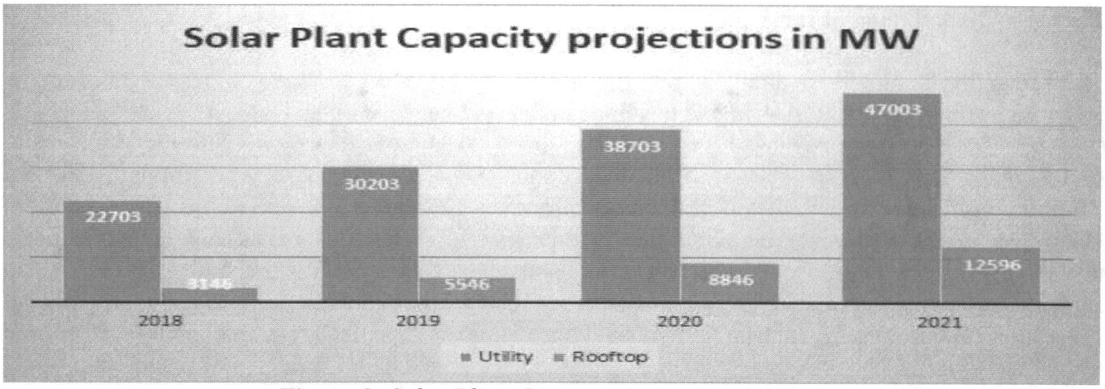

Figure 2: *Solar Plant Capacity projections in MW*

Table: 1 year wise and technology wise capacity additional targets to 2022 (Grid connected RE only)

Year	Rooftop Solar	Ground-Mounted Solar	Solar	Wind	Small Hydro	Biomass	Total
	(GW)	(GW)	(GW)	(GW)	(GW)	(GW)	(GW)
Cumulative installed capacity by 2014-15			3	24	4.1	4.4	35.5
2015-16	0.2	1.8	2	3.2	0.14	0	5.3
2016-17	4.8	7.2	12	3.6	0.14	0.9	16.7
2017-18	5	10	15	4.1	0.14	0.9	20.2
2018-19	6	10	16	4.7	0.14	0.9	21.8
2019-20	7	10	17	5.4	0.14	0.9	23.5
2020-21	8	9.5	17.5	6.1	0.14	0.9	24.7
2021-22	9	8.5	17.5	8.9	0.14	0.9	27.5
Total	**40**	**60**	**100**	**60**	**5.08**	**9.98**	**175**

Table 1 represent a brief data about 7 years targets in energy sector by Niti Aayog in sector like solar energy, wind energy, Hydro energy and biomass energy. Above table data represent that in India green energy is continuously increasing 27.5 GB in 2022.Major contributor in energy is solar and wind. These sources are sustainable and can contribute more significantly is country's energy demand. Least contributor in energy is Hydro sector in India. Biomass energy have full potential to meet future energy demand [2].

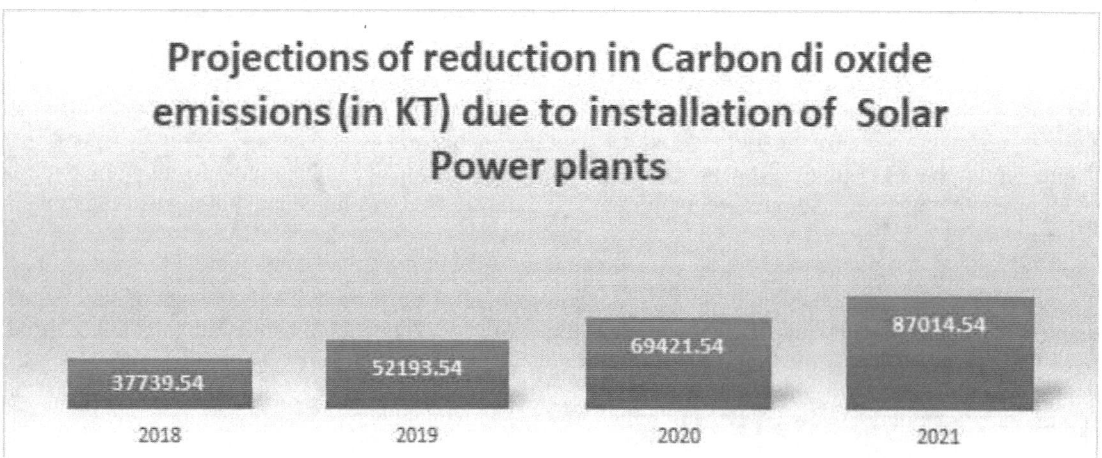

Figure 3. Projections in reduction in Co2 emission (in KT)

Figure: 3 shows Projections in reduction in Co2 emission year wise. It is showing a trend that year wise as solar energy use is increasing carbon dioxide emission will also decrease. The power requirements will always be on the rise and the power generation capacity will also be increasing year on year to meet the demands of ever-increasing power. It is wise to build capacities in a sustainable way so that the unfavorable climatic changes can be mitigated drastically. Hence it could be concluded that usage of solar energy as an alternative source would help us to leave a clean, safe and healthier world for our future generations [15].

6. Renewable Energy contribution in sustainable development

Green energy or renewable energy most important feature is that it is accessible in abundant supply. It is unlimited. RE sources are natural and sustainable that have no harmful impact on nature. RE can be used in household and industrial functioning [16].

- Eco friendly environment: Green energy sources and technologies are clean source of energy than conventional energy technologies.
- Sustainable Source of Energy: Solar energy, thermal energy, winds energy and ocean energy all can be created by natural sources which are available vast quantity rather than nonrenewable sources which are finite and will sometime be depleted.
- Employment opportunity: Government is investing lots of funds on green energy R& D and infrastructure facilities which will create many jobs for people.
- Improved healthy life: green energy uses natural resources for energy making process like solar, thermal or ocean, which does not create any health issues or air pollution. Renewable energy is very clean energy which can save even cost health as well.

7. Recommendations for future renewable strategy for India

Renewable energy is best solution for environment protection and sustainable development. Best renewable source like Solar and wind energy are getting popularity nowadays. But it has also certain limitation exists such as: seasonal disparity as most RE resources are affected by seasons. So strategic planning with technological innovation is required. In main challenge lack of information to access the green energy comes [17]. So by the help of technology these shortcomings can be improved. Government should make Strategic future policy to make renewable energy plan successful. Main challenge include lack of infrastructure support, no financial viability and stringent regulatory approval by removing all these hurdles green energy for sustainability can be promoted [18].

8. Conclusion

In today's era there is a dire need of green energy or RE where challenges like environment pollution, carbon emission or health issues are increasing day by day. Every country is now working to develop technology based on RE to meet future energy demand. In this paper many issues have been discussed related with India's future possibilities of green and RE. India can be a hub of green energy supply if India will work on a strong strategy base plan. India can be self reliant in energy demand and supply by using proper geographical advantage in energy supply sources. Government should mandate for insurance companies to make better policy to invest in RE sector. The government should focus on Innovative latest techniques use in energy sector. Thus, to conclude it can be said that looking at the present situation, a self-sufficient better energy scheme in nation like India is vital for future growth & require fulfillment in energy consumption. RE sector can help to meet environment challenges and creating new ventures and job in green energy area.

9. Future Scope

Study can be done on issues like carbon emission or environment pollution solution by green energy or RE. More studies can be done like applications of RE and Innovation in green energy and their impact on people.RE have a huge potential to meet energy future need and it is a sustainable source so study can be done to develop a sustainable model on RE sources contribution in meeting energy future demand.

This research paper is prepared based on secondary data and limited data sources have been referred .RE or green energy is having vast potential for future of world so more and more primary data based research should be needed in this area. Government need to address implementation and supervision of green energy issue and model seriously in industry or household.

References

[1] Ministry of New and Renewable Energy, Government of India. [Online] Available: http://www.mnes.nic.in/ Accessed 10 Jan 2021.
[2] Ministry of Power, Government of India. [Online] Available: http://powermin.nic.in/ Accessed 19th Nov, 2020.
[3] Pachar S. 2020.Renewable & Non Renewable Energy sources in India: A Study on Future prospects**.** *Inter J. of Advanced Science and Technology* **29**, **03**.8956.
[4] Ministry of New and Renewable Energy (MNRE). 2019-2020 Annual report;
[5] *Garg S. & Pachar S. 2020,* Future of Electric vehicles: Empirical study in relations to Potential Factors *,Inter J. of Advanced Science and Technology* **29**, **03**, 5849.
[6] Kumar, A., Sudhakar, K., Baredar, P., &Mamat, R. 2018, Solar PV and BIPV system: Barrier, challenges and policy recommendation in India. *Renewable and Sustainable Energy Reviews,* **82** 3314–3315
[7] IEA (International Energy Agency) 2017a, Tracking Clean Energy Progress, OECD/IEA,
[8] IRENA 2017 International Renewable Energy Agency a, Accelerating the Energy Transition through Innovation, IRENA, Abu Dhabi, www.irena.org/DocumentDownloads/Publications/
[9] Hák, T., Janoušková, S., &Moldan, B. 2016. Sustainable development goals: A need for relevant indicators *Ecological Indicators, **60*** 571–572.
[10] Edenhofer, O., Pichs-Madruga, R., Sokona, Y., Seyboth, K., Matschoss, P., Kadner, S., von Stechow, C. 2011. *Renewable Energy Sources and Climate Change Mitigation*. Cambridge: Cambridge University Press.
[11] Twidell, J., & Weir, T 2015. *Renewable energy resources*
[12] Preeti H. Narnaware, Ramesh G. Surose& swati V. Gaikwad 2015. Current Status and the Future Potentials of Renewable Energy in India- *A Review. International Journal of Advances in Science Engineering and Technology,* **1** 1-3.

[13] Panwar, N.L. & Kaushik, S.C. & Kothari, Surendra, 2011. Role of renewable energy sources in environmental protection: A review, *Renewable and Sustainable Energy Reviews*, **15** 1519.
[14] Renewable Energy Policy Network for 21st Century (REN21). Renewable 2011: Global Status Report, 13–14
[15] Ajanovic, A. 2011. Biofuels versus food production: Does biofuels production increase food prices, *Energy, **36*** 2069–2071.
[16] Simon, C.A 2009. Cultural constraints on wind and solar energy in the U.S. context. Comp. Technol. Transf. Soc. **7**, 251
[17] IEA. International Energy Agency Database 2010, Release 01: (a) Energy Balances of Non-OECD Member Countries; (b) Energy Balances of OECD Member Countries;
[18] U.S. Department of Energy. www.energysavers.gov/renewable (2019). Accessed 10 march 2019.

Design and Development of All- Terrain Vehicle Roll Cage

K K Dubey [1*], B Pathak[1], B K Singh[1], P Rathore[1], S R S Yadav[1]

[1]Galgotias University, Noida - 203201, India

[*]**dubey.kaushalendra@gmail.com**

Abstract. ATV is an off-roading vehicle and one of its essential parts is its roll cage on which all parts are mounted. The roll cage not only protects the passenger inside the vehicle it also acts as a structural defense for the vehicle. Now because of the above-mentioned properties, the roll cage is required to be structurally robust and it should tolerate high Von mises stresses and should have the least possible deformation during a crash while protecting the motorist inside the vehicle. Hence due to these high concerns of the roll cage, it is necessary to undergo simulation of the design to detect points that are vulnerable to stresses and high deformation. High Impact forces are applied on the roll cage in analysis to achieve real world actions of the chassis, which in fact helped us in developing a design that have mandatory safety standards and adequate potency to withstand these loading.

Keywords: Baja ATV, Roll cage Design, static structural analysis, structural tests, impact force calculation.

1. Introduction

ATV stands for All-Terrain Vehicle, defined by ANSI (American Standard National Institute) as an "automobile which travels on low-pressure tires, with a seat that is straddled by the operator, along with handlebars for steering control". An ATV is a type of vehicle that can be used on any type of terrain; it is generally used for extreme off-roading. The proper material selection is an essential part of the design and development of the chassis of ATVs. Various papers and thesis have been written on this topic but the main focus of some 190 plus papers was based on the accidents and injuries that are caused during extreme off-roading on ATVs. [1]. Clinging on to the safety concerns and strength of the chassis, it should have the capability to withstand high stresses that are generated during ATV operation [2]. Material selection for roll cage manufacturing is among AISI 1018, AISI 4130, AISI 1020, E-Glass epoxy, Carbon fiber [3], and various other factors are also considered before selecting a material like cross-section, design ergonomics of the vehicle, and Factor of safety (FOS). Another important factor in material selection is the weldability of the material, in R. Soundararajan et al. [4] paper focuses on TIG welding, where they analyzed experimentally the weld made by TIG welding by the process of shot peening (a process in which a material is bombarded with small spherical balls). Now since we know that ATVs are used in extreme off-roading, therefore, driver's safety has to be focused on, this all roll cages must have Seat Belts, Shoulder Belts, Lap

Belts, Anti-submarine belts, Arm restraint, Head restraint and Driver Equipment like Arm restraint, Safety shoes, Knee guards, Tear off goggles etc. [5-6].

For designing, software like SolidWorks, CATIA, Creo Parametric 1.0, and CAD programs can be used were as Ansys and SimScale software can be used for analysis. Different tests like Rear Impact, Side Impact, Front Impact, and Rollover Analysis are being performed [7]. Design Failure Mode and Effective Analysis (DFMEA) is used for the investigation of the chassis components. Monocoque chassis, Ladder-Frame chassis, and Space frame chassis are used in today's industries. For vehicle dynamics control, there is a need for Global Chassis Control GCC [8-9]. Different parameters like Ground Clearance, driver's safety, and weight of the structure were considered while designing the Chassis and focus is to achieve a factor of safety above the given limits in all loading situations [10].

2. Methodology and Design

Roll cage is a base component of any vehicle; any other components like engine, CVT, etc. are all mounted on the roll cage. Since ATVs are mostly used for off-roading it is required that the roll cage of the vehicle must bear with various loads and it should also protect the driver in case of a mishappening. In the roll cage there are two types of members namely: -

A. Primary member

B. Secondary member

Primary members are better for higher stress endurance than secondary members. Secondary members are there for equal distribution of stresses [11]. For the material of the roll cage in the BAJA rulebook it is specified that the minimum carbon content of the material should be 0.18%.

2.1. Material Selection

As the design of the roll cage is finalized then we need to decide the material of the roll cage with the required 0.18% minimum carbon content as advised by the SAE Baja rulebook. Other important factors that have to be kept in mind during material selection are vehicle weight, the process of fabrication, the weight to strength ratio of the material, and the good weldability of the material selected [4]. The materials that are taken into consideration are AISI 1018, AISI 1020, and AISI 4130.

Properties like carbon content, tensile strength, yield strength, etc. of the materials AISI 1018, AISI 1020, and AISI 4130 that are considered for the manufacturing of the roll cage is given in Table 1.

Table 1. Properties of Materials that are considered for Roll cage Manufacture. [12]

Properties	AISI 1018	AISI 1020	AISI 4130
Carbon Content (%)	0.18	0.27	0.30
Tensile Strength (MPa)	440	463.7	731
Yield Strength (MPa)	370	341.3	460
Bending Strength (Nm)	392.68	477.62	662.78
Bending Stiffness (Nm2)	2763.1	4258.2	2786.1
Weight\meter (kg\m)	1.686	1.47	1.122
Cost (Rs. /meter)	325	325	920

Therefore, based on the above Table 1, availability and broad study in the market we decided that AISI 4130 should be best for our ATV.

2.2. Roll cage Tube Cross Section

The tube's size is also an important factor in ATV manufacturing. It is responsible for factors like weight of the roll cage, the strength of the roll cage, and the cost of the roll cage. Generally, there are two different cross-sectional pipes namely circular and square. Although it is said that circular cross-section pipes are a cut above their counterpart square-shaped pipes [13].

Based on the Baja SAE rulebook primary members of the roll cage should have a cross-section of 25.4mm outer diameter and 3mm inner diameter whereas the secondary members of the roll cage should have a cross-section of 25.4mm outer diameter and a minimum of 0.89mm inner diameter [14-15].

2.3. Ergonomics

The ergonomics deal with the following factors: -

- Safety of the Driver
- Comfort for all types driver (fat, slim) while driving
- Roof clearance
- Seated eye height – so that the driver can clearly see the road up ahead

Ergonomics of the vehicle like lap angle, ankle angle, seat angle etc. and the values that we have considered for our ATV is mentioned in Table 2.

Table 2. Ergonomics of the vehicle [11]

Parameter	Range	Value That We Have Incorporated In Our Atv
Lap Angle	110°-130°	120°
Ankle Angle	85°-110°	90°
Seat Angle	95°-100°	95°
Elbow Angle	80°-165°	105°
Upper Arm Angle	0°-35°	19°
Angle of vision	20°-40°	30°
Steering Column Angle	20°-50°	50°

2.4 Proposed dimensions Of ATV

The Solidworks based three dimensional model has been developed with the consideration of weight of the vehicle should be kept as minimum as possible because additional mass and incorrect geometry can disturb the constancy of the ATV and would affect its perfect execution in the dynamic rounds [16]. Also the design and material selection of the roll cage should be as such that it takes minimum time and cost during fabrication [17]. Isometric View is a method of representing a 3-D engineering or technical drawing in 2-D. Figure 1-3 represent the Isometric View, Side View and Front View obtained from CAD based software solidworks respectively.

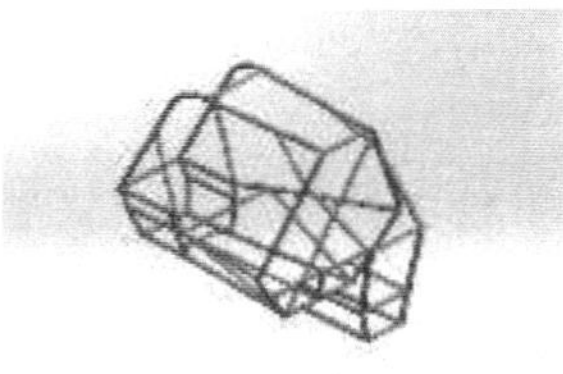

Figure 1. Isometric View of Rollcage

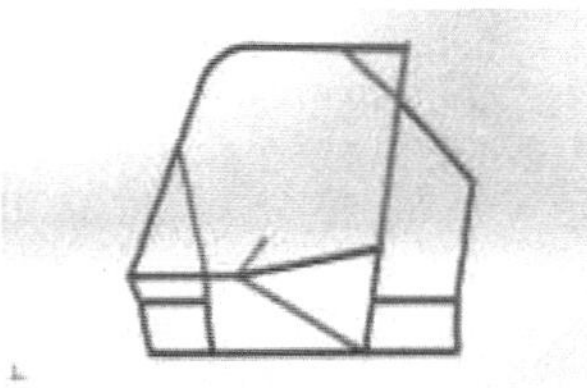

Figure 2. Side View of Rollcage

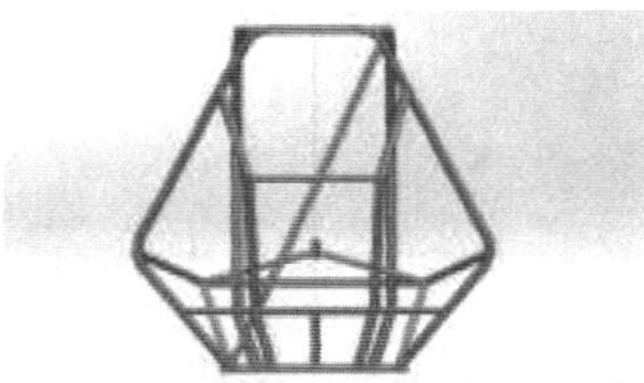

Figure 3. Front View of Rollcage

IOP Conf. Series: Materials Science and Engineering 1149 (2021) 012021 doi:10.1088/1757-899X/1149/1/012021

Length, Width and Height of our designed ATV is given in Table 3 below.

Table 3. Dimensions of our Designed ATV

Parameters	Value(mm)
Length of the ATV	1526.57
Width of the ATV	1620
Height of the ATV	1016

3. Static Analysis of ATV

After the design and material of the rollcage was finalized we moved on to analysis part of the rollcage. The best part of software based simulations is that it is economical and it takes less time and arrangement than analyzing with an actual model. [18]. The following test were performed on Ansys 19.2.

- Front Impact Test
- Side Impact Test
- Rollover Impact Test
- Rear Impact Test
- Dump Test
- Bump Test
- Torsional Rigidity
- Modal Analysis

Assumptions taken for Structural Analysis of ATV

- Roll cage act as distorted body.
- Impact time is considered 0.30 seconds in case of deformable body.
- Impact time is considered 0.15 seconds in case of rigid body.

3.1 Mechanical Impact & Strength Analysis

Calculations are obtained from mechanical system design handbook referred in reference no. [19-20].
Data that we have considered for our ATV:-
ATV Weight (m) = 210 kg
Initial velocity ($V_{initial}$) = 16.67m/s (60 Km/hr.)
Final Velocity (V_{final}) = 0 m/s
Time of Impact (t) = 0.30 seconds. (Since we have considered our rollcage as a deformable body)

Impact Force Calculations for Front, Side and Rear Impact Test:
From Principle of Work and Energy,
Work done= Change in Kinetic Energy,

$$|W| = |1/2 \times m \times (V_{final})^2 - 1/2 \times m \times (V_{initial})^2| \quad (1)$$

$$|W| = |1/2 \times m \times (V_{initial})^2| = 29178.33 \text{ Nm}$$

Displacement (S): -

$$S = (V_{initial}) \times t \quad (V_{initial} = \text{max. speed of vehicle at the time of impact}) \quad (2)$$

$$S = 16.67 \times 0.3 = 5.001 \text{ m}$$

Now we know that "Work is said to be done when an object moves (displaces) along the direction of application of force" hence by the formula:-

$$W = F \times S \rightarrow F = W/S \rightarrow F = 29178.33/5.001 = 5834.49\text{N} \approx 5900 \text{ N} \quad (3)$$

3.2 Front Impact Test:

This test is performed to know what will happen if the vehicle crashes into an object or into another vehicle head-on and what will be the stresses and deformation on the roll cage.

Impact Points - Front Bracing Members (FBM) and if there is no FBM then apply ¼ of the load on the front 4 nodes.

Fixed Support - Fire wall behind the driver.

Total deformation of the roll cage and Equivalent (von mises) stress acting on the roll cage for Front Impact Test, performed on Ansys software is given below in Figure 4 and Figure 5 respectively.

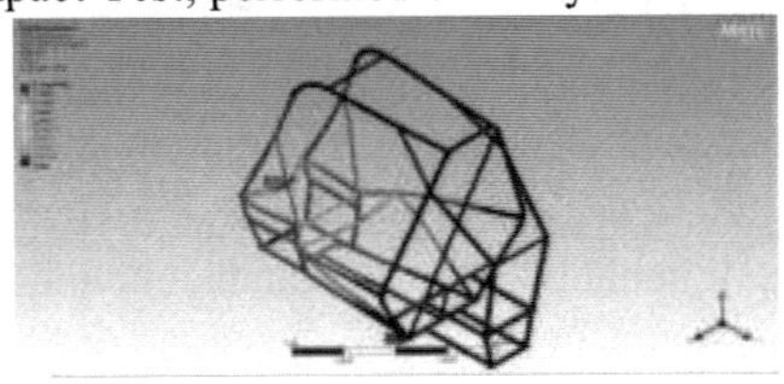

Figure 4. Total Deformation for Front Impact Test.

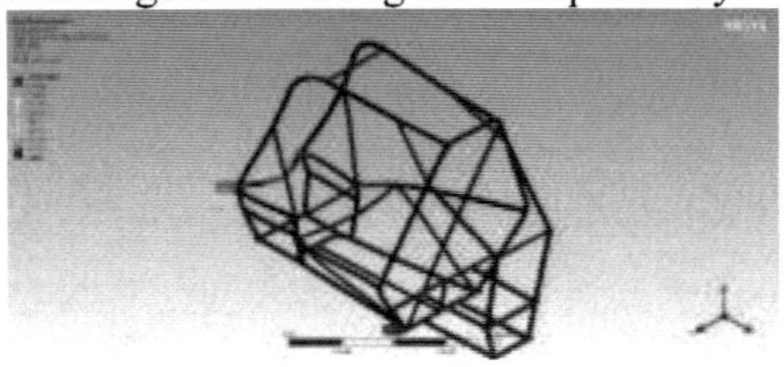

Figure 5. Equivalent (von mises) stress for Front Impact Test.

3.3 Side Impact Test:

This test is performed to know what will be the deformation and stress generated if two vehicles crash into each other making a 90-degree angle.

Impact Points - Side Impact Members (SBM) or the outer part of the side of the vehicle.

Fixed Support - opposite side's SBM.

Total deformation of the roll cage and Equivalent (von mises) stress acting on the roll cage for Side Impact Test, performed on Ansys software is given below in Figure 6 and Figure 7 respectively.

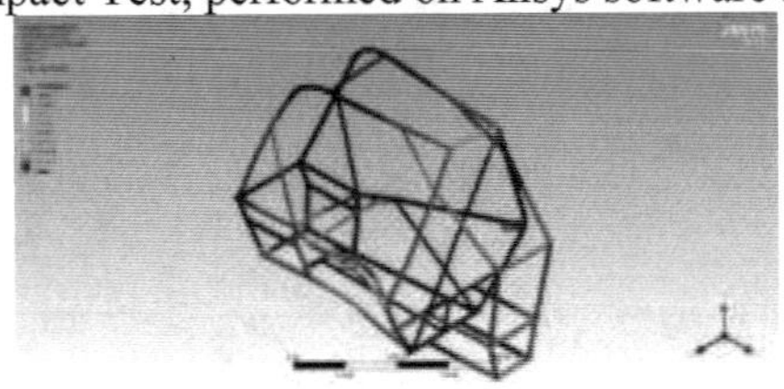

Figure 6. Total Deformation for Side Impact Test.

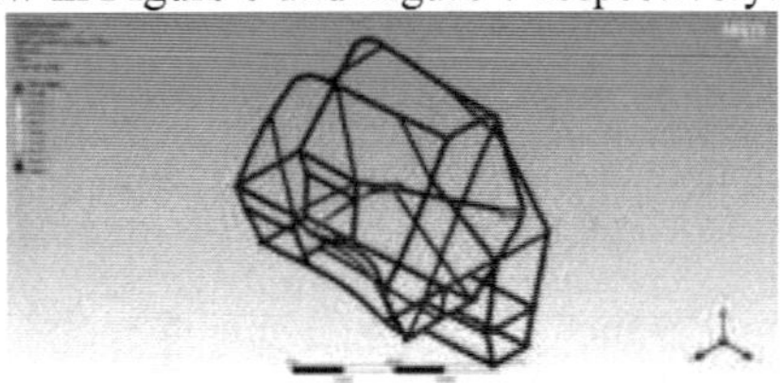

Figure 7. Equivalent (von mises) stress for Side Impact Test.

3.4 Rear Impact Test:

In this test, the vehicle is bumped head-on by another vehicle from behind.

Impact Points - Rear part of the vehicle.

Fixed Points - Front Bracing Members (FBM).

Total deformation of the roll cage and Equivalent (von mises) stress acting on the roll cage for Rear Impact Test, performed on Ansys software is given below in Figure 8 and Figure 9 respectively.

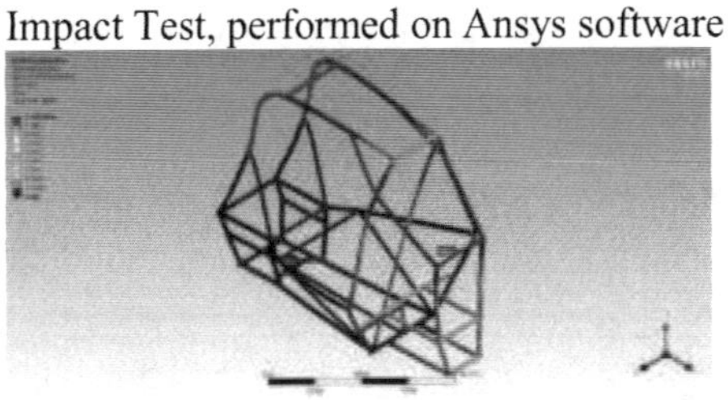

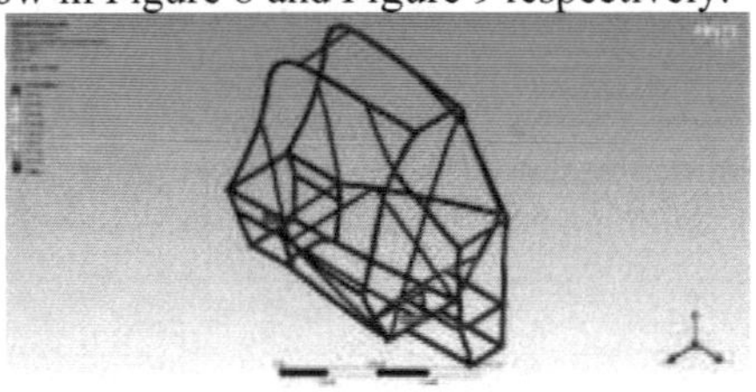

Figure 8. Total Deformation for Rear Impact Test

Figure 9. Equivalent (von mises) stress for Rear Impact Test

3.5 Dump Test:

This test is performed to know what will happen to the vehicle after it falls from a height or jumps from a ramp.

Impact Points - Front Lateral Cross Member (FLC), Aft Lateral Cross Member (ALC) and Rear part of the ATV.

Fixed Points – Pipes on which Suspension would be fixed.

Impact Force Calculations:

The distribution of weight of most ATVS is 55:45 ratio (Rear: Front) which is also our case.

ATV Weight (m) = 210 kg Time of impact (t) = 0.15seconds.

Height (h) = 1 m Velocity, $V = \sqrt{2gh} = 4.42$ m/s

Weight at the Front of ATV, m_f = Ratio Of weight at front axle × Actual weight of the ATV (8)

$m_f = 0.45 \times 210 = 94.5$ kg

Force at Front of the vehicle, $F_f = m_f \times V/t \rightarrow F_f = 94.5 \times 4.42/0.15 = 3213\ N \approx 3300N$ (9)

Weight at the Rear of ATV, m_r = Ratio Of weight at front axle × Actual weight of the ATV (10)

$m_r = 0.55 \times 210 = 115.5$ kg

Force at Rear of the vehicle, $F_r = m_r \times V/t \rightarrow F_r = 115.5 \times 4.42/0.15 = 3927\ N \approx 4000N$ (11)

Total deformation of the roll cage and Equivalent (von mises) stress acting on the roll cage for Dump Test, performed on Ansys software is given below in Figure 10 and Figure 11 respectively.

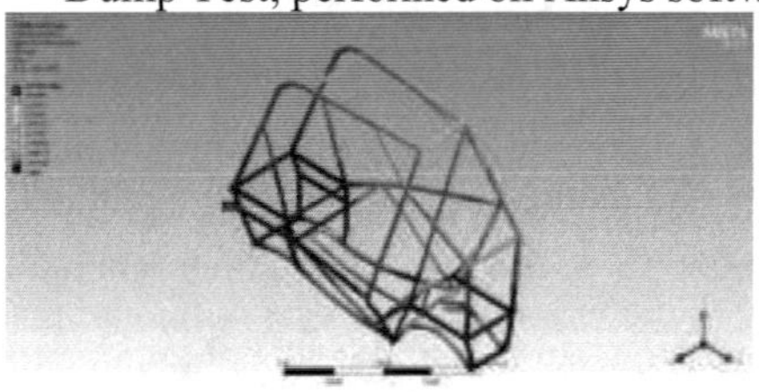

Figure 10. Total Deformation for Dump Test

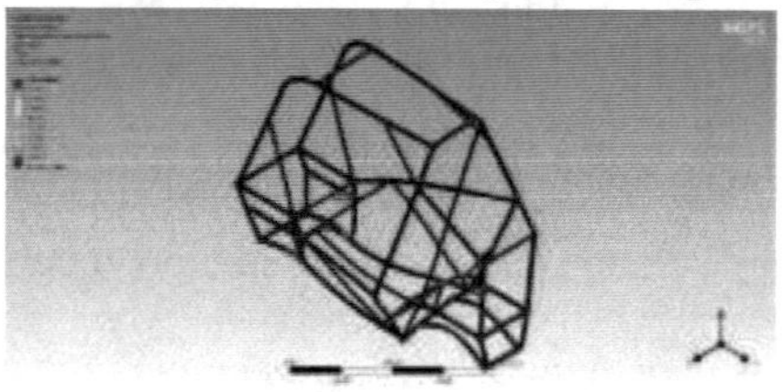

Figure 11. Equivalent (von mises) stress for Dump Test

4. Dynamic Analysis:

Since ATVs used for off-roading it is subjected to several uneven loading. So due to these uneven loadings, the vehicle vibrates this can cause resonance of forced and natural frequencies which is quite disastrous for the ATV's roll cage. Hence Modal Analysis is performed to check the natural frequencies of the ATV that are generated due to its self-weight.

Fixed Support – Fire wall of the ATV behind the driver's seat.

The Natural Frequencies that are obtained during modal analysis in Ansys 19.2 software is cited below in Table 4.

Table 4. Natural Frequencies of the Rollcage.

Serial No.	Frequency
1.	32.602
2.	51.031
3.	60.035
4.	77.41
5.	82.742

6.	93.095

The below Figure 14 show the Graph of Natural Frequencies that obtained during modal analysis performed on Ansys 19.2 software.

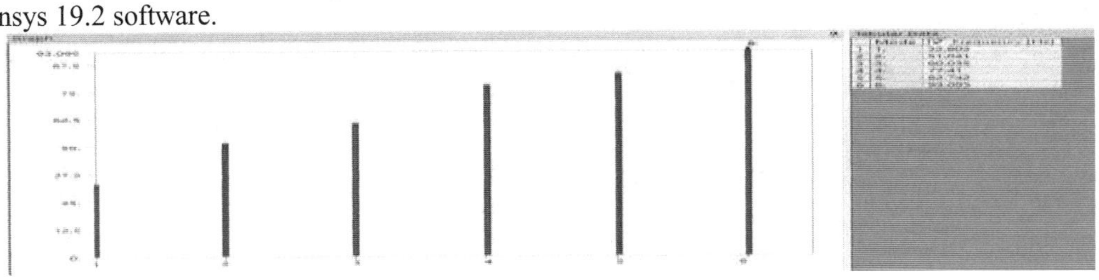

Figure 14. Graph of Natural Frequencies obtained from Modal Analysis (Taken from Ansys Software).

5. Results and Discussion

The acquired results Maximum Stress, Maximum Deformation and Factor of Safety is well within the range of what is given in the Baja SAE rulebook and previously published papers that are cited in this paper. Also the highest Factor of Safety 4.39 is obtained in Dump Test whereas the lowest Factor of safety is obtained in the Front impact test which is 1.93 and that is also above the mentioned lower limit of 1.5 Factor of safety. Henceforth it can be concluded that it is safe to begin the manufacturing of the roll cage. The Tabulated Data for the calculations of the proposed work is given in Table 5.

Table 5. Tabulated Data of Result of above Calculations.

	Force Applied (N)	Maximum Stress (MPa)	Maximum Deformation (mm)	Factor of Safety (FOS)
Front Impact Test	5900	378.67	0.72931	1.93
Side Impact Test	5900	345.25	2.0266	2.11
Rear Impact Test	5900	325.24	3.5552	2.24
Dump Test	Front – 3300 Rear – 4000	166.28	0.29041	4.39

6.Conclusion

This paper helped us in having better understanding of the Static Analysis of the ATV and the following conclusions are made:-

- Finite element analysis was the main key card in this entire project as it gave us critical knowledge about the design of the rollcage.
- The ergonomics of the vehicle is also kept in mind while design so that the driver can drive the ATV with absolute comfort and well-being.
- Greater the value of factor of safety greater will be the ability of the roll cage to sustain various loads in different terrains.
- All the impact test were performed and based on the maximum stress, factor of safety (FOS) was calculated. The factor of safety was well above the anticipated limit which is 1.5 and henceforth it is deduced that the modal of the ATV that was presented in this paper is safe to be fabricated.

- **References**

[1]. Alok D, Vedant K, Rajnish K and Manoj S 2015 Structural analysis of an atv frame *Int. J. Sci. Res.* **5**
[2]. Harvir Singh and Mayank K 2020 Design and optimization of an All-Terrain Vehicle roll cage *AJ* **8** pp 1735-41.
[3]. Penugula D R R, Naresh C and Phanisankar B S S 2020 Design and structural analysis of baja frame with conventional and composite materials *IJASRET* **5** pp 23-31.
[4]. R. Soundararajan et al. 2020 A novel approach for design and analysis of an all-terrain vehicle roll cage *Materials Today: Proceedings.*
[5]. Viktoria S, Ivan N and Sergey G 2020 Design and specifications of racing car chassis as passive safety feature *Transportation Research Procedia* **50** pp 591-607.
[6]. Shridhar A S, Tukkar A, Vernekar A, Badderu V, Patil A Y and Kotturshettar B B 2020 *Advances in Engineering Design and Simulation* Chandrasekhar U et al. (Singapore, Springer) pp 281- 292.
[7]. Kodavanti V R S N M, Kota Sai M and Nagireddy K P 2020 Design and analysis of baja rollcage *IRJET* **7** pp 4563-67.
[8]. Nipun S and Kaushik K 2020 Efficacy of Vehicle Chassis of Polymeric Composite *2nd International Conference on Materials Manufacturing and Modelling* **22** pp 2638-46.
[9]. Carlos A V L, Diana H A, Juan C T M and Ruben M M 2013 Review on global chassis control, *IFAC Joint Conference SSSC* pp 875-80.
[10]. Satputaley S, Ksheersagar K, Sankhari B, Kavishwar R and Waghdhare K 2021 *Advances in Mechanical Engineering* ed Kalamkar Vand Monkova K (Singapore, Springer) pp 317-327.
[11]. Prof. Gaffar M, Shital S, Aniket T, Rupesh T and Tejas G 2017 Design, analysis and fabrication of a rollcage for an all-terrain vehicle *IJTIMES* **3** pp 201-07.
[12]. Shiva K J, Ambesh S and Prabhudev M 2019 Design and analysis of chassis for sae baja vehicle *IOSR JEN* pp 51-57.
[13]. Anil B S, Ayub K A A and Krishnamurthy H S 2020 Buggy role cage – analysis and design *IJSSE* **10** pp 589-599
[14]. Ketha S 2020 Design and simulation of roll cage of an all-terrain vehicle *IRJET* **7** pp 450-57.
[15]. *SAE Baja Rulebook* 2013, 2019, 2020.
[16]. L. Rajeshkumar et al 2020 Design and optimization of static characteristics for a steering system in an atv *IOP Conf. Ser.: Mater. Sci. Eng.* **954**(2020)012009.
[17]. Dhruv V, Akash R, Amit W, Kaustubh M and Guruprasad Y S 2020 Design, analysis and fabrication of atv roll cage *IJAIR* **7** pp 65-69.
[18]. T. Safiuddeen, P. Balaji, S. Dinesh et al. 2020 Comparative design and analysis of roll cage for automobiles *Materials Today: Proceedings* **39** pp183-200.
[19]. David G U 2010 *The Mechanical Design Process* Bill Stenquist (New York: McGraw Hill).
[20]. Shubham P V and Shaikh A A R 2019 Static and modal analysis of all terrian vehicle roll-cage *IJERT* **8** pp 103-07

IOP Conf. Series: Materials Science and Engineering 1149 (2021) 012022 doi:10.1088/1757-899X/1149/1/012022

Heat Transfer Analysis of Recuperator for Waste Heat Recovery Purpose Using LMTD & NTU Method

Agraj Pandey[1] and Kaushalendra Kumar Dubey[1]

[1] School of Mechanical Engineering, Galgotias University, Greater Noida, India
***kaushalendra.dubey@galgotiasuniversity.edu.in**

Abstract: The dumped heat from captive industry like power generation plant, steel plant, refineries, etc. has tremendous potential for multiple energy effect like cooling, power and heating effect. The novel thermodynamic system has been invented for waste heat recovery (WHR) purpose as organic Rankine cycle, kalian thermal system, etc. The recuperator type heat exchanger (HEx) can support for WHR at category of medium and high-grade heat discharge. The design and operating parameters of recuperator is essential for complete development and its thermodynamic analysis. The proposed title of research deals with the heat transfer modeling using LMTD & NTU approach with the help of MATLAB R2017 analytical tool. And ANSYS is used for design work of mentioned heat exchanger. The major outcomes of this paper are maximizing the heat transfer rate, improve overall efficiency by operating parameters, and HEx effectiveness. The shell and tube material, diameter, number of turns of tubes.

1. Introduction

A Recuperator is a form of counter flow heat exchanger (as shown in Fig. 1) that is used for energy recovery in air handling systems or in exhaust gases from industry or plant processes [1]. The Recuperator is simple to make and can be used for a variety of purposes in the factory, including waste recovery in gaseous and liquid mediums at high pressures and temperatures. This form of Recuperator is used in thermal plants, food plants, refineries, and chemical plants, among other places [2-3]. In this paper, a standard method for designing the Recuperator is derived, which is coded with the aid of MATLAB in order to obtain the Recuperator's outlet temperature using the Epsilon-NTU method [4-5]. Waste energy is the energy released into the atmosphere by industry's flue gases, which accounts for 27% of total energy [6-8]. To recycle waste energy by reusing flue gases with the aid of a suitable heat exchanger for switching from one state to another. The first benefit of waste energy is that it improves the performance of the Recuperator by thermal efficiency. The second aspect involves minimizing emissions by reducing the size of the Recuperator [9-11]. The term "recuperator" refers to a heat exchanger that is used to recover waste energy [12].The Recuperator has a wide range of applications and is primarily used to recover waste energy in industries such as metallurgy, petrochemistry, mechanical engineering, chemical industry, and shipbuilding.The other type of Recuperator is Radiation Recuperator which is defined as indirect contact flow heat exchanger which is used for recovering the waste energy at high temperature application [12]

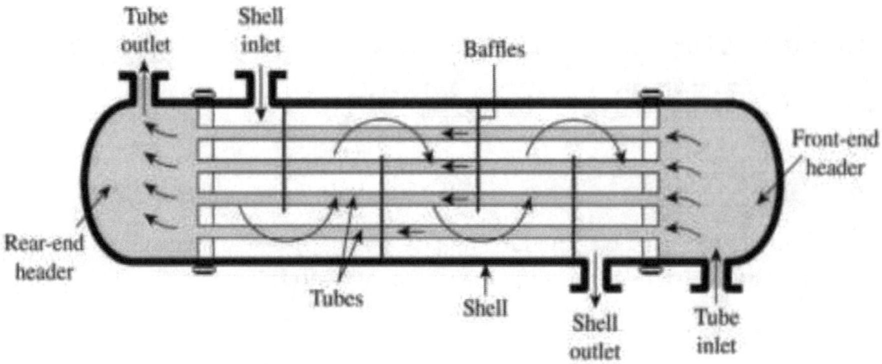

Figure.1 Tube and shell type Recuperator

2. Experimental Analysis

2.1 Heat Transfer Analysis of Recuperator

In this paper the design of recuperator with giving standard tube length and tube diameter dimension are followed by TEMA (tabular exchange manufactures association) and the code is written in MATLAB. Program is run for the obtaining maximum heat transfer by comparing the heat transfer is obtained by the LMTD method and heat transfer is obtained by the NTU method. The comprehensive heat transfer analysis is obtained from Fraas and Green D. model [2-3].

Let us assume the length and diameter of tube under the TEMA specification [2]

$$0.067 < (\text{shell diameter (S.D)/ tube length(l)}) < 0.201 \tag{1}$$

$$1.20 < (\text{diameter of tube}) < 1.499 \tag{2}$$

For laminar flow-

$$\text{Nusselt number} = \frac{h*D}{K1} = 0.023*(R_e^{0.8})*(Pr^{0.33})*(\eta^{1.25}) \tag{3}$$

Where-

R_e= Reynolds number

P_r= Prandtl number

Fixing the Recuperator's the inlet & outlet temperature for H.F. (hot fluid) as well as C.F. (cold fluid), considering the operating temperature values as per source-

$$T_{hfi}=353K;\ T_{hfo}=323K;\ T_{hci}=293K;\ T_{hco}=283K \tag{4}$$

$$U=20W/m^2K \tag{5}$$

$$LMTD=\frac{T1-T2}{\ln(T1/T2)} \tag{6}$$

$T1=T_{hfi}-T_{hci}$ (7)

$T2=T_{hfo}-T_{hco}$ (8)

Q=U*A*LMTD (9)

The above equation is coded in MATLAB in this way for obtaining the value of heat transfer as well as the graph of heat transfer also by variation in the size of tube.

Now taking the other parameter of Recuperator for finding the maximum heat-transfer effectiveness of Recuperator also.

300kg/s< hot fluid's mass flow rate (m_h) <500kg/s (10)

C_{ph}=1.005kJ/kg-K (11)

C_{pc}=4.18kJ/kg-K (12)

30kg/s< cold fluid's mass flow rate (m_c) <130kg/s (13)

$X=\frac{Cmax}{Cmin}$ (14)

$NTU=\frac{U*A}{Cmin}$ (15)

$$Q1 = mh * Cph * (Thfi - Thfo) = mc * Cpc * (Thci - Thco) \quad (16)$$

$$Cmin = \min(mh * Cph, mc * Cpc) \quad (17)$$

$$Qmax = Cmin * (Thfi - Thci) \quad (18)$$

$$\text{Effectiveness } (€) = \frac{1-e^{-NTU(1-X)}}{1-X*e^{-NTU(1-X)}} \quad (19)$$

These above equations are coded with the help of MATLAB in this way to find out maximum heat transfer and effectiveness of Recuperator which is essential for any type of Recuperator.From the Eq^n (19) this is clear that Effectiveness is the function of Capacity Ratio(R) and NTU. €=f (NTU, R).

3. Result and Discussion

The resultant graphs generated by using the LMTD equation with the help of MATLAB. So, from Fig. 2 and 3 this is clear that heat transfer from the Recuperator depend upon the diameter of tube which means that as slow as the diameter of tube increases then heat transfer from Recuperator also increases.The maximum value of heat transfer (Q) is 8.56×10^5kW at the value of tube diameter(D)=3m.

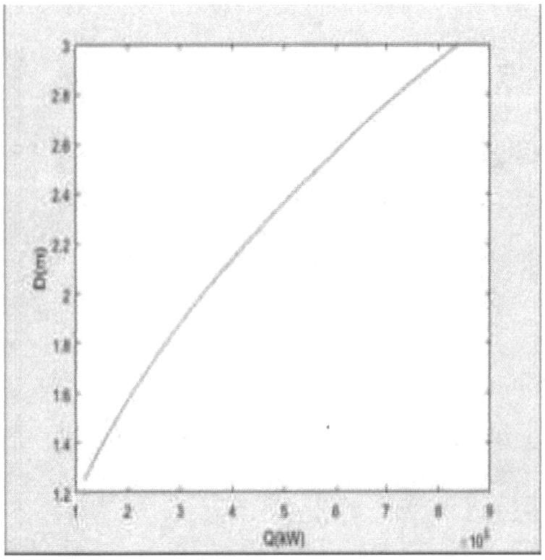

Figure 2. Tube diameter vs heat transfer

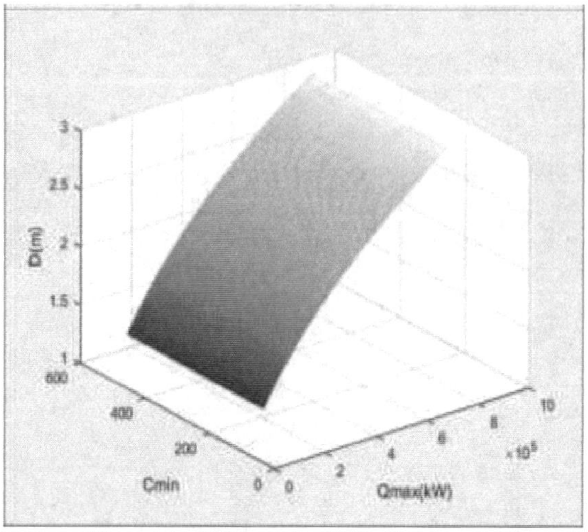

Figure 3. 3D graph of D, C min and Q max

The Figure 2 is obtained by using the NTU method which is coded in the MATLAB in this way to obtained the maximum value of HT (heat-transfer) to satisfy the value of HT (heat-transfer) which is obtained by using LMTD equation. From Figure. 3 & 4 this is clear that the maximum value of heat transfer (Q)=8.56×10^5kW at the value of tube diameter (D) = 3m

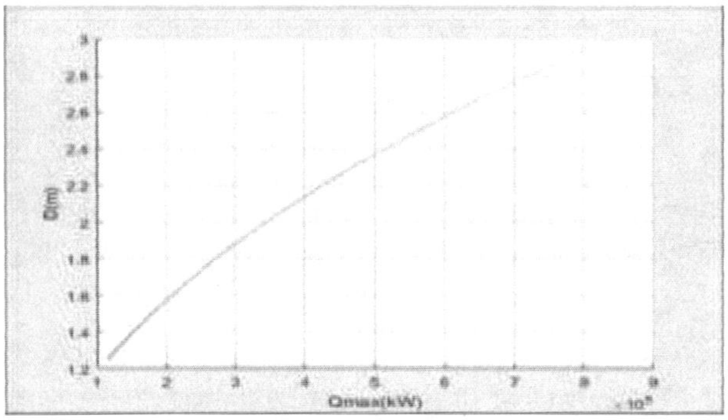

Figure 4. Tube diameter vs heat transfer (max)

Fig 5 is shown below which indicate the graph between X and Effectiveness, and from the eq^n (19) which shows the relationship between effectiveness, NTU and X (capacity ratio) or from the equation (19).

From the figure it is clear that effectiveness for counter flow depends on the NTU and X. From Fig. 5 effectiveness ramps on with decreasing the capacity ratio(X) and from the Figure 5 value of effectiveness is 1.21. Figure 6 indicate 3-D graph of effectiveness which is generated by the MATLAB

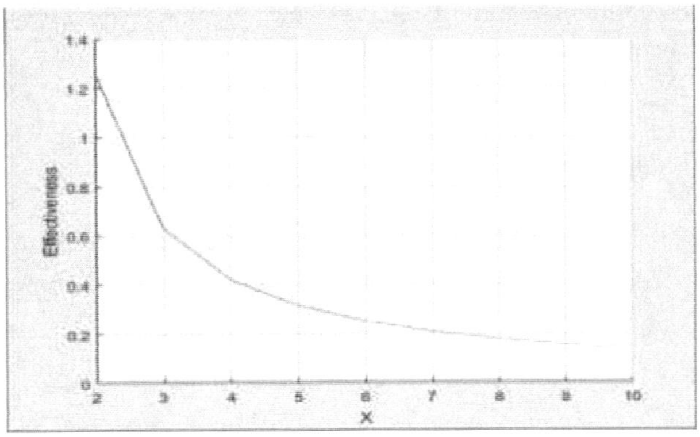

Figure 5. Effectiveness vs capacity ratio

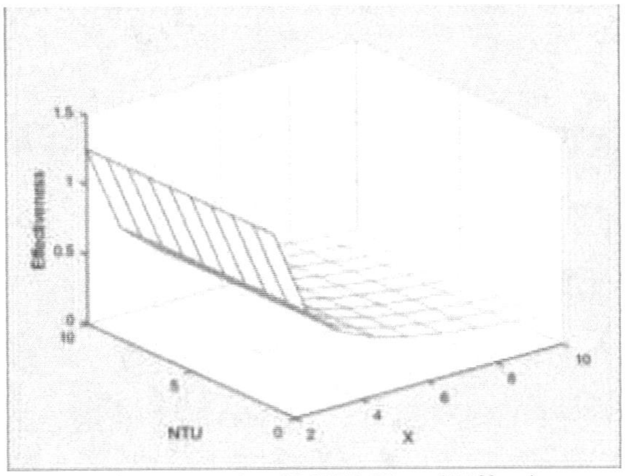

Figure 6. 3D graph of NTU, X and Effectiveness

- From the above graph the value of tube diameter, maximum value of heat transfer and effectiveness are mentioned in table which is shown below.

4. Conclusion

The proposed recuperator type heat exchanger is basically designed for waste heat recovery purpose. The selective diameter, material, and other essential parameters have been considered as per available heat source for heat recovery. The measure highlights have been concluded-

- A MATLAB code is written in such a way to find out the value of heat transfer which is essential for the Recuperator and taking the tube diameter under the consideration of TEMA specification.
- The method to find out maximum value of heat transfer for shell and tube type Recuperator by coding in MATLAB and solving the LMTD Equation and NTU method Equation by analyzing the

graph which has shown in above figure. That graph represents the heat transfer on x-axis and Tube diameter on y-axis.

- In case of Fig.2, the value of C min (mass flow ratex heat capacity) on y-axis, heat transfer on x-axis and tube diameter on z-axis which appears to be stationary.
- Effectiveness of Recuperator measures the performance of Recuperator. Effectiveness and Heat transfer play important role in the designing of Recuperator.
- The final value of tube diameter, heat transfer and effectiveness are obtained with the help of Matrix Laboratory which is D=3meter, Q=8.56×10^5kw and €=1.21 respectively.

Nomenclature

Dynamic viscosity= ŋ(Ns/m2)

Q = heat transfer(W)

Thfi = hot fluid temperature at inlet(K)

Thfo hot fluid temperature at outlet(K)

Thci = cold fluid temperature at inlet(K)

Thco = cold fluid temperature at outlet(K)

m.h = hot fluid mass flow rate (kg/s)

m.c = cold fluid mass flow rate(kg/s)

Cph = hot fluid's heat capacity(kJ/kg-K)

Cpc = cold fluid's heat capacity(kJ/kg-K)

U= overall heat transfer coefficient (W/m2K)

A = Recuperator's area(m2)

h = Heat-transfer coefficient, (W/m2K)

D = diameter(m)

K1 = (W/mK) material's thermal conductivity

X = (C.R.-Capacity Ratio)

C.R. = Capacity Ratio

NTU = Number of Thermal Unit

HX = Heat Exchanger

MATLAB = Matrix Laboratory

HF = Hot Fluid

CF = Cold Fluid

TEMA: = Tabular Exchange Manufacturing Association

LMTD: Logarithmic Mean Temperature Difference

References

[1] Mohammed Rabeeh, V., & Vysakh, S. (2014). Design of Shell and Tube Heat Exchanger Using MATLAB and Finding the Steady State Time Using Energy Balance Equation. *International Journal of Advanced Mechanical Engineering*, *4*(1), 95-100.

[2] Fraas, A. P. (1989). *Heat exchanger design*. John Wiley & Sons, ISBN: 978-0-471-62868-2

[3] Green, D. W., & Southard, M. Z. (2019). Perry's chemical engineers' handbook. McGraw-Hill Education, ISBN: 9780071834087.

[4] Bergman, T. L., Incropera, F. P., Lavine, A. S., & Dewitt, D. P. (2011). *Introduction to heat transfer*. John Wiley & Sons, ASIN : B00APYBOB0.

[5] Peters, M. S., Timmerhaus, K. D., & West, R. E. (2003). *Plant design and economics for chemical engineers* (Vol. 4). New York: McGraw-Hill, ISBN-10 : 0072392665.

[6] Salahuddin, U., Bilal, M., & Ejaz, H. (2015). A review of the advancements made in helical baffles used in shell and tube heat exchangers. *International Communications in Heat and Mass Transfer*, *67*, 104-108.

[7] Ayala, H. V. H., Keller, P., de Fátima Morais, M., Mariani, V. C., dos Santos Coelho, L., & Rao, R. V. (2016). Design of heat exchangers using a novel multiobjective free search differential evolution paradigm. *Applied Thermal Engineering*, *94*, 170-177.

[8] Choo, P. L. (2015). *Generating Electricity Power From Solar Energy: Preview From First Solar Malaysia Sdn Bhd* (Doctoral dissertation).

[9] Hatami, M., Ganji, D. D., & Gorji-Bandpy, M. (2014). A review of different heat exchangers designs for increasing the diesel exhaust waste heat recovery. *Renewable and sustainable energy reviews*, *37*, 168-181.

[10] Kayabasi, E., Kılınç, E., Rahman Çalhan, A. C., Tabak, A., Furtun, F., Arslan, K., ... & Özalp, M. DESIGN SIMULATION AND ECONOMIC EVALUATION OF A HEAT EXCHANGER FOR HOT STOVES.

[11] Sharma, H., Kumar, A., & Khurana, S. (2014). A review of metallic radiation recuperators for thermal exhaust heat recovery. *Journal of Mechanical Science and Technology*, *28*(3), 1099-1111.

[12] Asadi, M., Song, Y., Sunden, B., & Xie, G. (2014). Economic optimization design of shell-and-tube heat exchangers by a cuckoo-search-algorithm. *Applied thermal engineering*, *73*(1), 1032-1040.

IOP Conf. Series: Materials Science and Engineering 1149 (2021) 012023 doi:10.1088/1757-899X/1149/1/012023

Optimizing Agricultural to Hydro power water transfer

Maya Kurulekar[1*], K Kumar[2], Shardul Joshi[3], M Kurulekar[4] and Neha Shinde[5]

[1]Research Scholar & Assistant Professor, Vishwakarma University, Pune, India
[2]Scientist 'C', Central Water and Power Research Station, Pune, India
[3]Professor, Vishwakarma Institute of Information Technology, Pune, India
[4]Research Scholar, Vishwakarma Institute of Information Technology, Pune, India
[5]B Tech, Electrical Student, AISSMS College of Engineering, Pune, India
***maya.kurulekar@vupune.ac.in**

Abstract. Hydro energy is one of the richest and most useful renewable energy sources in the world. It contributes 16-20% of total electricity generation in the world. A trade-off between irrigation and power generation is essential in order to achieve optimal social benefits. Proper allocation of water to irrigation and optimal utilization of water for power generation is one of the most important area of concern. The study is mainly concentred on using educators, auxiliary services and inline axial turbines. It helps to identify the energy recovering potential in terms of closed conduit. About 90% excess released through irrigation gates can be used for inline turbines. The inline turbine will also be exploited during the lower release where the main turbine could not be used during lower releases of water to irrigation. The remaining 10% water can be used for educator pumping and auxiliary services. The system of retrieving energy potential can be implemented in an agricultural to hydro power dependent irrigation projects. A pilot study on Dhom (2X1 MW) irrigation dependent hydroelectric plant, MAHAGENCO, Satara, Maharashtra has been conducted in order to tap the water during the lower release from the reservoirs. This paper emphasizes the methods of utilizing untapped energy during the lower & higher water release in irrigation dependent hydro power plant.

1. Introduction

In view of the global renewable energy production is steadily increasing to meet demands for clean and reliable energy. Presently, Hydropower contributes 16-20% of total electricity generation in world and turbine is the power conversion component in hydropower plant.

The International Hydropower Association (IHA) reports that renewable comprise 23% of the global electricity mix as of 2014, with 16% of the world's energy production coming from hydropower [1]. A study by the World Energy Council (WEC) shows the hydro power generation accounts about 71% of all renewable energy [2]. Hence, hydro power plays a vital energy player in a world scenario with its outstanding characteristics of sustainable, clean and environmental source of energy. Water resources and their management aims for two major economic activities which are irrigation crop production and hydro energy generation. These two activities are often referred as water for crop irrigation as a static asset and hydro power generation as a dynamic asset [3]. Thus, dynamic management approach is critical for multipurpose reservoirs in order to achieve optimal social benefits. A study conducted at International Water Management Institute (IWMI) has shown that the demand for water is to increase by 50% by the year 2025 [4]. The importance of sustainable water – use efficiency is vital in a water resources sector [5].

Among the renewable energy sources, hydropower generation is one of the primary sources which date back to 1770s. Presently, The International Hydropower Association (IHA) reports that renewable comprise 23% of the global electricity mix as of 2014, with 16% of the world's energy production coming from hydropower [1].

1.1 Literature Review

Hydraulic turbines are rotary machine which utilizes the potential and kinetic energy that converts into a useful mechanical energy [6]. Hydropower turbines areoftwo types namely impulse and reaction type. Impulse Turbines are high head, low discharge, and low specific speed and operate at atmospheric pressure, while the reaction turbinesoperate under low and medium head with high specific speed and operate under variable pressure. Especially Kaplan turbine is a reaction type mainly used to create the power from the low head and to provide loading at large flow rates. Kaplan turbines are always pre dominantly engaged in an agricultural to hydro power water transfer projects [7]. Most of the low head reservoirs call for a dynamic allocation of irrigation crop production and hydro power generation. A tradeoff between irrigation and power generation is essential in order to achieve optimal the social benefits. In this scenario, all hydro power generating companies bound to respect pre-defined monthly release targets reflecting the agricultural demands [8].

A storage type hydroelectric irrigation power plant possesses a high capacity factor of around 50% as compared to 10% and 30% of solar and wind power [9]. A plant capacity is selected based on the annual rain fall pattern, storage capacity, gross head, flow rate available, capacity factor etc. On the operation point of view, a hydro turbine is subjected to cavitation, erosion, fatigue which dictates operational conditions in turn invariablyaffects the exploitation of the units [10]. Most of the irrigation dependent hydro plants are of low/medium head which employs the reaction turbines. Hence, reaction turbines are subjected to trailing edge cavitation, leading edge cavitation, draft tube swirl cavitation, inner blade cavitation and travelling bubble cavitation [11]. In hydraulic turbines, due to cavitation: the machine performance drop, high and low pressure is developing, the noise and vibration level increases. Hence, standard guidelines are issued by the regulatory authorities for safe running of plants and that limits minimum discharge in which the plant tobe operated. A study [12] has shown that in hydro power generation projects involving impounding of water, adequate water is required to be released round the year to meet the needs of the downstream users and flora and fauna to maintain the eco system in that area. However, during the minimum flow released to the canal systems could not be utilized for power generation. In view of the above, some of the energy potentials at irrigation dependent power plants could not be tapped because of the limitation imposed on the operation of the power units. In addition to the above, a study [13] suggested that the hydro plants are not immune from unexpected severe incidents during the operation periods especially floods related damages. Some of the references [14] show that the flood related damages in hydro plants are steadily increasing trade due to the economic activities in flood plains and global warming. The work [15] elaborates on the mathematical model generated for the optimization of power generation by minimizing the water consumption and also the dynamic programming approach is explained. A paper [16] focussed on the different ways of retrieving energy potential in an irrigation dependent power plant. In order to minimize the damages, disturbances in generation various guidelines are issued time to time by the regulatory authorities like National Disaster Management Authorities (NDMA), Central Electricity Authority (CEA) and Central Water Commission (CWC). In this area, limited research has been attempted on the irrigation dependent power units towards optimal use of energy. Recent flood related incidents in hydro power plants like Kopili hydroelectric plant, Assam and Bhatghar hydro power plant, Maharashtra have opened up for discussion for safe operation of power plants. During the flood related incidents in a typical hydro power plant, the existing pumping system becomes inactive due to the chances of short circuit and operational limitations. In this connection, the use of standby generator is also inactive. In such a catastrophic situation, the use of educator jet pump will be appreciable in order to avoid the flooding or the minimal damage to the plant equipments and systems. Hence, the present paper addresses the models and methods for utilizing the untapped energy potential at irrigation to hydro power water transfer schemes especially to utilise the

IOP Conf. Series: Materials Science and Engineering 1149 (2021) 012023 doi:10.1088/1757-899X/1149/1/012023

Environmental Flow (EF) and unscheduled minimum flow for irrigation through closed conduits by exploiting educator and inline hydro turbines.

2.Problem Statement and Methodology

Proper allocation of water to irrigation and optimal utilization of water for power generation is one of the most important areas of concern. Most of the irrigation based hydro turbines employs the reaction type turbines due to the advantages of low/medium head and high discharge through the irrigation canals. The capacity of the turbines based on the water release and head available in the reservoirs. However, during the lower release of water for irrigation and regulation of mandatory water release for flora and fauna, the water release in the irrigation gates are been utilized. The power units of reaction type could not able to tap the energy due to the restriction on the lower discharge conditions, the turbine may subject to cavitation and leads to noise and vibration and subsequent catastrophic failure of units. This is a case of universal in nature for irrigation storage dependent power generating units. This condition leads to reasonable generation loss to the power generating units. A pilot study on Dhom (2X1 MW) irrigation dependent hydroelectric plant, Mahagenco, Satara, Maharashtra has been conducted in order to tap the water during the lower release from the reservoirs. This paper emphasizes the methods of utilizing untapped energy during the lower & higher water release in irrigation dependent hydro power plant.

2.1Salient details of Dhom Hydroelectric plant

DHOM 2 X 1 MW Capacity Hydro Power Station has been constructed at Vyahali dam near Wai, Satara district, Maharashtra.The dam lies right bank of river Krishna and the two separate outlets, one for Irrigation Discharge (ID),other for power have been embedded in the right bank of the dam. Dhom left bank canal bifurcates from right bank canal via embedded on downstream side is shown in Figure 1. The rain fall data and the energy utilized and the excess water release in irrigation gates are given in table 1 & figures 2. The features of power plants are,

1. Installed capacity = 2 X 1MW **2.** Annual design Energy generation = 10.56Millio units (MUs)
3. Average Rain fall in catchment area1500 – 6250 mm **4.** Gross dam storage 382 MCM
5. Average inflow of water 423.91 Million Cubic Meters (MCM)
6. Live storage 331MCM 7. Full reservoir level 747.70 m **7.** Design head 18 m
8. Minimum draw down level MDDL 733.08 m **9.** Design discharge 14 m^3/sec
10. Diameter of penstock 2.5 m

From the data collected, the average power generated by the two units from the year 2005 to 2018 (Table 1) is 7.247 million units per year. The capacity factor of generating unit is,

$$\text{Capacity Factor} = \frac{\text{Average energy generated per year in kWh}}{\text{Installed capacity in W*24*365}} \quad (1) \quad = 7247000 \;/\; (2000*24*365) = 0.463$$

So, it has been observed that the plant is working at its 40% of capacity. The maximum discharge through the power plant is 500 cusec and irrigation requirement is 900 cusec at a time. The total release of water for irrigation by left and right bank canal at a time is 650 cusec and 235 cusec respectively. Due to simultaneous discharge of water from both the canal, 400 cusec water releases continuously causes generation loss. It has also remarked that the project report design working table schedule is not followed. Due to which no yearly design generation of 10.597 million units is achieved since from commissioning. It has been also observed that average energy loss per year is 2.6919 MU. Thus total units lost per year are 2691900kWh. Therefore the power generated from this loss would be,

$$\text{Power loss}=2.6919*\frac{10000000}{12*30*24}= 311.56 \text{ kW} \qquad (2)$$

So, the additional power of around 300 kW can be generated through the losses.

The data from year 1991 to 2018 in Table 1, Figure 2 shows that the water released through the irrigation gates at minimum discharge during the high rain fall period and maximum discharge during the low rain fall period.

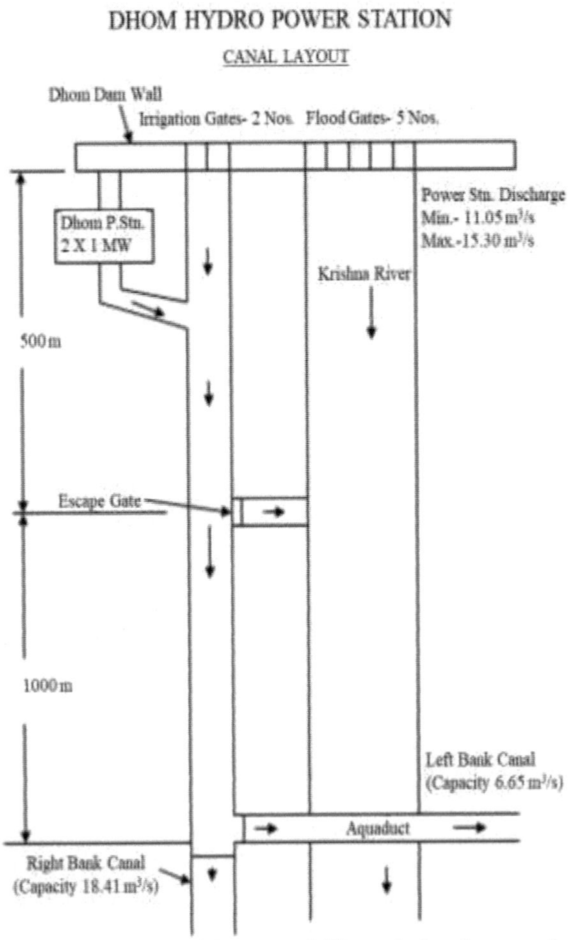

Figure .1 Layout of Dhom Irrigation canal

Table 1. Irrigation and Machine Discharge

Year	Average Rainfall	Irrigation Discharge	M/c Discharge
	(mm)	Million Cubic Meter (MCM)	Million Cubic Meter (MCM)
91-92	967.60	98.54	0.77
92-93	910.20	54.35	19.40
93-94	1129.30	58.24	106.11
94-95	1170.90	37.42	90.23
95-96	660.60	92.36	21.60
96-97	1114.40	61.44	85.85
97-98	1086.60	165.29	212.33
98-99	858.70	88.54	158.20
99-2000	852.90	27.65	236.56
2000-01	565.50	59.56	148.77
2001-02	702.30	86.46	155.55
2002-03	532.50	163.44	214.51
2003-04	661.20	56.35	128.30
2004-05	1164.20	48.48	191.58
2005-06	1454.80	30.32	303.90
2006-07	1504.00	57.35	342.24
2007-08	1200.00	27.85	284.60
2008-09	1009.00	154.24	211.01
2009-10	1163.00	29.84	160.53
2010-11	984.00	52.21	155.42
2011-12	953.00	90.87	194.20
2012-13	811.00	58.51	158.56
2013-14	1082.00	161.32	198.73
2014-15	1106.00	58.76	174.71
2015-16	580.00	58.90	75.24
2016-17	888.00	165.39	185.68
2017-18	897.00	46.24	20.19

Also, the behavior of machine during high and low rainfall period is seen. Hence, a proportionate energy loss at the given head and discharge due to which the power units could not able to operate having constraint of low discharge and also the period when high releases, the units are operating at maximum load. Also there is a change in rainfall pattern, frequency and quantity as one of the effect of the global warming which may cause the flooding over any. Similarly sudden power failure may also disrupt the routine functioning of plant due to improper working of auxiliary circulating pumps. This cascade of low and high discharge has been accommodated by adopting following components of energy recovery model as shown in Figure. 3.

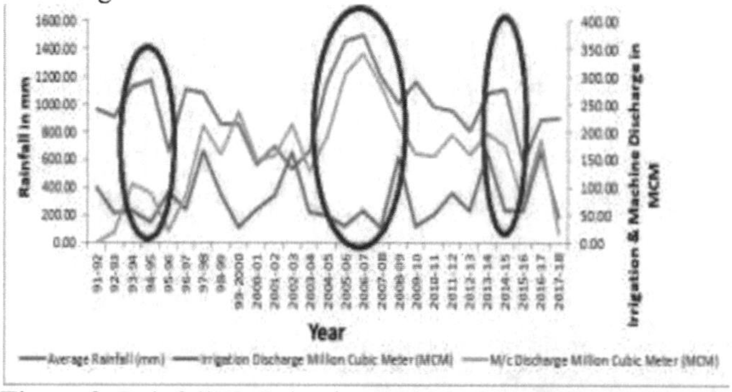

Figure 2. Rainfall Vs Irrigation and Machine Releases

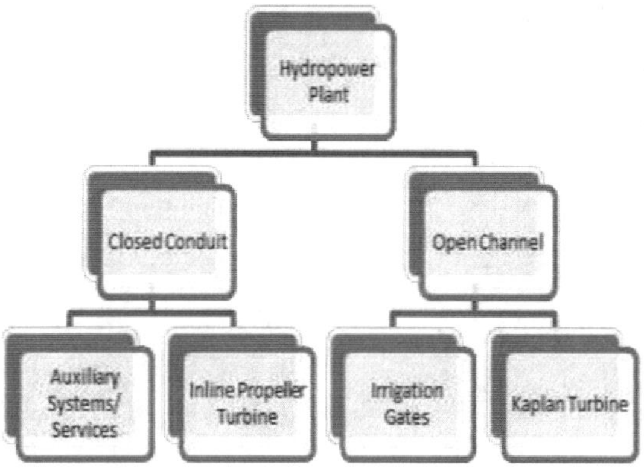

Figure 3 Components of Energy Recovery Model

Figure 4. A Typical section of a penstock section at main inlet valve of a hydro turbine

1. Dewatering component – an Eductor
2. Closed conduit inline propeller turbine in the existing suction line of the circulating pumps for cooling and auxiliary services
3. Arrangement of split type Kaplan/propeller turbines installed in the structure of the irrigation gates
4. Arrangement of very low head Kaplan turbines in the canal in which the energy developed through the water release from the turbines and the release through irrigation gates.

A typical reaction turbine plant is provided with number of pressure tapings such as pressure equalizing line connecting the upstream and downstream of the main inlet valve, drain lines on penstock at inlet and outlet of the main inlet valve and cooling line tapping for circulating pumps for generator cooling, bearing cooling, lubricating oil cooling and other auxiliary services. The figure 4 shows the existing layout of the water conduit at main inlet section of a hydro turbine. Sudden power failure, overflowing of existing area of power plant may create circumstances of improper and discontinuous working of hydro power plant. This paper emphasizes on how the excess water in the hydro power plant can be restored by employing suitable methods or devices. Additional dewatering system has been suggested supporting the existing auxiliary circulating pumps. Figure.5 shows the existing layout of hydro power plant.

3. Proposed methodology for eductor pumps

The penstock of the existing hydro power units is of 2500 mm diameter each in which flow to the turbine passes. Further, the auxiliary circulating pumps of capacity 15 kw each is provided to govern oil cooling, generator air cooler and other auxiliary services. The suction of auxiliary circulating pumps draw water from the tapping provided in the penstock of the turbines. Usual circulating pumps require electricity for their operations.

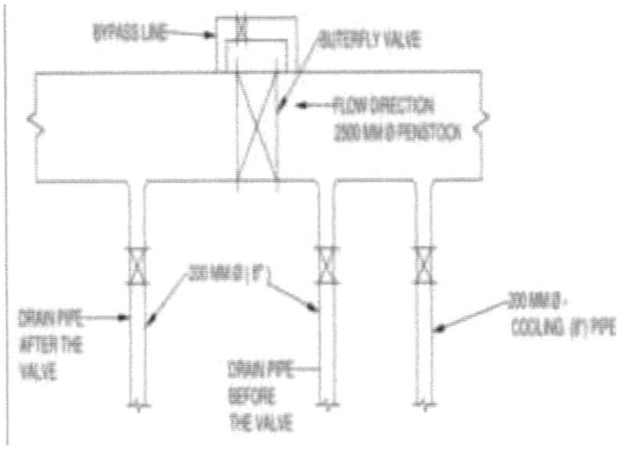

Figure 5: Existing layout of Hydropower Plant

In case of unavailability of power and the brimming in the power plant, make these pumps redundant. Thus, the discharge and velocity of flowing fluid has been effectively used in solving such real time problem without using the external driving force as electricity. The Venturi based pumps which are worked on fundamental principle of flow dynamics known as Bernoulli's principle and follows the basic Continuity equation known as an Eductor has been suggested. Eductor (Figure. 6) is functioning without any electricity and rotation of any parts such as impeller, motor. Any type of educator is having three basic portions; Motive portion, Suction portion and Outlet portion. Motive portion takes in motive (primary) fluid under high pressure and allows it to flow through the nozzle created thereafter. Suction portion insist the suction of suction (secondary) fluid. The suction fluid gets in and transported by motive fluid with the increased velocity of motive fluid after passing through nozzle. There is a kinetic energy transfer takes place in the Outlet portion due to the friction between these two fluids and increased velocity. This will make two fluids to leave outlet portion having diverging end with increased pressure and reduced velocity. The various parts diagram of eductor is shown in and Figure. 7 show the modified layout of energy recovery model in a hydropower plant.

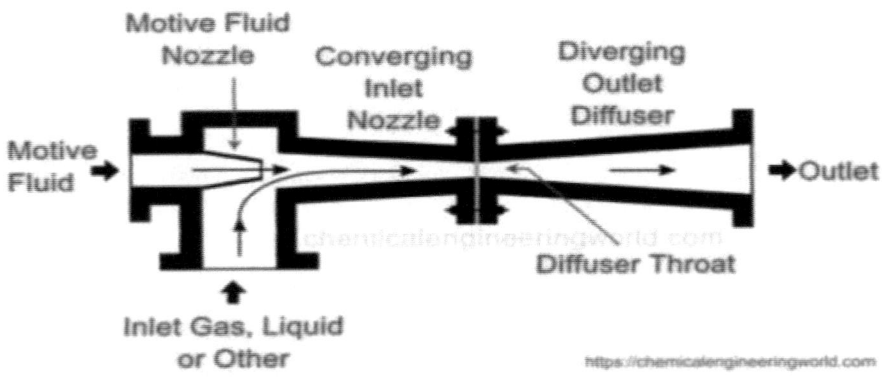

Figure. 6 Concept of Venturi based Eductor

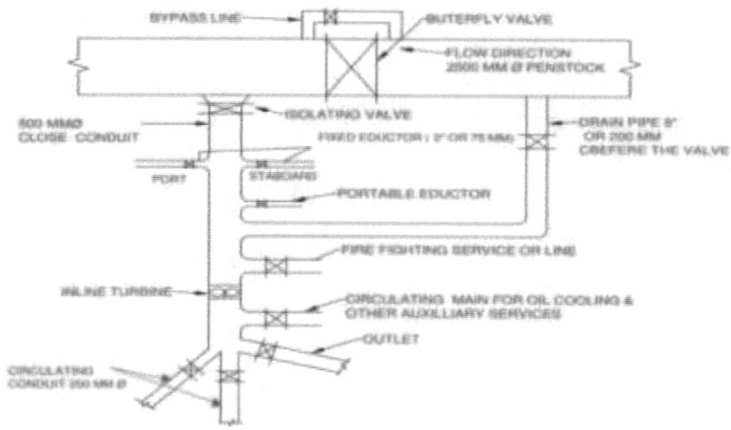

Figure. 7 Modified layout of Energy Recovery model

The existing layout of services to turbine units along with penstock can be modified to address the problem of reducing the loss of water or converting this loss into a potential use by modifying the layout. Alternative for dewatering the excess water can be used in the new arrangement to improve the conditions which may arise due to sudden power failure. Eductors can be used as fixed or portable installation. This ability of educator of working without power and rotating parts reduce the cost of operation and maintenance, portable eductors increase its flexibility as per requirement of time and place. This will certainly minimizes the chances of risk during the flooding of plant. Eductors are available according to the purpose of work like mixing eductors used in chemical industries, sand and mud eductors useful at the site of reservoirs to pump accumulated material. Also solids handling eductors are dealing with dry and wet solids, multinozzle eductors in operation for suction fluid containing solids and semi solids. Deep well, recirculating and priming eductors are useful in lifting water with few variations. The discharge from the eductor will be calculated to know the performance of this Bernoulli's based pumps over the submersible or borehole pump. The discharge is calculated on the basis of following assumptions:

1. The water vapour pressure is assumed = 3.023m of water column **2.** Required Net Positive Suction Head (NPSH) =10.13m of water column **3.** Performance ratio = 1.1 (if both motive and suction fluid is water) **4.** The atmospheric or barometric pressure required for NPSH is calculated on the basis of altitude of a dam.

The sample calculation to determine the size of eductor in case of a Dhom dam is shown as follows.

1. Design head of dam (H_{ds}) = 18m **2.** Assume suction head (H_s) = -3m **3.**Discharge head(H_d) = 10m **4.** Total Discharge Head (TH_d) = 13m **5.** Operating head (H_o) = H_{ds}-H_s = 21m **6.** Available NPSH ($NPSH_a$) = 3.47m **7.** Suction flow (Q_s) = 5kW of monoblock pump running for 6 hours for Discharges of 12 cumec **8.** Motive flow (Qm) = Qs/1.1 = 10.9 cumec **9.** Total discharge (Q) = Qs+Qm = 22.90 cumec **10.** The size of eductor required is 0.0635 m with velocity of 2.438 m/sec

4. Result

Eductors can replace the conventional pumps on many factorssuch as saving on electricity, continuous discharge without electricity, portability. The calculation of discharge required through eductor for other dams based on net head of respective dams and energy recovery is given in Table 2. This data can be used to know the size of eductor. By using eductor the saving in electricity can be possible since NO electricity required for this pump to run. The necessary calculation from Table 2 shows that if conventional pump of 5kW would run for 6 hrs at the cost of Rs.5/- per hour, it would require Rs. 54000/- per year in case of normal operation. In addition to the normal operation, during the catastrophic situation, the use of educator jet pump will be appreciable in order to avoid the flooding or the minimal damage to the plant equipment and systems. Further, the average energy loss due to low discharge is,

21.535/8 = 2.6919 million unit/year (3)

Table 2. Potential for utilizing water for auxiliary services

Sr. No.	Name of Dam	Design Head in m	Suction head in m	Discharge Head in m	Total Discharge Head in m	Operating Head in m	NPSHa in m	Discharge through Monoblock pump in cumec	Motive Flow in cumec	Total Discharge in cumec	Power of Pump in kW	Running hours of pump	Units consumed by pump per month	Cost of running the pump in Rs./year
1	Dhom	18	-3	10	13	21	3.47	12	10.90	22.90	5	6	900	54000
2	Kanher	21.2	-4	11	15	25.2	2.50	18	16.36	34.36	6	8	1440	86400
3	Bhatghar	26	-5	12	17	31	1.56	24	21.81	45.81	7	7	1470	88200
4	Dudhganga	42	-4	13	17	46	1.51	30	27.27	57.27	7.5	6	1350	81000
5	Panchet	45	-3	14	17	48	3.49	33	30	63	8	8	1920	115200
6	Varasgaon	47	-5	15	20	52	1.66	42	38.18	80.18	9	7	1890	113400
7	Bhira	48	-5	15	20	53	1.54	45	40.90	85.90	10	6	1800	108000

The power lost through the irrigation release of an irrigation dependent power plant is about 311.56 kW derived from the equation 2. An additional power generation can be tapped by providing closed conduit inline axial turbine in the existing penstock section by modifying pressure line to accommodate a conventional inline axial flow turbine. An inline axial flow propeller turbine installed in the modified piping of 500 mm nominal bore can able to generate 150 kW of power with design head of 18 m with water velocity of 6.5 m/s at flow 1.3 m^3/sec considering the efficiency between 80 – 85%. The power produced through the inline turbine can be used for internal load of power house and it can be optimally used to retrieve the lower releases and high releases at the irrigation canals. The installation of inline hydro turbine in the leading pipe from existing penstock will generate an additional energy of 300 units per hour will yield 2.592 million units per year.

5.Conclusion

A pilot study on Dhom (2X1 MW) irrigation dependent hydroelectric plant, Mahagenco, Satara, Maharashtra has been conducted in order to tap the water during the lower & higher release from the reservoirs. The study is mainly concentred on using educators, auxiliary services and inline axial turbines. It helps to identify the energy recovering potential in terms of closed conduit power utilization by adopting inline axial flow turbine as well as by utilizing auxiliary services. About 90% excess released through irrigation gates can be used for inline turbines. The inline turbine will also be exploited during the lower release where the main turbine could not be used during lower releases of agricultural to irrigation. The remaining 10% water can be used for educator pumping and auxiliary services. The system of retrieving energy potential can be implemented in an agricultural to hydro power dependent irrigation projects.

6.Acknowledgement

The authors are grateful to Shri A K Agarwal, Director, CWPRS, Pune, for giving the guidance throughout the studies. The authors are also thankful to Dr. R G Patil, Scientist ‘E’ for his continuous support and guidance during the period of investigations at various stages.

References

[1] Greg Schellenberg et al. 2017, *Sedimentation and Hydropower: Impacts and Solutions, Hatch.*
[2] World energy Council, *https://www.worlgenergy.org/data/resources/hydropower*
[3] A. Tilmant, Q. Goor, and D. Pinto2009, *Agricultural to hydro power water transfers: Sharing water and benefits in a hydro power- Irrigation Systems, Hydrology and Earth System Sciences*, 13, pp. 1091 – 1101.
[4] Seckler. David, Amarasinghe, Upali, Molden. David, Radhika and Barker. Randolph1990 to 2025, *Worldwide Demand and Supply, Scenerio and Issues*, Research Report 19. International Water Management Institute, Colombo, Sri Lanka.
[5] D. Suresh Kumar and K.Palanisami, Impact of drip irrigation on farming System2010, *Evidence from Southern India, Agricultural Economics and Research Review*, 23, , pp.265-272.

[6] Ece Ayli2019, *Cavitation in Hydraulic Turbines, International Journal of heat and Technology*,23, , pp. 334-344
[7] Munendra Kumar and A Nikhade 2015,*Cavitation in Kaplan Turbines, Journal of Material Science and mechanical Engineering*, 2(5), , 425-429.
[8] A Tilmint, Van der Zaag, and P. Fortemps 2007*Modelling and Analysis of Collective management of water resources, Hydrology and Earth System Sciences*, 11, pp. 711 – 720.
[9] Bhilal A. Nasir 2014*Design Considerations of Micro Hydro Electric Power plants, Energy Procedia*, 50, pp. 19-29.
[10] Doina. Frunzaverde, V. Campion, D. Nedelcu, G. A. Gillichand G.Marginean, *Failure Analysis of Kaplan Turbine Blade by Metallographic and Numerical Analysis, Continuum Mechanics*, ISSN No. 1790 – 5095, pp. 60-66.
[11] M. K. Padhy and R P Saini2008*A review of silt erosion in Hydro turbines, Renewable and Sustainable energy reviews*, 12(7), 1974-1987.
[12] Sharad K. Jain and Pradeep Kumar2014*Environmental Flows in India: Towards Sustainable WaterManagement, Hydrological Sciences Journal*, 59(3-4), pp. 751-769.
[13] US Patent No. 8884454B2, Method and apparatus for improved hydropower generation at existing impoundments, 2009
[14] US7972108B22003*Turbine and hydroelectric power plant for very low head.*
[15] Jiqing Li, May Myat Moe Saw, Siyu Chen and Hongjie Yu 2020 *Short-Term Optimal Operation of Baluchaung II Hydropower Plant in Myanmar*, Water.
[16] Maya Kurulekar et.al. *2020A Pilot Study on Retrieving Energy Potentials during Minimum Discharge Through Irrigation Dependent Hydro Power Plants*; Accepted for Publication. International Conference on Fluid Machinery and Fluid Power (FMFP 2020); IIT Guwahati

3D FEM simulation on Deformation and Strain analysis for grain refinement of Mg Alloy passed through ECAP

Arshit Kapoor[1*], Mohd Atif Wahid[2], Pankul Goel[3], Krishna Mohan Agarwal[1] and Utkarsh Mohan[1]

[1]Mechanical Engineering Department, Amity University, Uttar Pradesh, Noida, India
[2]Delhi Technical Campus, Greater Noida, India
[3]IMS Engineering College, Ghaziabad, India
[*] **arshitgemini@gmail.com**

Abstract. Equal Channel Angular Pressing (ECAP) is a cutting-edge Severe Plastic Deformation (SPD) process for grain refining of materials up to ultra-fine or nanostructured standards, resulting in superior mechanical properties. To consider the effect of each processing parameter on the end results, computer simulation is one of the better alternatives to real ECAP deformation. The alloys used in this study are AZ61 magnesium alloy, which is a difficult material to work with in automotive applications. The ECAP process is subjected to 3D FEM simulation using ANSYS software in order to better understand the process/die design parameters. 90° channel angles and corner angles 20° i.e the best combination suggested by experimental studies were used in the simulation. For a single ECAP pass, the ECAP phase was simulated. Extrusion load, established stresses, effective strain, and temperature variation on the processed material were all noted in the results. The minimum channel angle produces the highest temperature profile on the material, developing maximum stress and strain at the deformation zone, according to the observed results. Furthermore, the results obtained using such computational techniques show that the shear strain decreases as the channel and corner angle increase.

Keywords- Severe Plastic Deformation, Equal Channel Angular Pressing, AZ61 magnesium Alloy, ANSYS, Element method simulation

1. Introduction

Nowadays, the manufacturing industry needs products that are high in strength, have a broad tolerance range, and are affordable to consumers. Magnesium is widely used and has a large scope of applications. For example, AZ61 magnesium is used in the automotive industry because it is a lightweight material with a high load capacity [1]. The ECAP technique is one of the most well-known methods for meeting the needs of the industry in today's world. It is a technique in which a material is subjected to high stress, resulting in a positive strain in the material. That strain is what helps to improve the properties and create an ultra-fine microstructure [2]. Equal channel angular pressing is the safest technique for extreme plastic deformation in ductile metals. In a number of ways, equivalent channel angular pressing has been found to be superior to other techniques [3]. The key benefit of ECAP is that the material's cross-sectional area remains unchanged. Figure 1 depicts the ECAP basic sketch. The two angles visible in the above figure are angle of channel (Φ) and angle of corner, which are two significant ECAP parameters. The angle of channel, which ranges from 90° to 135° is the angle at which two die channels converge. Corner angle, on the other hand, is the outer arc angle at the point of intersection, which can range from 0° to 30° degrees [4]. The software simulation technique was discovered to be an effective method for saving time, energy, and material while determining the impact of each processing parameter on material behaviour. For investigating the ECAP operation, the finite element method of software simulation is a well-simplified method. More specifically, by changing the formation of die gas corner and geometry of mould, the material strain hardening rate was investigated. The impact of angle of channel variations, angle of corner, strain intensity, and friction force that existed between channel and billet has been visualized using parametric studies.

According to the available research, the strain softening behaviour must be used to accelerate the strain localization in various regions. Furthermore, in most cases FEM studies does not include convergence analysis for depicting the results in their models, and automated re-meshing functions have been used in some cases to achieve stable simulation results. However, in the most recent research available, there is only a small amount of work for the simulation of ECAP on AZ61 magnesium Alloy. This research is performed on a AZ61 magnesium alloy for various channel angles and corner angles in order to evaluate the strain rate deformation and total deformation that occurs after a plunger enforces a load. [5-7]

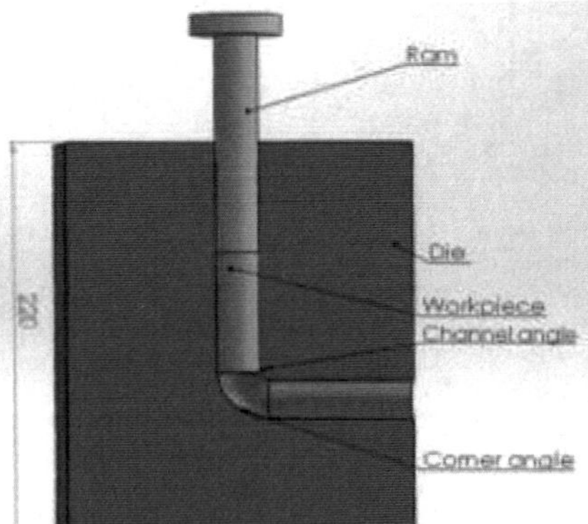

Figure 1. Setup for analysis for grain refinement of Mg-Alloy with angle of channel as Φ and angle of corner as Ψ

2. Material and Methods

AZ61 magnesium Alloy have a lot of promise in research and development because of their wide variety of applications in the machining and automotive industries. As magnesium is alloyed with aluminium, it has a higher strength and ductility. Mg proves to be the best source for overcoming iron adulteration. Table 1 and Table 2 showing the AZ61 magnesium Alloy composition and properties of material without ECAP [8].

Table 1 Composition of Material used (in %)

Composition of Material used (in %)							
Mg	**Al**	**Zn**	**Mn**	**Fe**	**Si**	**Cu**	**Ni**
93.1	5.95	0.64	0.26	0.005	0.009	0.0008	0.007

Table 2 Properties of material without ECAP

Properties	**METRICS (MPa)**
Tensile strength	310
Yield strength for value of strain as 0.200%	230
Compressive yield strength 0.2% offset	130
Ultimate bearing stress	470
Bearing yield stress	285
Shear strength	140

Simulation 3D FEM is used to investigate the variation in the extrusion load, and also amount of effective stress and strain of AZ61 magnesium alloy developed by ECAP method of simulation. The benefits and merits of using this specific method, namely software simulation, can be attributed as study of the changes and characteristics on experimental test material, and the outcomes can be predicted or visualized on the actual process [8-10]. One of the most prominent disadvantage of this kind of technique is that one single simulation is insufficient for the large size billets, so some method for large billets is needed to be created. The ANSYS Workbench programme was used to run 3D FEM simulations on rigid visco-plastic materials [6]. The APDL Known as ANSYS Parametric Design Language solver was used to perform static structural analysis. During process of severe plastic deformation the conclusion of ductile, fracture damage was incorporated by the means of ANSYS WORKBENCH device .For the purpose of depicting the boundary conditions during the simulation, input was considered from the tensile test data, displacement control medium and free extending length on Y axis with one end free and other fixed. With a quadrilateral/triangular element shape and a 1 mm element dimension, a hex-dominant method was used to mesh. Finally, from ANSYS WORKBENCH 18 analysis, Von-Mises stress, principal stresses, equivalent plastic strain, and results concluded from the strain simulation technique were extracted for further theoretical and numerical modelling [11-12].

3. Result and Discussion

Incorporating true stress-strain gained from experimental setup with the Hollomon's flow law, the uniform deformation regions were identified. Table 3 summarizes the (K) i.e. coefficient of strength and (n) i.e. exponent of strain hardening values for every ECAP passes. The 0.2 offset yield force, ultimate tensile strength, elongation, and modulus of elasticity were all evaluated using experimental tensile test results. The Poisson's ratios of each ECAP pass were estimated using ANSYS WORKBENCH 18.0 and a trial and error approach with reverse simulation analysis [13]. One pass of technique of ECAP with inconstancy of design of die parameters is proceeded using ANSYS computer simulation.

3.1 Analysis of the Extrusion load

The impacts of various angles such as channel angle and corner angle on extruder load is depicted using a single pass ECAP modelling using FEM-based simulation software. Figure 2 illustrates the simulation effects in terms of an extruder force vs. time graph for the optimal channel and corner angle values [14-15]. The increase in load during the initial process indicates that the billet has reached the die channel's corner, and the load has stabilized since reaching its full value. For the purposes of this study, Fig 2 shows that the maximum load for a die with a channel angle of 90 and a corner angle of 20 is approximately 861.83 KN. Similar results have also been documented by Liu el. The maximum load is reached as the material moves through the Deformation region, according to their brass ECAP simulation analysis.

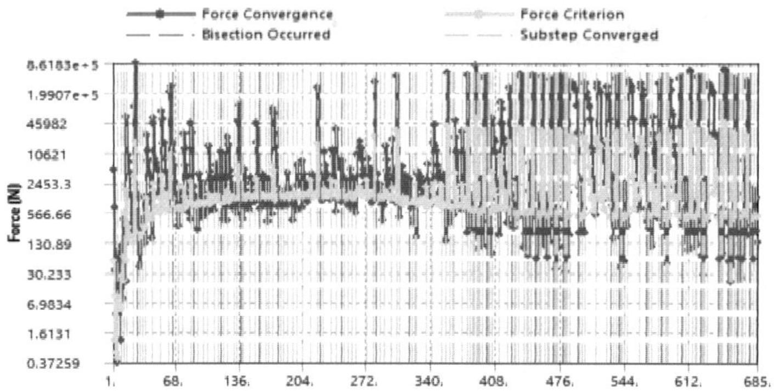

Figure 2. Graph between extrusion force (in N) vs. Time (in sec)

3.2 Deformation analysis

Two-stage deformations were considered in this paper: The uniform deformation stage was defined as the damage nucleation site, including the GTN parameter ranges (nucleation strain and the corresponding nucleation void volume fraction); The non-uniform deformation stage, which includes the second stage GTN impact parameter ranges, was defined as the non-uniform deformation stage. [16-17]

3.2.1 Total Deformation By passing the material piece through the bent ECAP channel and using FEM-based simulation tools, the complete deformation of Mg alloy is investigated. During the extrusion process, the Successful total deformation of a Mg alloy billet is shown in Fig 3(a). The curve representing the variance of this deformation over time is shown in Fig 3(b). The bent zone of the channel or the corner area of the channel, known as the main deformation zone, is where the Mg alloy experiences the most deformation. The maximum deformation measured is 0.021m, while the minimum deformation measured is 0.009m. The scale is from 0.009m to 0.021m.

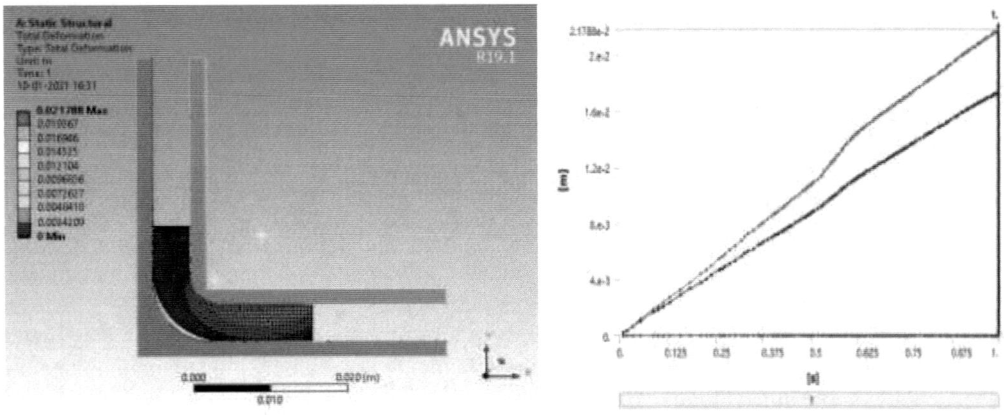

Figure 3. (a) Setup for Calculation of total deformation (b).Variation of Total deformation (in m) with Time (in sec)

3.3 Stress analysis

The maximum stress in the billet during the extrusion phase is shown in Fig.4(a). The highest stress is observed in the channel's corner area, which is also known as the key deformation zone. The overall stress is reduced and the stress distribution is widened when an outer radius is present. [nine] The added inner radius reduces the overall stress and further spreads the stress distribution. On all simulation scenarios, the overall stress that exists on a 105° channel angle is slightly higher than on a 90° channel angle on every corner radius. Furthermore, for channel angles greater than 120°, the stress is significantly reduced.

3.3.1 Equivalent Stress Fig. 4(a) illustrates the Effective Equivalent Stress distribution in billet during extrusion process and Fig 4(b) illustrates the Equivalent stress vs Time curve during simulation in the billet during the period of extrusion. The key deformation zone is located inside the channel's corner area, which has the highest stress. The maximum value of effective equivalent stress can be shown to range from nearly zero to 14.4 GPa. Similar patterns in stress growth have been established by Niu et al. In their numerical simulation of ECAP processing, they discovered that the highest stress generation occurs at Deformation [18].

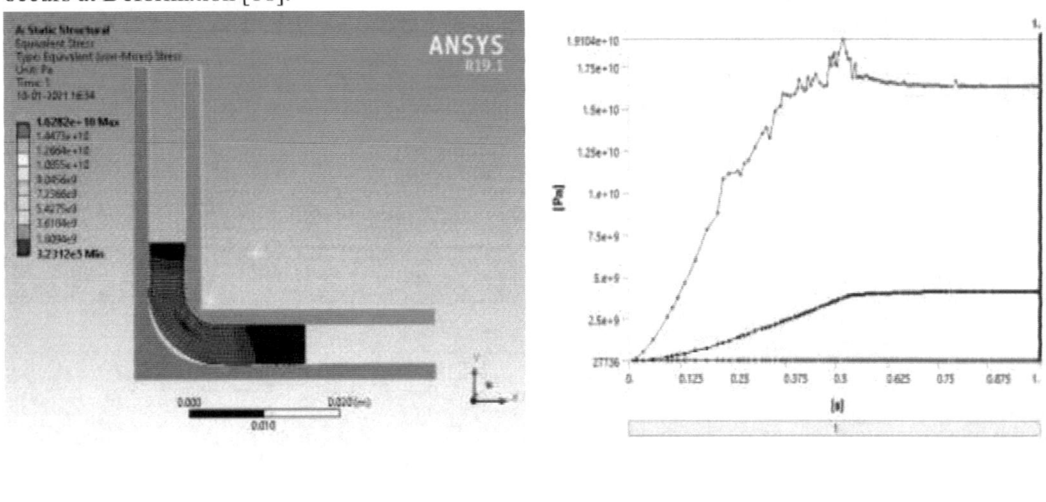

Figure 4. (a) Setup for calculation of Equivalent stress (b)Variation of Equivalent Stress (in Pa) with Time(in sec)

3.3.2 Maximum principal stress Simulations of ECAP plastic deformation in the AZ61 magnesium alloy and the resulting overall principal stress distribution ranges. The maximum value of the effective principal stress is 152.43 GPa at the corner or bent region of the channel. This very high value increase the strength of the Mg alloy billet thus making the worth of ECAP method. Maximum Principal Stress is concentrated to a region known as main deformation zone. However the minimum value is -4.331.

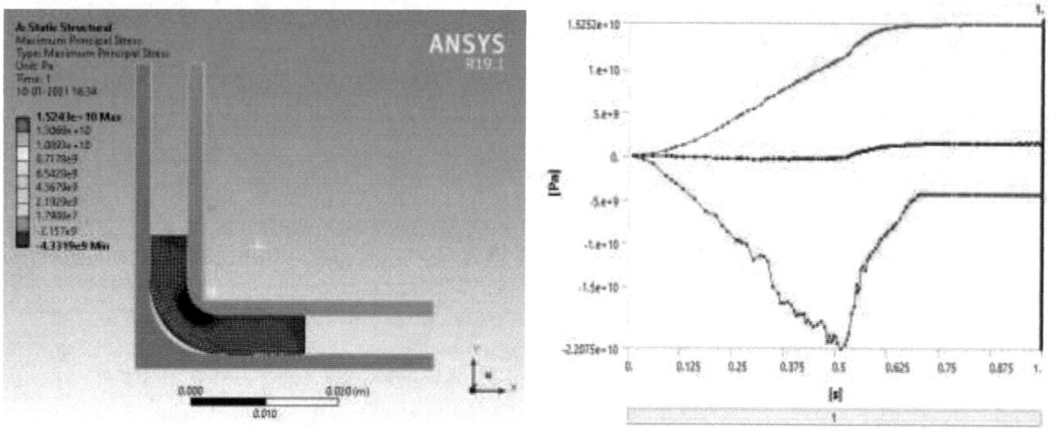

Figure 5 (a) Setup for calculation of Maximum Principal Stress (b) Variation of Maximum Principal Stress (in Pa) with Time(in sec)

3.3.3 Shear stress Fig 6 (a) illustrates process of finding shear stress in billet during extrusion process .The Maximum value of shear stress is found to be 9182.9 MPa which is as usual at the inner bent region of the channel or inner corner region .However the magnitude of minimum shear stress obtained is also significant and turns out to be -7613.3 MPa by the ANSYS software. Negative sign symbolize that shear stress is on negative side on the axis. Inner bent region is also called as main deformation zone

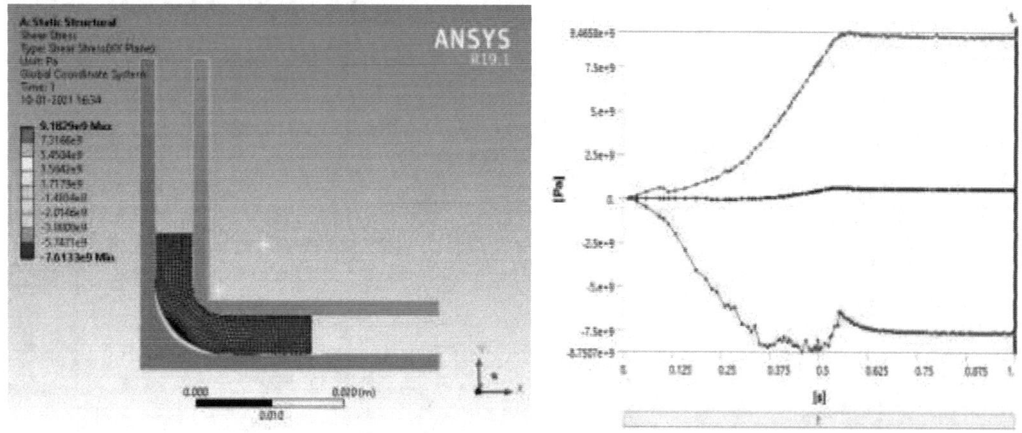

Figure 6 (a) Setup for calculation of Shear Stress (b)Variation of Shear stress (in Pa) with Time (in sec)

IOP Conf. Series: Materials Science and Engineering 1149 (2021) 012024 doi:10.1088/1757-899X/1149/1/012024

3.4 Strain analysis

The strain distribution in the billet tends to be highly non-uniform, with significantly different levels of residual strains in different regions. As a result, we used the strained comparative fraction of the sample amount in the total volume versus the strain strength to compare the amounts of cumulative strains in the material after various modes of ECAP. Volume of sample referring to a strained proportional fraction and strain strength coordinate for a given limit.

3.4.1 Maximum Principal Elastic Strain Fig. 7(a) illustrates the Effective Equivalent Maximum Principal Elastic strain distribution in Mg alloy billet during extrusion process and Fig.7(b) illustrates the Equivalent Maximum Principal Elastic strain vs Time curve during simulation of Mg alloy billet during the period of extrusion. The highest elastic strain occurs inside the corner region of the channel with the value of 0.2972m/m and is also known as the main deformation zone. The minimum elastic strain can be attributed to be less than 0.033m/m .It can be observed that the maximum value of effective elastic strain varies from almost 0.033 to 0.2972 m/m.

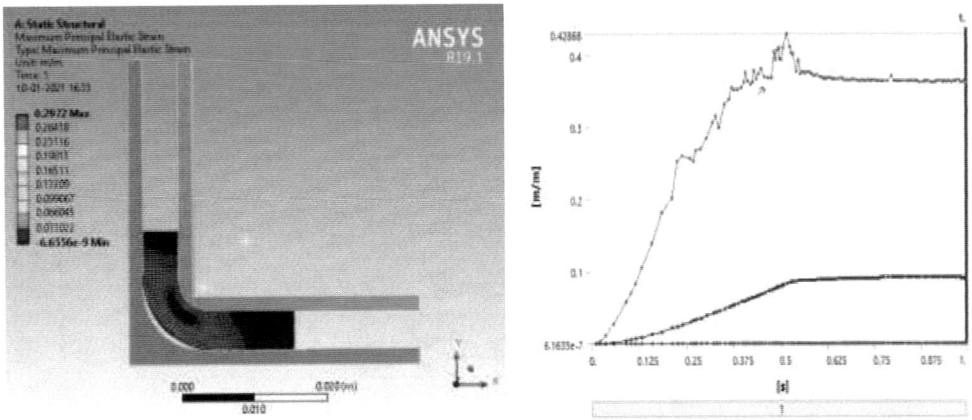

Figure 7 (a) Setup for calculation of Maximum Principal Elastic Strain (b) Variation of Maximum Principal Elastic Strain (in m/m) with Time (in sec)

3.4.2 Shear elastic strain Fig.8(a) shows the effective shear elastic strain in Mg alloy billet during extrusion process of analysis Fig 8(b) shows the curve for the variation of shear elastic strain which is obtain by the means of ANSYS simulation software. It can be easily seen that maximum value of the shear strain is concentrated to very small region at the bend of the channel. The maximum value of shear strain is limited to 0.5509 m/m and minimum value correspond to -0.4568.We get the negative value of the shear strain where the shear strain is acting in the opposite direction of an axis.

3.4.3 Equivalent total strain Fig.9(a) illustrates the setup for calculation of effective total strain in a Mg alloy billet during the extrusion process, Fig 9(b) provide the curve for the variation of Total Strain with time. This curve is obtained by the means of simulation software when the analysis is done for Mg alloy. Maximum value of the Total strain can be seen in corner region of the channel known as main deformation zone. 0.362m/m is the maximum value of the total strain produced and minimum value of total strain is approximately equal to zero.

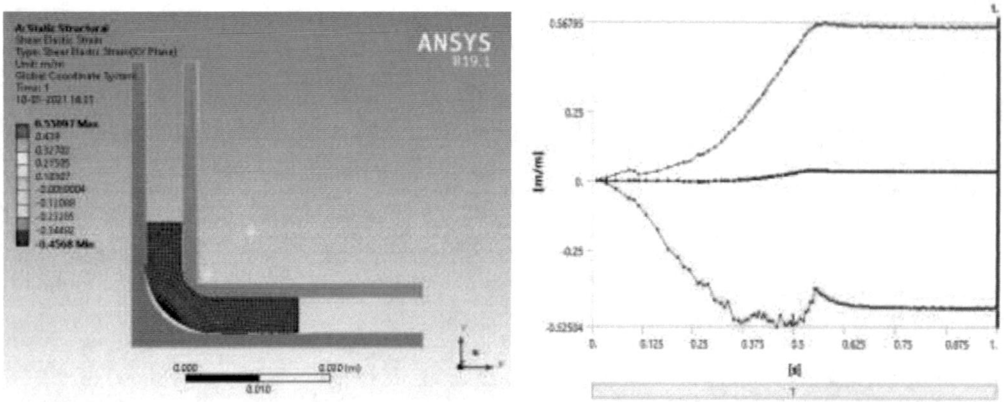

Figure 8 (a) Setup for calculation of Shear Elastic Strain (b) Variation of Shear Elastic Strain (in m/m) with Time (in sec)

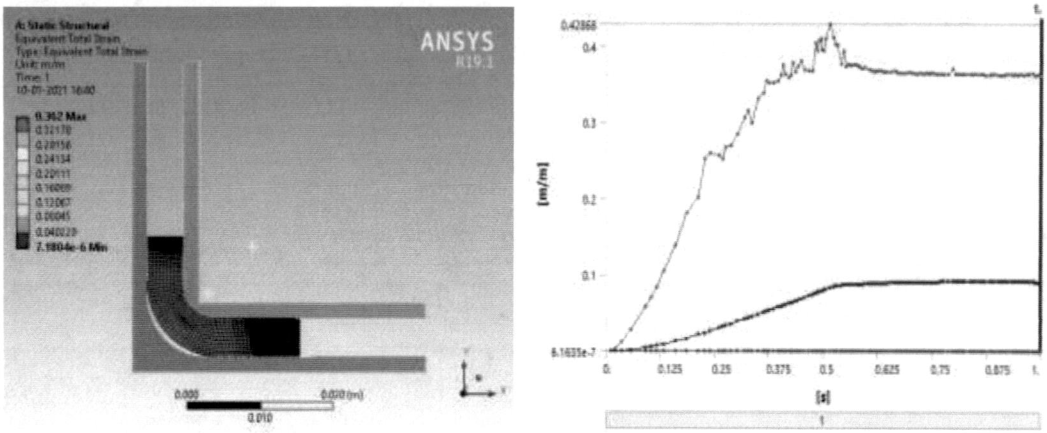

Figure 9 (a) Setup for calculation of Equivalent Total Strain (b)Variation of Equivalent Total Strain(m/m) with Time (in sec)

4. Conclusion

From automotive to aerospace and from industrial to commercial application Mg based alloy can be linked to great use in manufacturing and machining sector. Extreme plastic deformation (SPD) is a useful technique to achieve the desirable properties in a material by going through the grain refinement analysis. These grain refinement analysis can be done by Finite Element Method based Software (FEM), by this means the working condition, mechanical performance , tensile strength of a material can be determined. Therefore to observe these characteristics simulation of the ECAP method of the Magnesium alloy using Ansys software was used. The following conclusion can be drawn from the simulation data:

- Effective Total deformation was studied with the help of simulator. The value of total deformation comes in the range of 0.009m to 0.021m. With 0.021m is the maximum value at the main deformation zone. Also the minimum value is 0.009m

- Effective Maximum Principal strain in the Mg alloy billet was simulated and observed. Maximum value during the extrusion of the billet is 0.2972m/m while the minimum value is nearly zero. The main deformation zone has highest value of the effective principal strain.
- Effective Shear elastic strain in the Mg alloy was studied by the means of FEM based analysis software ANSYS. We get the maximum value and minimum value of effective shear elastic strain as 0.5509 m/m and minimum value correspond to -0.4568. Maximum value is found near the main deformation zone
- Effective total strain in the billet ranges from approximately zero at the straight region of the channel to reaching a maximum value of 0.362 m/m at the inner corner region of the channel. These results was observed during analysis of the Mg alloy using ANSYS software

5. References.

[1] Petch, N. The cleavage strength of polycrystals. J. Iron Steel Inst. 1953, 174, 25–28.
[2] Valiev, R.; Alexandrov, I.; Zhu, Y.; Lowe, T. Paradox of strength and ductility in metals processed by severe plastic Deformation. J. Mater. Res. 2002, 17, 5–8.
[3] Wang, Y.; Chen, M.; Zhou, F.; Ma, E. High tensile ductility in a nanostructured metal. Nature 2002, 419, 912–915.
[4] Ma, A.; Jiang, J.; Saito, N.; Shigematsu, I.; Yuan, Y.; Yang, D.; Nishida, Y. Improving both strength and ductility of a Mg alloy through a large number of ECAP passes. Mater. Sci. Eng.A 2009, 513, 122–127.
[5] Valiev, R.Z.; Langdon, T.G. Principles of equal-channel angular pressing as a processing tool for grain refinement. Prog. Mater. Sci. 2006, 51, 881–981.
[6] Kim, W.; Chung, C.; Ma, D.; Hong, S.; Kim, H. Optimization of strength and ductility of 2024 Al by equal channel angular pressing (ECAP) and post-ECAP aging. Scr. Mater. 2003, 49,333–338.
[7] Frint, P.; Localization Phenomena after Combined Highly Plastic Forming by Extrusion and ECAP of A 6000 Aluminum Alloy. Ph.D. Thesis, Chemnitz University of Technology, Chemnitz, Germany, 2015.
[8] Lapovok, R.; Tóth, L.S.; Molinari, A.; Estrin, Y. Strain localisation patterns under equal- channel angular pressing. J. Mech. Phys. Solids 2009, 57, 122–136.
[9] Agarwal, K. M.; Tyagi, R. K.; Chaubey, V. K.; Dixit, A. Comparison of Different Methods of Severe Plastic Deformation for Grain Refinement. IOP Conference Series: Materials Science and Engineering, 2019, 691 (1).
[10] Agarwal, K. M.; R. T. Investigation of Mechanical Properties of Metals and Alloys Processed After Equal Channel Angular Pressing. AGU International Journal of Engineering & Technology, 2017, 5 (5), 53–65.
[11] Agarwal, K. M.; Tyagi, R. K.; Dixit, A. Theoretical Analysis of Equal Channel Angular Pressing Method for Grain Refinement of Metals and Alloys. Materials Today: Proceedings, 2020, 25 (4), 668-673
[12] Kim, H. S.; Seo, M. H.; Hong, S. I. Finite Element Analysis of Equal Channel Angular Pressing of Strain Rate Sensitive Metals. Journal of Materials Processing Technology, 2002, 130–131, 497–503.
[13] Yoon, S. C.; Kim, H. S. Finite Element Analysis of the Effect of the Inner Corner Angle in Equal Channel Angular Pressing. Materials Science and Engineering A, 2008, 490 (1–2), 438–444.
[14] Kim, H. S.; Seo, M. H.; Hong, S. I. Plastic Deformation Analysis of Metals during Equal Channel Angular Pressing. Journal of Materials Processing Technology, 2001, 113 (1–3),622–626.
[15] K. M. Agarwal, R. K. Tyagi, and K. K. Saxena, "Deformation analysis of Al Alloy AA2024 through equal channel angular pressing for aircraft structures," Advances in Materials and Processing Technologies, vol. 00, no. 00, pp. 1–15, 2020.

[16] Bowen, J.; Gholinia, A.; Roberts, S.; Prangnell, P. Analysis of the billet Deformation behavior in equal channel angular extrusion. Mater. Sci. Eng. A,2000, 287, 87–99.
[17] Agarwal, K. M.; Tyagi, R. K.; Gupta, B.; Singhal, A. Evolution of Microstructure Through Various Techniques of Severe Plastic Deformation. In Smart Innovation, Systems and Technologies; Springer Smart Innovation, Systems and Technologies, 2020; pp 251–260.
[18] Agarwal, K. M.; Tyagi, R. K.; Kapoor, A. Deformation and Strain Analysis for Grain Refinement of Materials Processed through Equal Channel Angular Pressing. Materials Today: Proceedings, 2020, 21 (3), 1513–151

Aluminium Metal Matrix Composites: Manufacturing and Applications

Lavepreet Singh[1*], Sandeep Kumar[1], Shivam Raj[1], Shivam[1] and Piyush Badhani[1]

[1]Department of Mechanical Engineering, Galgotias University, Greater Noida, Uttar Pradesh, India
[*]**punstu@gmail.com**

Abstract: Aluminium metal matrix composites(AMCs) is widely used in the industrial applications right now. Aluminum metal matrix composites have properties that no other monolithic material can match. Due to their superior strength to conventional materials, aluminium matrix composites (AMCs) have a broad variety of industrial applications. The nature of reinforcing, that can take the form of constant or undefined fibres, has a big influence on the properties of aluminum metal matrix composites. Thus it depends on the fabrication methods for aluminium matrix composites, which are influenced by a number of factors including the type of reinforcement and matrix used, its required degree with surface morphology integrity, as well as physical, mechanical, electro-chemical, and thermal properties. This article provides an overview of the manufacturing processes and different reinforcing elements used during the synthesis of Al-MMCs. Generally, the reinforced particles like carbides, nitrides, and compounds of oxides are used. This paper gives a brief overview on various methods that are beung used to manufacture aluminium metal matrix composites. The present study offers a description of the synthesis, mechanical behaviour, and utilisation of aluminium metal matrix composites. The main processing methods for making or production of aluminium metal matrix composites(AMCs) are thoroughly discussed. Finally, questions of commercialization as well as business issues are also discussed.

Keywords: AMC; Matrix; Commercialization; Industrial Aspects; Mechanical Behavior

1. Introduction:

Aluminium metal matrix composites have a wide variety of applications in the automotive sector. Aluminium matrix composites are a high-demand commodity in the aerospace, vehicle, and other engineering applications. A composite is a matter made up of components formed with the help of physically combining already present monolithic materials for generating a single material with diffrent properties from the proposed prototype. Every composite, in general, has two phases: (a)metal matrix ;(b) reinforcements In particular, the matrix is consistent and accompanies the discrete reinforcement process of the composite material. The composites have been distributed into four categories: 1) matrix (ceramic, metal, and carbon), 2) reinforcements (oxide, carbide, and nitride), 3) structure (continuous fibre, short fibre, whisker, and particulate matter), and 4) processing directions, and 5) orientation [1-2]. The form of matrix and reinforcement, its mechanical and thermal properties, and the degree for micro - structural integrity needed are all factors influencing the processing technique selected. The type of reinforcement, the variance for reinforcement materials, as well as the relationship of both the matrix with the reinforcement are all important factors in deciding the composite's properties of the resulting [3].

Nonmetallic ceramics like Silicon carbide, Carbon, alumunium-di-oxide, Silicon-di-oxide, Boron, are commonly used[4]. Magnesium reinforced composites have received much interest because of their better mechanical and corrosion properties. The majority of Magnese-Aluminium alloys contains about 8% to9% Al and a trace of Zn is used to increase its tensile strength of material formed. Resistance to corrosion is improved by adding 0.1–0.3 wt% Mn. Magnesium alloy composites reinforced with silicon carbide (SiC) dispersion particles also have low population density of 2.1–2.2 g/cm3 and have 25–35 percent excellent mechanical properties and quality than without reinforced magnesium alloys. The greater reactivity of Mg makes the production of Magnesuim-based composite extremely difficult. Incorrect manufacturing techniques can degrade rather than improve mechanical properties. As a consequence, the reactive species at the interaction amongst SiC particles and the Mg matrix should be closely controlled. Powder metallurgy is used to make Fe-based composite materials because it produces a homogeneous phase with no contact between the matrix and the reinforcement state. Various properties like hardness, strength, fatigue, fracture, and degradation improve in Fe-Al2O3 metal matrix composites, according to Gupta et al. The properties are improved by the creation of iron aluminate (Fe-Al2O4). The above process is generated by the active sintered metal of iron and alumina particles [5–9].In aluminum metal matrix composites, higher level Al2O3 particulates are widely used, and there is some literature on the use of nanometric Al2O3 particulates. [10,11]

Jamwal et al. [12] used a stir casting method to make Al/Al_2O_3-TiC composites and found that perhaps the wear rate of the composites falls as the reinforcement content increases. Caused by heavy interfacial bonding of matrix material including Al_2O_3 and TiC, composite compressive modulus has increased about 149.3 MPa.

Bandil et al. [13] study the impact of Silicon Carbide reinforcing it on Al-Si alloy property. The composites has been made using stir casting. As the % of SiC in composite increases, the density of this is said to decrease. As the SiC content of composites increases, the wearing rate was decreased. The uniform dispersion with SiC particles also helps with corrosion resistance. The maximum corrosion resistance efficiency is 56.58 percent at 20 wt% SiC.

Vinod et al. [14] used a double stir casting method to fabricate A356 alloy RHA-Flyash reinforced hybrid matrix composites and examined the mechanical and physical properties. The incorporation from both organic and inorganic particles increases the mechanical properties, according to the findings.

This paper describes the basic production techniques in order to make Matrix Composites. It also goes over the mechanical properties, corporatization, and implementation of different Al_2O_3 strengthened aluminium based composites manufacturing methods. This review is indeed a comprehensive review of the various manufacturing processes used for the production of matrix composites throughout period. Aluminium-based composite materials are expected to positively influence in the development and design of components for lighter industries, mainly aviation and transportation.

2. Primary production process for AMCs:

The way of manufacturing path used to fabricate any composites is critical in deciding the composite's final properties [15]. The three major classes of primary processes for producing AMCs on an industrial scale are:

1) Solid-state production processes.
2) Production that take place in a liquid.
3) Methods of deposition production process.

2.1. Solid - state production process for AMCs:

2.1.1. Powder blending & consolidation process (P/M processing): In solid state processes, this procedure, also known as powder metallurgy, seems to be the most commonly used technique. Powder metallurgy processing is appealing because it uses lower temperatures, allowing for more influence about system kinetic model.The use of composite material compounds and micro structural improvements who are only achievable with rapidly solidified pigments are also possible with this processing techniques. As given below, powder metallurgy composites involve following number of processing steps.

1) The first step involves determining the composition and preparing the pigments which will be components including its combination to be treated throughout subsequent stages, such as metal/metal alloy powders and reinforcement powders.
2) The blended pigments get compressed into some kind of "green form" utilizing uni - axial press or cold isostatic applying pressure. The resulting green form, also known as the green body, is around 80percent thick.
3) Sintering is achieved with in final stage via warming that green body to the desired processing temperature, which is normally performed in an inert atmosphere [16]. A sintered body's theoretical density is usually 100 percent.

2.1.2. Diffusion bonding process:

Heat-sensitive substances, substances with such a feisty oxide film, and composite materials that are difficult to join using traditional methods were the primary targets of diffusion bonding methods [17]. Diffusion formation is a process that uses mechanical pressure and high temperatures to create a phase that tightly connects the matrix and also fibre [18]. Diffusion bonding is a compression weld technique that requires atoms inter-diffusion through the welding line, which is typically divided by a liquid or solid phase inter -layer. It mainly focuses three interconnected process parameter: a) joining temperature, b) joining power, and c) retention or hold time. The joining force ought to be sufficient to ensure direct contact among both the meeting surfaces and to help in the bending of substrate abrasive particles that fill most empty spaces in the weld area.

2.2. Liquid state processes for AMC production:

2.2.1. Stir casting process:

Within this casting procedure of composite production, a reinforcement is stirred into melted metal. The stir casting method typically starts with the formation of the melt with the desired matrix content, quickly by addition of a reinforcing material and mixing to achieve the desired dispersion. The stir casting of Magnesium-oxide reinforced Al metal matrix composite is shown in Figure 1.This is the process which requires traditional stir mixing and then casting, is among the mainly cost-effective technologies for obtaining greater near-net form parts using metal matrix composites [19].

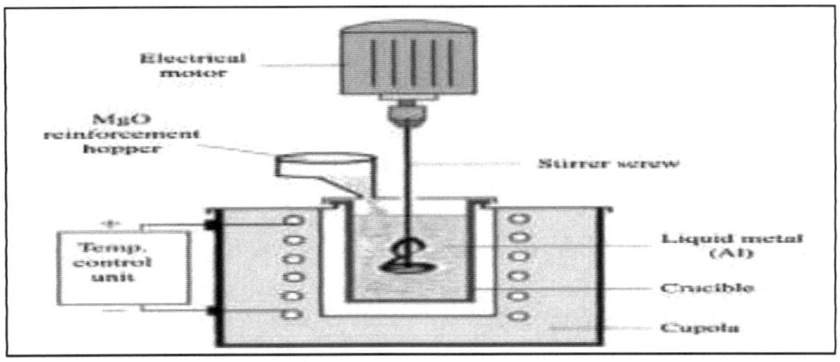

Figure 1. Mg oxide reinforced with (Al)matrix composite by stir casting.

The vortex method is the most basic and widely used of all stir casting methods. The vortex technique [20] entails injecting ceramic particles into a vortex of molten material formed by a spinning impeller prior to treatment. The introduction of ceramic particles into the vortex created by stirring molten metal is the first step in the vortex process. After that particles have been fed into the melt, swirl the

IOP Conf. Series: Materials Science and Engineering 1149 (2021) 012025 doi:10.1088/1757-899X/1149/1/012025

mixture to ensure that the particles are uniformly distributed, and eventually, cast the molten mix to receive the ceramic content.

2.2.2. The method of infiltration:

Within this pressing process, the melt is injected into a preform created during the reinforcing step to completely fill all of that open porosity. The friction effect caused by the viscosity, and its melted matrix, which occupies the composite preform, determine the pressure needed to combine the matrix and reinforcement. Various factors, such as alloy composition, ceramic preform content and crystal structure, temperature, and time, affect the wetting of the ceramic preform due to the liquid alloy. Figure 2 shows method of infiltration:

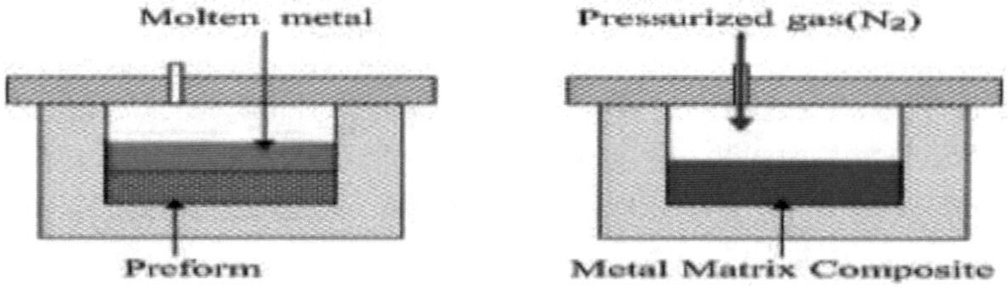

Figure 2. Using pressurised N2 gas, mechanical force has been showed to infiltrate the preform.

2.2.3. Forced Infiltration process:

The outer pressure is used to direct the penetration of melted material within permeable reinforcement in the forced infiltration process. This procedure can be divided among a number of groups, as shown in figure 3: (a) The penetration mechanism using a pressure die. (b) The infiltration of gas pressure. c) Penetration by squeeze casting. (d) The vacuum infiltration method. (e) Method of using Lorentz force infiltration. (f) Method of using ultrasonic infiltration. (g) Phase of using centrifugal infiltration.

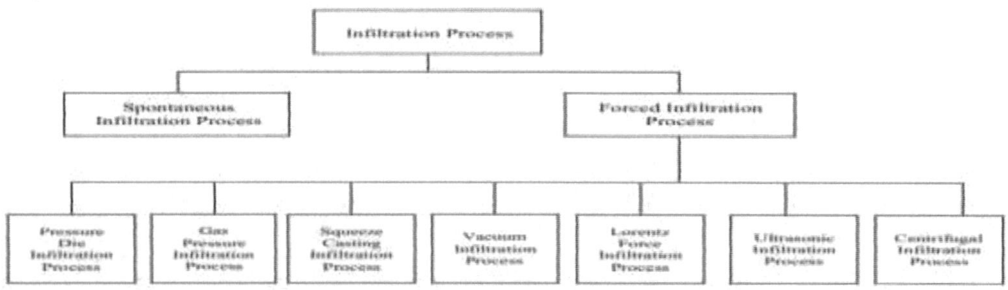

Figure 3. The method of infiltration and the various processes of infiltration.

2.2.4. Processing by In-situ (reactive processing):

This category includes processes such as liquid with gas, liquid with solid, liquid with liquid, and combined salts reaction. In this process chemical reactions among components or even among element and compound lead to formation of reinforcing states in the matrix material during the composite fabrication process [21].The in-situ formed reinforcing processes become thermally stable, independent of foreign material contamination, include smaller dimensions, and more evenly distributed in the matrix, resulting in enhanced particle-matrix bonding in in-situ manufactured composite [22]. Figure 4 shows in-situ mechanism:

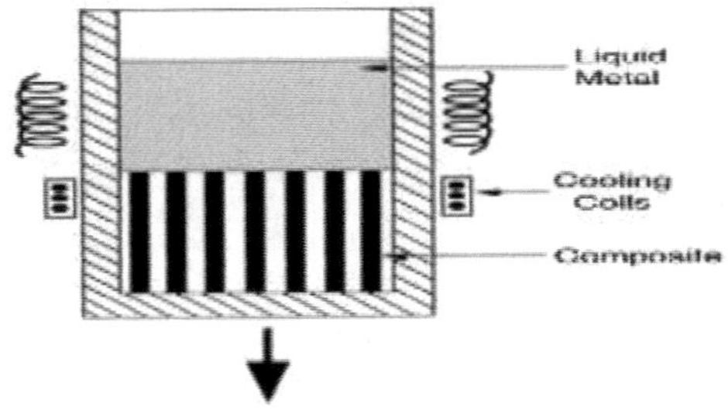

Figure 4.A eutectic alloy is processed by in-situ controlled one directional solidification process [22].

2.3. Method of deposition:

2.3.1. Physical Vapour Deposition process (PVD): This deposition is being utilized to create matrix composites, which is a time-consuming procedure. The whole process entails moving a fibre via a slight positive pressure region of the metal to still be collected, where condensation occurs, resulting in a relatively thick coating upon this material [23-24].A powerful electrons light is directed it onto edge of the a trusted platform feed stock to create fumes.. The coated fibres are packed or arrayed as well as assembled in a pressing process or we can say HIP process to achieve the best composite content. PVD process that exist categorized as:

1. Vaporization also characterization techniques.
2. Sputtering techniques.

In the above stated first method, a gun generates a large electric field (EB) that vaporises the composite materials and causes the metallic moisture to encapsulate upon this fiber. In the sputtering method, ions from the manufacturing gas (such as argon) strike a piece of coating, breaking away molecules first from workpiece material and leaving a residue upon this material.

2.3.2. Spray Deposition Procedure:

This techniques are classified among two categories based on how much the given droplet flow is generated by either the melted baths (osprey procedure) or through continuous flow with cold hand into some kind of area of high temperature infusion (thermal spray procedure)[25]. The spray deposition method is visualized in figure 5.

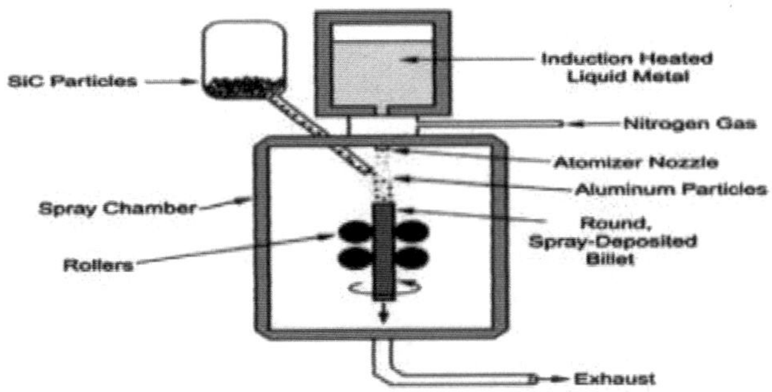

Figure 5. Process of spray deposition [25].

3. Mechanical behaviour of aluminium matrix composites as a feature of processing parameters:

Composite materials (MMCs) have a higher elastic modulus, fracture resistance, and strength properties. Also they are perhaps more heat, wear, also resistant to corrosion.

Rahimian et al. used powder metallurgy to build Al-Al2O3 composites in order to examine the effect of alumina crystal structure, heating rate, and sintering temperature on the reinforcement material's specific entities [26]. The size of the particles of the aluminium being used was 3, 12, and 48m, with temperature of sintering ranging from 500 to 600 C including time of sintering ranging from 30 to 90 minutes. When the time of sintering was raised from 45 to 90 minutes, the amount of residential construction increased. Furthermore, higher sintering temperature changes contribute to the emergence of a structural performance, which results in higher membrane permeability.

In an experimental environment, Rabiei et al. [27] investigated the crack durability of aluminum metal matrix composites with particles of different reinforces. To compare the findings to fracture toughness, the model of Hahn–Rosenfield was used. This Hahn–Rosenfield design being discovered to have a credibility range of 5–10m for reinforcing particles. The design has been tweaked for calculate the break resilience among matrix composites including higher particle reinforcement.

4. Corporatization of aluminium metal matrix composites:

Aluminium matrix compounds (AMCs) are being tested for validation as an alternative to industrial applications due to their low weight and high strength. One of the international markets for matrix composites products and technologies is construction, building materials, and other industrial sectors. Knowing the need of markets and industry research, the worldwide aluminum-based composite materials marketplace is forecast to attain a new high point with in coming years. The cost/performance ratio of ceramic composite materials, such as that of other materials, is the most important factor for companies when deciding whether to use ceramic composite materials in their goods. Pretty much across the board, massive commercial need for ceramic composite materials is expected to happen very soon [28-29].

The demand for aluminium composites is being increasing by the use of lightweight and high tensile strength components. Although most manufacturers have begun to use aluminum-based composites in their manufacturing processes, from very small units to large products. Industrial production, construction equipment, and other manufacturing industries are among the global markets for aluminum-based composite products and technologies. Because of their multifunctional properties such as stiffness, strength, and wear resistance, their demand will rise.

5. Commercial uses of aluminum metal matrix composites in industries:

Because of its high strength to its composite materials, many manufacturers were forced to prefer ceramic composites over aluminium alloys. Following are some applications:

- Aviation or aerospace sector: Airplane wing and supporting structure, fuel tank, fighter aircraft, and transport aircraft.
- Design & construction material: Ceramic materials give construction materials more strength and stiffness, making it a great choice for material selection in development.
- Athletics and wellness: The production of athletic apparel with boron(B) and silicon carbide(SiC) strengthening in aluminium gives higher strength while leading to the possibility.

Conclusion:

Each of the methods mentioned above has its own set of benefits and drawbacks. This article evaluates how a different reinforcement combination is used in AMCs and how it influences their performance. Several factors can affect the manufacturing methods chosen, including cost of production, process performance, preferred quality of products, and applications, among others.

- Powder metallurgical processing is appealing for a multitude of reasons, such as the ability to better monitor system metabolic pathways.
- The diffusion bonding method is simple to use, highly efficient, adaptable to a variety of situations, and produces an uniform joint product. Complex parts and structures are not appropriate for this method.
- Among the most cost-effective methods for manufacturing large relatively close formed parts from metal matrix composites is stir mixing accompanied through casting. The above process cannot easily be said to produce nanocomposites because nano-sized

ceramic particle reinforced with metal matrix composites materials manufactured using traditional stir casting methods generally include poor nanoparticle distribution with high porosity.

- In metal matrix composites manufactured using the infiltration technique, some porosity of localized variation with in volume fraction of a reinforcement are frequently observed. The whole method can be seen as an extremely versatile, nearby net shape fabrication method that provides excellent micro structural control.
- Some voids and localized variations in the particle size including its reinforcing are usually found in MMCs manufactured using the infiltration technique.
- The in-situ developed contamination-free reinforcing states spread more evenly in the matrix, resulting in greater particle –matrix attachment.
- The matrix surface morphology has extremely fine sizes and low separation, which is a key advantage of the deposition process.
- Spray diffusion becomes incredibly beneficial also for processing of difficult-to-prepare materials (alloying elements metal matrix composites).

References:

[1] Miracle DB. Metal matrix composites — from science to technological significance. Comp Sci Technol2005;65:2526–40.

[2] Rosso M. Ceramic and metal matrix composites: routes and properties. J Mater Proc Technol 2006;175:364–75.

[3] Singh N, Banerjee S, Parkash O, Kumar D. Tribological and corrosion behavior of (100-x)(Fe70Ni30)-(x) ZrO2composites synthesized by powder metallurgy. Mater Chem Phys2018;205:261–8.

[4] Kaczmar JW, Pietrzak K, WøosinÂski W. The production and application of metal matrix composite materials. J Mater Proc Technol 2000;106:58–67.

[5] Gupta P, Kumar D, Parkash O, Jha AK. Structural and mechanical behaviour of 5% Al2O3-reinforced Fe metal matrix composites (MMCs) produced by powder metallurgy(P/M) route. Bull Mater Sci 2013;36(5):859–68.

[6] Gupta P, Kumar D, Parkash O, Jha AK. Effect of sintering on wear characteristics of Fe-Al2O3metal matrix composites.Proc Inst Mech Eng J: J Eng Trib 2013;228(3):362–8.

[7] Gupta P, Kumar D, Quraishi MA, Parkash O. Corrosion behavior of Al2O3reinforced Fe metal matrix nano composites produced by powder metallurgy technique. Adv Sci Eng Med 2013;5(4):366–70.

[8] Gupta P, Kumar D, Quraishi MA, Parkash O. Effect of sintering parameters on the corrosion characteristics of iron-alumina metal matrix nano composites. J Mater Env Sci2015;6(1):155–67.

[9] Kumar UJP, Gupta P, Jha AK, Kumar D. Closed die deformation behavior of cylindrical iron–alumina metal matrix composites during cold sinter forging. J Inst Eng India Series D 2016;97(2):135–51,http://dx.doi.org/10.1007/s40033-015-0089-1.

[10] Surappa MK, Rohatgi PK. Preparation and properties of cast aluminium-ceramic particle composites. J Mater Sci1981;16(4):983–93.

[11] Rahmana MH, Al Rashed HMM. Characterization of carbide reinforced aluminium matrix composites. Procedia Eng2014;90:103–9.

[12] Jamwal A, Vates UK, Gupta P, Aggarwal A, Sharma BP.Fabrication and characterization of Al2O3–TiC-reinforcedaluminum matrix composites. In: Advances in industrial and production engineering. Singapore: Springer; 2019. p. 349–56.

[13] Bandil K, Vashisth H, Kumar S, Verma L, Jamwal A, Kumar D,et al. Microstructural, mechanical and corrosion behavior of Al–Si alloy reinforced with SiC metal matrix composite. JComp Mater 2019,http://dx.doi.org/10.1177/0021998319856679.

[14] Vinod B, Ramanathan S, Ananthi V, Selvakumar N. Fabrication and characterization of organic and in-organic reinforced A356 Aluminium matrix hybrid composite by improved double-stir casting. Silicon 2019;11(2):817–29.

[15] Olszówka-Myalska A, Szala J, Cwajna J. Characterization of reinforcement distribution in Al/(Al2O3)p composites obtained from composite powder. Materials characterization2001;46(2–3):189–95.

[16] Garg P, Gupta P, Kumar D, Parkash O. Structural and mechanical properties of graphene reinforced aluminum matrix composites. J Mater Environ Sci 2016;7(5):1461–73.

[17] Feest EA. Metal matrix composites for industrial application. Mater Des 1986;7(2):58–64.

[18] Casati R, Vedani M. Metal matrix composites reinforced by nano-particles — a review. Metals 2014;4(1):65–83.
[19] Su H, Gao W, Feng Z, Lu Z. Processing, microstructure and tensile properties of nano-sized Al2O3 particle reinforced aluminum matrix composites. Materials & Design2012;36:590–6, 1980-2015.
[20] Contreras A, Lopez VH, Bedolla E. Mg/TiC composites manufactured by pressure less melt infiltration. Scr Mater2004;51(3):249–53.
[21] Sahin Y, Acılar M. Production and properties of SiCp-reinforced aluminium alloy composites. Compos AAppl Sci Manuf 2003;34(8):709–18.
[22] Chawla KK, Chawla N. Metal-matrix composites. Universityof Alabama at Birmingham and Arizona State University.2013.
[23] Mubarak A, Hamzah E, Toff MRM. Review of physical vapour deposition (PVD) techniques for hard coating. J Mekanikal2005;20:42–51.
[24] Chaudhury SK, Panigrahi SC. Role of processing parameters on microstructural evolution of spray formed Al–2Mg alloyand Al–2Mg–TiO2composite. J Mater Process Technol2007;182(1-3):343–51.
[25] Agarwal A, McKechnie T, Seal S. Net shape nano structured aluminum oxide structures fabricated by plasma spray forming. J Therm Spray Technol 2003;12(3):350–9.
[26] Rahimian M, Ehsani N, Parvin N, Baharvandi HR. The effect of particle size, sintering temperature and sintering time on the properties of Al–Al2O3composites, made by powder metallurgy. J Mater Proc Technol 2009;209:5387–93.
[27] Rabiei A, Vendra L, Kishi T. Fracture behavior of particle reinforced metal matrix composites. Compos A Appl SciManuf 2008;39:294–300.
[28] Gupta P, Kumar D, Quraishi MA, Parkash O. Influence of processing parameters on corrosion behavior of metal matrix nano composites. J Mater Environ Sci 2016;7(7):2505–12.
[29] Gupta P, Kumar D, Parkash O, Jha AK, Sadasivuni KK. Dependence of wear behavior on sintering mechanism foriron-alumina metal matrix nano composites. J Mater Chem Phys C 2018;220:441–8.

Validation of Valve Coefficients Using Experimentation and Numerical Analysis

K. Kumar[1*], S. Komble[2], S. Thorat[3], P. Korgaonkar[3], V. Bartakke[3], P. Sonje[3], O. Chavan[3] and K. Kavire[3]

[1]Scientist 'C' Central Water & Power Research Station, Pune, 411024, India
[2]Assistant Professor, Mechanical Engineering department, Vishwakarma Institute of Technology, Pune, 411037, India
[3]Final Year B. Tech Mechanical Engineering Student, Vishwakarma Institute of Technology, Pune, 411037, India
***krishnaswamy_kumar@yahoo.co.in**

Abstract. Today's demanding projects calls for a quality valve that is effective in controlling the flow in the closed conduit with optimum flow velocity and least energy losses in the system. Proper selection of valve requires information on maximum and minimum flows and the head loss for wide range of applications. Hence a comprehensive research study was carried out on an improvised tricentric butterfly valve of 100 mm NB over a conventional type of butterfly valve and globe valve at different percentages of openings to determine the flow performance characteristics at CWPRS, Pune. These studies highlight the improved performance of the tricentric butterfly valve over the conventional valve. The findings exhibit the exceptional performance and advantage of equal percentage flow control loop characteristics of a new tricentric valve over the conventional type S-shape flow performance characteristic curve and the linear characteristics of the globe valve. The tricentric valve has demonstrated in the opening range between 15° and 75° showing an equal percentage characteristic. The minimal pressure drop in these valves, causes significant energy savings over a broad range of control applications like water distribution systems, power generation, refineries, chemical, and desalination. The validation by experimental data with CFD is obtained with reasonable near-optimum results for valve coefficients.

1 Introduction

An essential function of a control valve in a closed conduit is to open, throttle and close the rate of flow. The throttle characteristics of a valve may be quick opening, linear or equal percentage controlling of flow. As such, above inherent characteristic of control valve plays a vital role in process control engineering [1]. Proper selection of a control valve for a particular application may impact on economic consequences and efficiency of the system [2, 3]. Proper size and control characteristic of a valve will have impact on the optimum flow passes through the closed conduits irrespective of operation on pneumatic or electrical [4]. Butterfly valves, ball valves and plug valves are frequently used as on/off and throttling services with negligible pressure and energy losses [5]. Butterfly valves and globe valves are used in large sizes for handling larger flow of liquids, liquids with suspended solids in water treatment, waste water, chemical applications and fire protection systems [6, 7]. Additional advantages of the butterfly valve are comparatively high coefficient of flow (Cv), standard face to face dimensions and their availability in chemically resistant materials. The butterfly valve is considered a high recovery valve with less head loss across the valve. Hence, the size of the required pump is reduces and considerably implies cost saving on the project. Hence various researchers [8, 9] have carried out studies on butterfly valve characteristics. However, the disadvantages of the butterfly valve are not bubble tight due to the erosion and pitting when used for slurry applications. Butterfly valves are good control limited to 60^0 opening and it is subjected to frequent wearing out of seals. Noting the number of advantages of butterfly valves, R&D efforts have been made recently to overcome most of the disadvantages of these valves by introduction of "Tricentric butterfly valves". The paper highlights the improvised performance of the new tricentric butterfly valve which was established in high precision test facility at CWPRS, Pune. Comparison of flow performance of this new valve with conventional valves indicates the superiority achieved by the new valve design in respect of flow characteristics. The study is to verify and validate CFD analysis [10, 11] to obtain reasonable results for control valve coefficients. The steady and incompressible Navier-Stokes equations are solved numerically to predict the flow behavior and compute the pressure loss and flow coefficients. A limited research has been focused on the validation of experimentation and CFD analysis. Hence, a comprehensive 3D simulation study is carried on 4" (100 DN) butterfly valves and globe valves is conducted to establish a trusted and a calibrated numerical solution model after comparing with experimental data.

1.1 Valve Performance Coefficients

The principal use of valve performance coefficients is to assist within the selection of appreciated valve size for a specific application. All the relevant sizing factors must be known at different valve disk angles (α), valve performance coefficients include pressure loss and flow coefficients. The characteristics of the valves are generally obtained experimentally or alternatively by using CFD.

1.2 Flow coefficient

The flow coefficient Cv flow per hour passes through a valve with a head loss of 1 bar and can be calculated by:

$$\text{Flow coefficient } C_v = Q/\sqrt{\Delta P/SG} \qquad (1)$$

1.3 Pressure loss coefficient

The pressure loss coefficient, K, is a dimensionless and generally used to predict the minor head loss. Thanks to the presence of valve in fluid flow field, it is important to get the valve pressure loss coefficient as a function of valve disk angle (α). Two different methods are used numerically to analyze.

$$\text{Loss coefficient } K = 2\Delta P/\rho V^2 \qquad (2)$$

ΔP- Pressure difference in (bar)
Q - Flow rate in (m^3/hr)
SG - Specific gravity
ρ - Water density (kg/m^3)
V – Fluid velocity (m/sec)

2 Experimental Test Circuit

The high precision test facility at CWPRS is used for assessment of hydraulic performance of various flow elements (Fig.1). The circuit equipped with volumetric tank for accurate measurement of flow volume. The test valve is installed in piping circuit by providing required upstream and downstream straight lengths. The service pumps of the circuit pump the stabilized flow passes through the test specimen. The differential head across the test valve is measured by pre-calibrated pressure transducers whereas each set of flow rate is measured volumetrically.

IOP Conf. Series: Materials Science and Engineering 1149 (2021) 012026 doi:10.1088/1757-899X/1149/1/012026

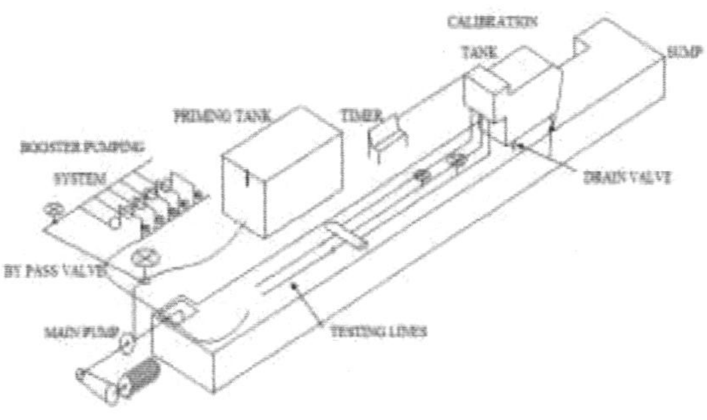

Figure 1. Volumetric Calibration Circuit

2.1 Instrumentation

The details of instruments used, and their accuracies are as follows:

Flow: ±0.3 % by volumetric method

Head: ±0.1% by manometer

Time: Digital timer with 0.001 second least-count

Flow stability: Set flow rate of 0.001 % accuracy

3. Experimental Findings

The head loss values were measured for different flow rates with different percentage openings of valves viz., 100 mm NB (tricentric) and 100 mm NB (conventional) butterfly valves and 100 mm NB globe valve. The values of inlet velocity through the test valves were ranging from 0.25 m/sec to 3 m/sec with the various valve opening positions. The loss coefficients (K) and flow coefficient (Cv) were computed from the observation made during the test. Table: 1-3 illustrate the experimental findings and CFD in respect of the valves. To compare the relative performance of these valves, graphs indicating flow characteristics VIS percentage of valve openings are plotted.

4.Computational Fluid Dynamics

For doing the flow analysis of this model, there is a requirement of a CFD tool. For this work, ANSYS Fluent CFD tool was used to model the valve. In CFD analysis, numerical methods and algorithms are used to solve and analyse problems with fluid flow with governing equations and the boundary conditions as specified.

4.1 Methodology

The assembled file from CAD software is modeled into IGES and into design modeler. The fluid domain created, took the form of the whole void. The meshed volume of fluid domain is as shown in Figure 2-4. The optimum gap between the trim surface and valve seat had a very fine mesh which was necessary to capture accurately the flow through that narrow region.

The CFD code utilized in Fluent is usually based on Finite Volume technique wherein a complex geometry can be divided into a finite region of control volume and the mass and momentum conservation equations are applied to each control volume.

4.2 Mesh Generation

The steps in CFD simulation are pre-processing, solving, and post-processing. The studied flow volume including the valve disc and the connected pipes has meshed via ANSYS. The generated mesh has been repeated for alternate mesh types and sizes, the efficient mesh method for converging solution is executed using unstructured (tetrahedral). An illustration of the geometry and mesh is shown in (Figs. 2-4) with a locally re-fined numerical grid of high-density ranged from $0.8x10^6$ to $1.2x10^6$ elements. The model previously described is implemented directly into Ansys18.2 Partial differential equations are discretized into a system of algebraic equations and these algebraic equations are then solved numerically over each elemental discrete volume.

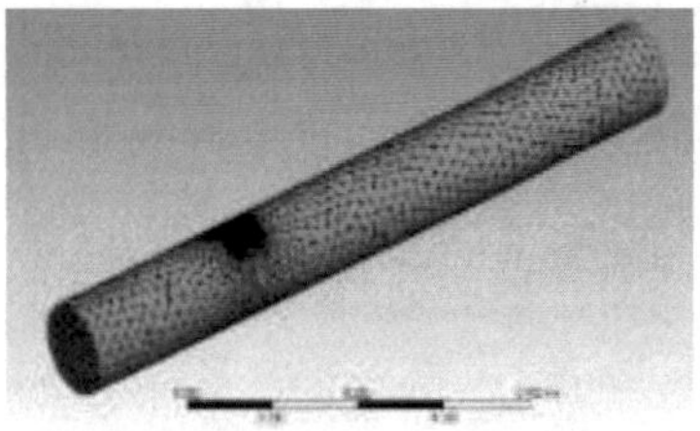

Fig 2. Meshing of Conventional BV

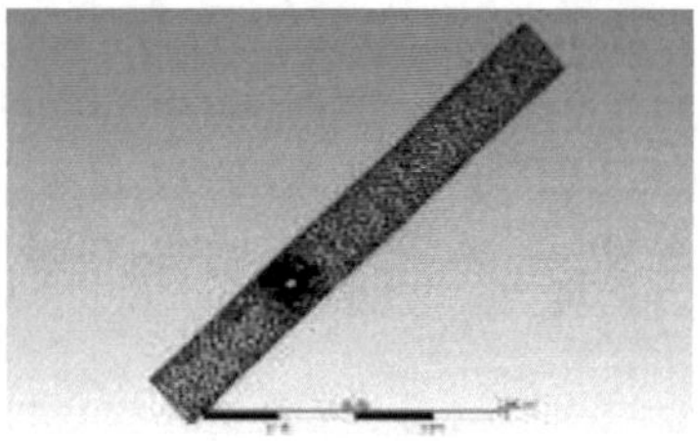

Fig 3. Meshing of GV

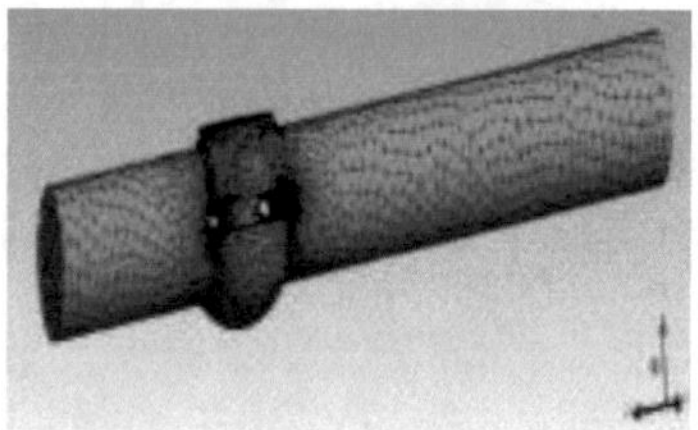

Figure 4. Meshing of TBV

4.3 Boundary Conditions
Sample boundary conditions for conventional butterfly valve at 11% opening are as follows.
Set up of the case (Common):

1. Time: Steady state
2. Pressure Based solver
3. Gravity: 9.81 m/s^2 in -z direction
4. Viscous Model: K epsilon, Realizable with Scalable Wall functions.
5. Material: Water
6. B.C: Pressure inlet: 200 kPa
   Pressure outlet: 100.46 kPa
7. Standard initialization: Computed from inlet.

4.4 CFD Modeling and Simulation

A 100 NB conventional butterfly valve, globe valve, and tricentric valve, along with other internal parts and alternate types of trim shapes were modeled in Solidworks as shown in figure 5-7. Flow analyses are carried out on the stem and bush of each plug.

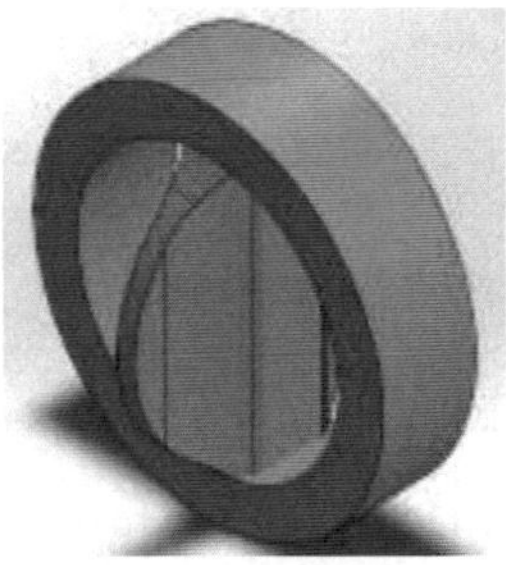

Fig 5. Conventional Butterfly Valve

Fig 6. Globe Valve

Fig 7. Tricentric Butterfly Valve

5 Results and Discussions

The performance of CFD results for flow coefficients and pressure loss coefficients are given in table 4-6. The velocity contours are given in figure 8. The comparative results of flow coefficients, pressure loss coefficients of butterfly valve, globe valve and tricentric valve are given in graphs.

IOP Conf. Series: Materials Science and Engineering 1149 (2021) 012026 doi:10.1088/1757-899X/1149/1/012026

5.1 Velocity Contours

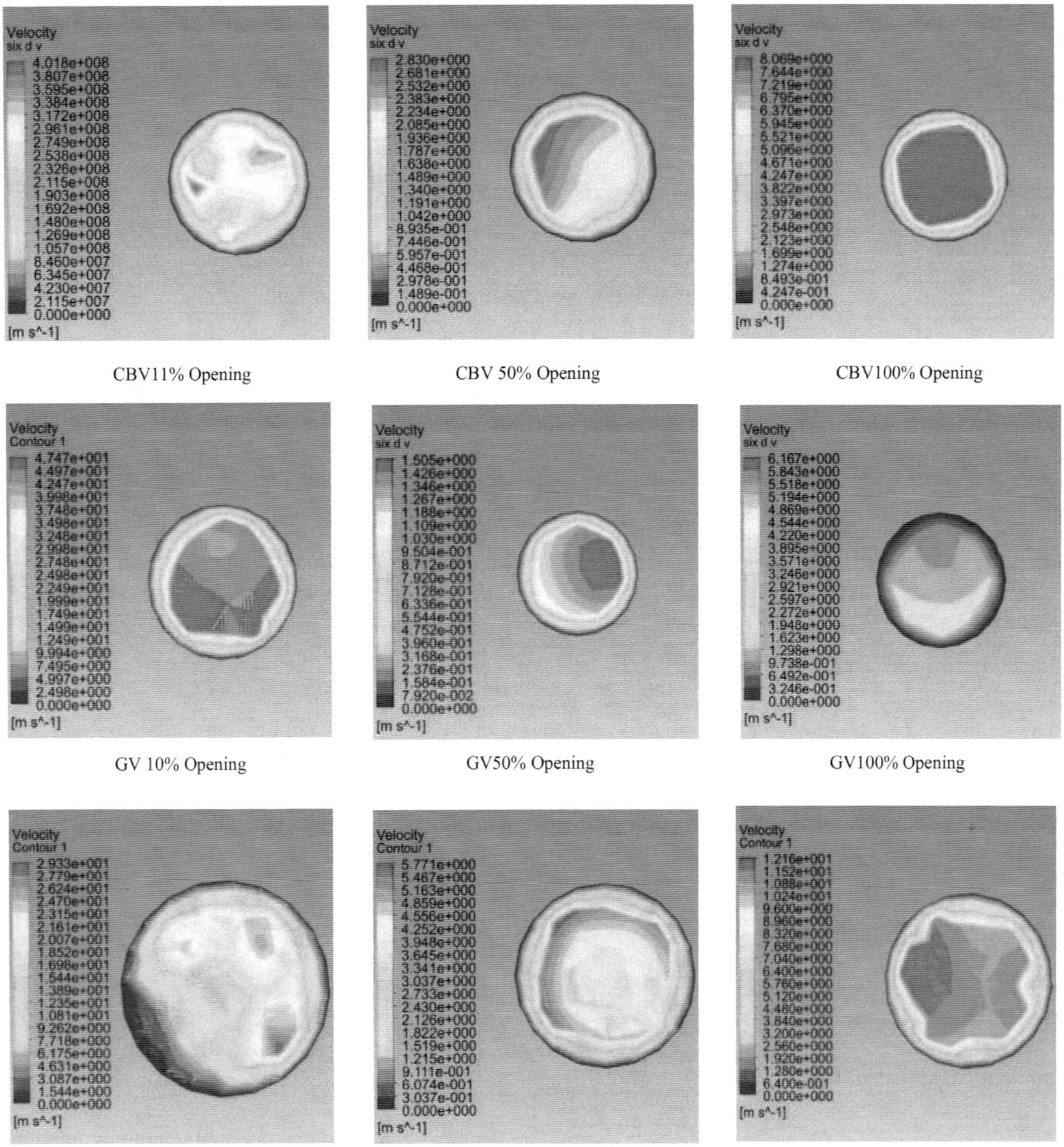

CBV11% Opening CBV 50% Opening CBV100% Opening

GV 10% Opening GV50% Opening GV100% Opening

TBV10% Opening TBV55% Opening TBV100% Opening

Fig 8. Velocity contours of various positions of different valves

CBV – Conventional Butterfly Valve
GV – Globe Valve
TBV – Tricentric Butterfly Valve

Table: 1-3 illustrate the experimental findings and CFD in respect of the valves. To compare the relative performance of these valves, graphs Indicating flow characteristics VIS percentage of valve openings are plotted.

Table 1. Characteristic coefficients of 100 mm NB tricentric butterfly valve Exp. and CFD

Flow rate	Differential Pressure	Valve Opening	Flow Coefficient	Loss Coefficient	% Flow Coefficient
(m3/hr)	(bar)	(%)	(C_v)	(K_v)	(%C_v/C_{vMax})
10.558	1.00109	10	10.578	1456. 839	5.122
20.506	1.00165	33	20.546	386. 113	9.949
53.779	1.00186	55	53,879	28.074	26.090
82.194	1. 00186	66	82.347	12.019	39.875
173.836	1.00106	80	174.160	2.6869	84.334
206.126	1.00186	100	206.512	1.9110	100.000

Flow rate	**Differential Pressure**	**Valve Opening**	**Flow Coefficient**	**Loss Coefficient**	**%Flow Coefficient**
(m^3/hr)	(bar)	(%)	(C_v)	(K_v)	(C_v/C_{vMax})
2.40295	1.00186	10	2.377802	3216	2.104012
6.883745	0.2583	25	13.4152	871.2774	11.87051
26.0084	0.2346	50	53.18438	55.43478	47.0605
35.6202	0.1816	75	82.78906	22.8773	73.25637
43.30964	0.1583	90	107.8149	13.48942	95.40059
45.7974	0.1611	100	113.0128	12.27709	100

Table 2. Characteristic coefficients of 100 mm NB conventional butterfly valve Exp. and CFD

Flowrate	**Differential Pressure**	**Valve Opening**	**Flow Coefficient**	**Loss Coefficient**	**% Flow Coefficient**
(m3/hr)	(bar)	(%)	(Cv)	(Kv)	(%Cv/Cv Max)
9.683	0.9954	11	9.703	1721 758	4.722
16.785	0.9450	16	17.246	544.667	8.391
42.467	0.7812	33	48.047	70.058	23.377
74.338	0.5972	50	90.195	17.478	46.804
144.579	0.8316	63	158.543	6.434	77.139
206.348	1.0080	100	205.528	3.830	100.000

Flow rate	**Differential Pressure**	**Valve Opening**	**Flow Coefficient**	**Loss Coefficient**	**%Flow Coefficient**
(m^3/hr)	(bar)	(%)	(C_v)	(K_v)	(C_v/C_{vMax})
7.9156	0.9954	11	7.858136	2539.286	3.423652
14.9831	0.945	16	15.26582	672.8373	6.651053
40.9915	0.7812	33	45.93537	74.31153	20.01324
68.4134	0.5972	50	87.68308	20.39478	38.20198
121.561	0.8316	66	132.0297	8.995133	57.52302
232.6621	1.008	100	229.5249	2.976397	100

Table 3: Characteristic coefficients of 100 mm NB globe valve Exp. and CFD

Flow rate	**Differential Pressure**	**Valve Opening**	**Flow Coefficient**	**Loss Coefficient**	**% Flow Coefficient**
(m3/hr)	(bar)	(%)	(Cv)	(Kv)	(%Cv/CvMax)
5.538	1.00186	10	5.538	5285.103	8.750
8.133	0.2583	25	14.2573	632. 964	22.530
16.285	0.2348	50	31.564	162.763	49.871
19.402	0.1816	75	45.529	78.196	71. 932
22.195	0.1583	90	55.777	52.102	88.125
25.405	0.1611	100	63.295	40.459	100.000

Flow rate	**Differential Pressure**	**Valve Opening**	**Flow Coefficient**	**Loss Coefficient**	**%Flow Coefficient**
(m^3/hr)	(bar)	(%)	(C_v)	(K_v)	(C_v/C_{vMax})
28.27	1.00189	18	27.97372	200.378	10.41651
39.578	1.00185	33	39.164	102.2296	14.58341
93.291	1.00186	55	92.31467	18.39963	34.375
155.485	1.00186	66	153.8578	6.623868	57.29167
203.544	1.00186	88	201.4138	3.865201	75
271.392	1.00186	100	268.5518	2.174175	100

The comparative valve coefficients with experimental and CFD results are given in graphs below

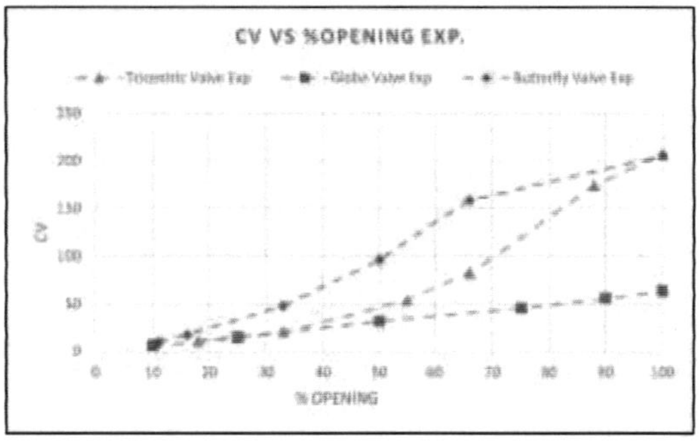

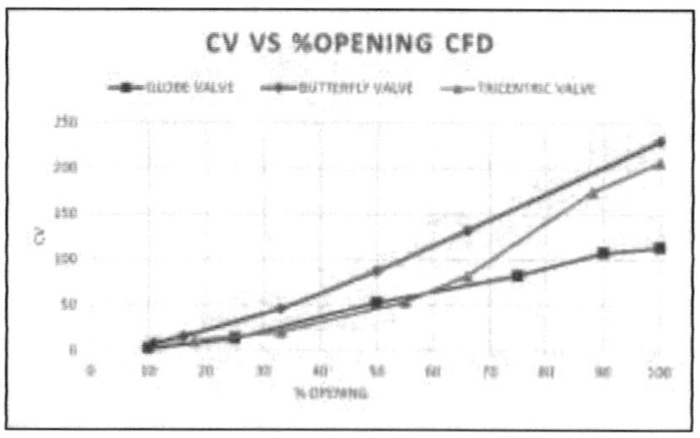

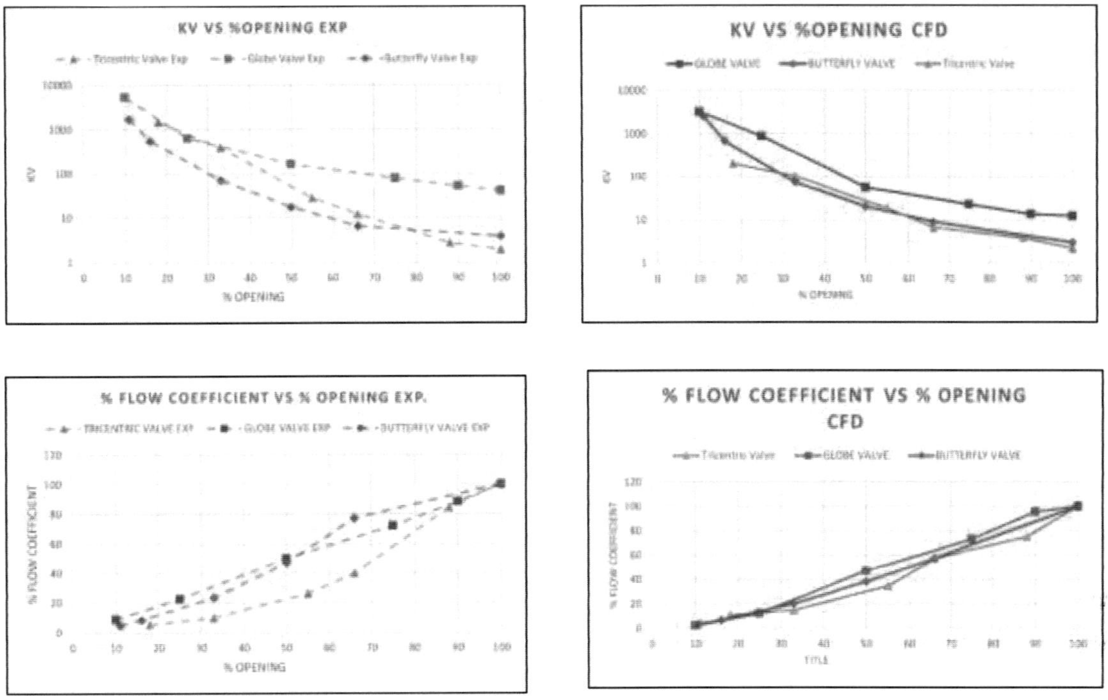

Figure 9 Graphs showing Comparison of experimental and CFD results

A comparison of flow performance curves of these valves indicates that the tricentric valve besides the advantage of perfect sealing also exhibits an equal percentage of flow control loop between openings 15^0 and 75^0, Therefore this valve provides a precise throttling through the lower range and can be used effectively for pressure control purposes. From the performance curve of a conventional butterfly valve, the flow reacts to small positioning steps however there is no proportion flow variation after the further opening of the valve due to the S-shape performance curve. Considering the behaviour of the conventional butterfly valve, its use for flow control purposes is considered inefficient. This drawback of the conventional butterfly valve has been eliminated in tricentric design. The loss coefficient for the tricentric design is less than the conventional butterfly valve and globe valve for the same percentage of valve opening. Similarly, the flow coefficient of tricentric design is better than the conventional butterfly valve and globe valve for the same percentage of valve opening. Therefore, to achieve the same flow rate the differential pressure resulting across the tricentric valve is less than the conventional type i.e., the flow control is achieved with less energy loss resulting in energy conservation. The study reveals that the characteristics of valves established by CFD with conventional experimental results which are near optimum results. The CFD analysis is a substitute for experimentation in certain conditions.

6 Conclusion

The study reveals that the tricentric butterfly valve has the merits over the conventional butterfly valve and globe valve in terms of construction, ease of installation, operation, and also overcoming its limitations. The triple offset metal seat design of tricentric valve provides superior performance and advantages over a broad range of applications. The equal percentages of tricentric valve characteristic provide good results and are very tolerant of over-sizing. It will further offer a more constant gain in control as the load changes. The study reveals that the CFD can be an alternative solution and complement to experimental validations in certain simulated conditions.

The future scope of the research has opened up for experimenting plug, piston type valves for gaseous application with CFD analysis.

Acknowledgement
The authors are grateful to Shri A. K. Agrawal Director, CWPRS, Pune, for giving the guidance and motivation throughout the studies. The authors are also thankful to Shri P M Abdul Rahiman Scientist ‘E’ for his continuous support and guidance during the period of investigations at various stages.

References

[1] Lyons, J.V. Valve Designer's Handbook. Nostrand Reinhold Company, New York, 1983.
[2] Tullis, J.P. Hydraulics of Pipelines - Pumps, Valves, Cavitation and Transients. John Wiley and Sons, New York, 1989.
[3] Fisher Controls International. Control Valve Handbook. Fisher Controls. Marshall Town, 1977.
[4] Zappe, R.W. Valve Selection Handbook. Gulf Publishing Company, London, 1981.
[5] HolgerSiemers. Notes from Valve World. www.valve-world.net, 2007.
[6] Guan Song Xue and Young Chul Park. WCECS 2007. Numerical Analysis of Butterfly Valve-Prediction of Flow Coefficient and
Hydrodynamic Torque Coefficient. 24-26 Oct 2007, San Francisco, USA.
[7] Weir Technical Bulletin, Weir Valves and Controls, USA, 2010.
[8] Ogawa, K and Kimura, T. Hydrodynamic Characteristics of a Butterfly Valve — Prediction of Torque Characteristic, ISA Transactions, 1995, 34, 327-333.
[9] Huang, C.D. and Kim, R.H. Three-Dimensional Analysis of Partially Open Butterfly Valves Flows. Transactions of the ASME, 1996, 118, 562-568.
[10] M.-J. Chern, C.-C. Wang, Control of Volumetric Flow-rate of Ball Valve Using V-Port, Journal of Fluids Engineering, 126(3) (2004) 471- 481.
[11] Kirik, M.J. and Driskell, L.R. Flow Manual for Quarter- turn Valves, Rockwell International Co, London, 1986.

IOP Conf. Series: Materials Science and Engineering 1149 (2021) 012027 doi:10.1088/1757-899X/1149/1/012027

Aerodynamics Characteristics of Compound Delta Wing at Sea Level

Gaurav Gupta[1,*], Pranav Tiwari[1], Bhanu Pratap Vatsa[1], Aashish Anand Sahay[1], K S Srikanth[1] and Shrikant Vidya[1]

[1]School of Mechanical Engineering, Galgotias University, Greater Noida, Uttar Pradesh, India
***gtech1499@gmail.com**

Abstract: Compound delta-wing aircraft are in the top tier for great maneuverability and satisfactory take-off and landing speeds and low distance as delta wing have a property of vortex generation at different speeds, during flight the property of vortex generated depends on factors like the coefficient of lift, coefficient of drag and shock wave impact absorption upon varying speed from subsonic speed and supersonic speed with changing attack angle is studied by flow visualization of the wing in Ansys Fluent using 'k-Ɛ 2nd equation model' to simulate mean flow characteristics for turbulent flow conditions. Vortex generation is studied for flow physics at Subsonic speed and Supersonic speed at sea level over Pressure and Density based respectively. The flow velocity at Subsonic is 260m/s and 686m/s for Supersonic. During the experiment, we observed that the vortex flow generation of observable output starts forming at 5° angle of attack following a powerful vortex at 10° leading to a shortfall powerful vortex at 15° for both subsonic and supersonic speeds but as velocity is more in the supersonic state the vortex is denser and more stable and provides better shock absorption while transitioning on higher attack angle and lift force is very sensitive for the supersonic state as drag is substantially increased with speed.

Nomenclature
C_L: Coefficient of lift
C_D: Coefficient of drag
α: Angle of attack
L: Lift force
D: Drag force

1.Introduction

Compound delta wing is mainly known for the unique ability to create vortex which can benefit the flight performance and stealth operational improvement for conventional and non-conventional aircraft. The vortex can be affected by drag and lift variation by a great deal. The problem associated with delta wing is that more wing area causes more drag increase rather than lift, better aspect ratio of L/D improvement has been the point of concern of most research teams. The other most important aspect of aeronautical engineering is to shorten the landing and take-off distance and speed, have more armament and cargo weight, and most important aerodynamics efficiency.

The flow in leading-edge delta wing is described as a movement in a spiral motion flow from the lower to the upper surface [1]. The wing causes flow separation that results in the pressure difference between the top and bottom of the wing [1].

T. Lee and his team studied the reverse delta wing to study the vortex breakdown further reported that, in addition to the lowered lift and drag forces, the 65°-sweep reverse delta wing also exhibited a delayed stall compared to its delta wing [2].

Flow separation will start at the leading edge from the apex a distortion of flow will be observed from the leading edge which wraps up in a spiral vortex. These vortices produce a pair of large primary vortices along the chord of the wing whose nature depends on the angle of attack, Mach number, and wing configuration for certain frequencies of undergoing conditions [2]. In compound delta wing at lower attack angle air flows is similar to a conventional wing but does not produce much lift but as the angle of attack increases the property of vortex generation upon the delta wing starts to appear, the vortex starts forming and the airflow starts separating as the air around spins very fast which creates a low-pressure region that provides an extra lift [3].

A multi-performance aircraft that needs to partake in many different operations can take advantage of this effect and maneuver this property by changing the angle of attack with speed to achieve better maneuverability and stable flight at heavy load conditions.

This study aims to computationally find out the angle of attack for aircraft with compound delta wings to produce a steady and uniform flow separation point to increase the maneuverability of the aircraft and improve the future scope for transportation.

2.Experimental Design Modelling and Methodology

The geometry of the delta wing is modeled in SOLIDWORKS the profile of the wing is shown in Figure 1 [4]. The geometry of the wing as shown in Figure 1 is derived from the model counterpart of military aircraft, with a leading-edge angle of 50° (first sweep) and 62.5° second sweep angle and 4° forward sweep trailing angle compound delta wing model with edge bevelled at an angle of 14° for 53 cm to study the vortex.

The flow fields are assumed to be symmetric on the centerline of the wing [5]. Therefore, the computational domain covers only half of the wing [5].

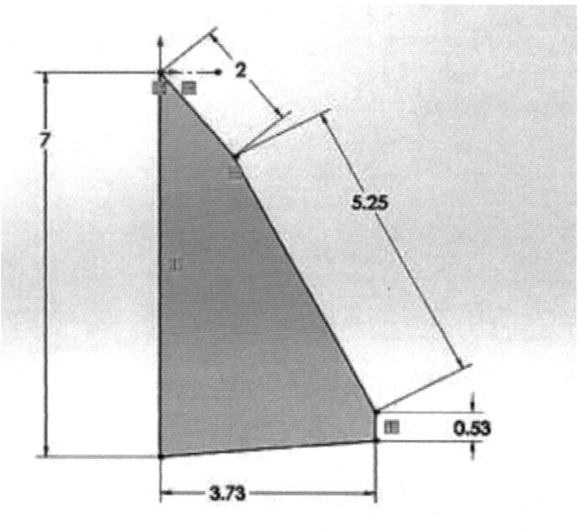

Figure 1. Geometry of wing

To simulate mean flow conditions over compound delta wing 'k-Ɛ 2nd Equation Model' is used in Ansys Fluent to perform CFD flow experiment and numerical calculations [6]. The traditional meshing of rectangular domain is used with Boolean tetrahedral meshing using with the cell count of 2317894, the simulations were run on subsonic speed(260m/s) condition using pressure-based model since at low speed the air instead of compressing just slides away around the edges and density-based model for Supersonic speed(686m/s) as in the high-velocity phase the sudden shock leads to the compression in the air and insight over the flow structures of delta wing pressure distribution for attack angle of 5°, 10° and 15° is studied.

3.Result and Discussion

Pressure distribution of surface contours was obtained which are shown by Figure 2 to Figure 7, coefficient of lift and coefficient of drag was acquired to discuss characteristics of delta wing in Figure 8 to Figure17. The flow visualization result of the leading-edge wing was obtained at a different angle of attack for both subsonic speed and supersonic speed and the results are discussed below:

3.1 Flow Visualization for Subsonic

In Figure 2, at 5° angle of attack faint profile of the vortex is observed at edges which formed the primary vortex although the lift was not quite improved and the density of the primary vortex was quite low [7]. Due to the pressure drag which caused the air to separate fast which made a delay in the formation of the vortex. In Figure 4, at a 10° angle of attack, as the attack angle increased the flow separated faster, and a powerful vortex is observed over the edge tailing to the primary vortex which has a higher density, and the lift was increased. In Figure 6, as the attack angle was high the pressure drag was increased and caused the flow to separate faster but due to high wave impact, at 15° a powerful vortex is observed but the vortex diminished near the middle of the wing showing the low density and the lift increase was quite significant but drag was also increased in multiples which compromised the stability of the vortex.

3.2 Flow Visualization for Supersonic

In Figure 3, at 5° a slight powerful vortex is formed from the edge towards the chord line and a lift increase of 7 times was observed with very high drag due to higher speed and pressure drag and faster flow separation but was lower than the subsonic counterpart. In Figure 5, as the flow separated faster and shock lead to compression of air, at 10° attack angle a very powerful vortex is originated from the edge creating a powerful primary vortex with high density and the wing is found to have a lower lift to drag ratio than the subsonic counterpart with a lift increase of about 6 times. In Figure 7, at 15° due to higher angle of attack, the compression of air around the wing was increased and pressure drag also increased which helped in early vortex generation to create a powerful but short vortex in contact with leading-edge and diminished in the middle of the wing but the drag increase was quite large which resulted in a lower L/D ratio.

3.3 Pressure Distribution of Wing

The pressure distribution of the wing shows quite a similarity between subsonic speed and supersonic speeds for the same angle of attack but at supersonic speed, the pressure is observed to be denser than the subsonic speed [8], the difference in pressure density can be observed upon the wings. Figure 2 to Figure 7 shows the pressure distribution on the contour of the wing as the speed increases for the various angle of attacks as the angle of attack increases vortex starts forming and starts to become denser and after the attack angle increases above 13° the vortex starts to distort due to shock separation [9].
The pressure distribution on the wing is shown below:

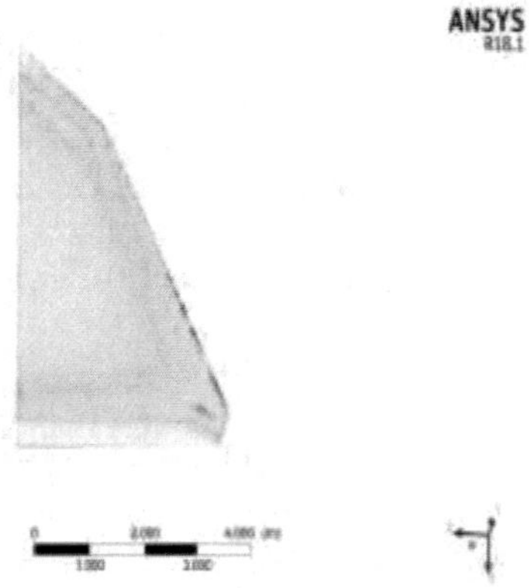

Figure 2. α = 5° Subsonic

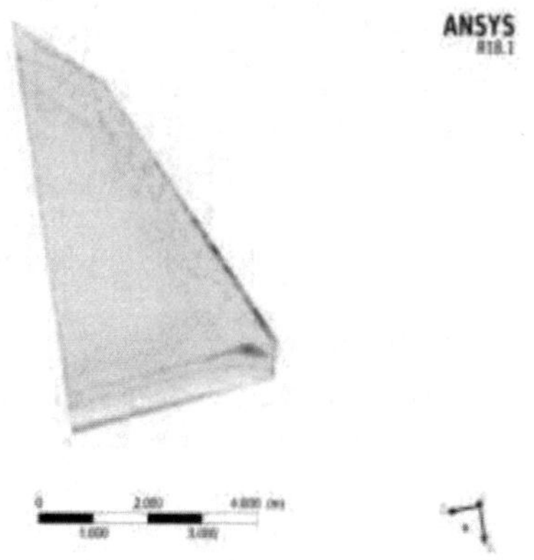

Figure 3. α = 5° Supersonic

In Figure 2, at 5° attack angle pressure is seen at the leading edge with density along the chord with most density at the trailing edge, and vortex is seen to be faint in the case of subsonic speed, and in Figure 3, a slight powerful and sustaining vortex is seen in supersonic speed with little lesser pressure on edge as compared to subsonic speed but higher pressure on trailing edge [10].

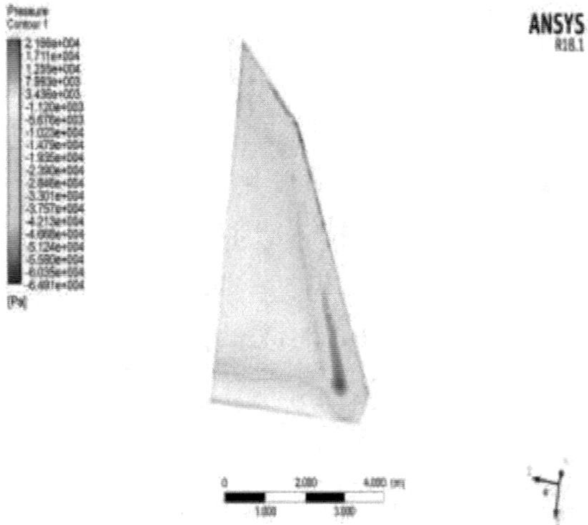

Figure 4. α = 10° Subsonic

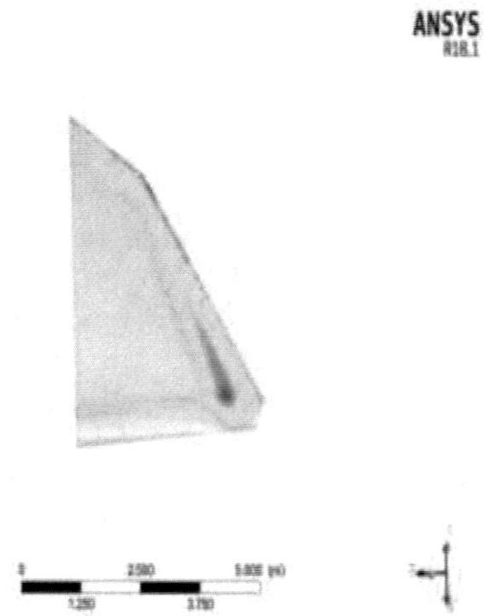

Figure 5. α = 10° Supersonic

In Figure 4, at 10° attack angle showed a stable and dense vortex formation at subsonic speed, and in Figure 5, denser vortex at supersonic speed with lesser pressure density at leading-edge than for the attack angle of 5° and more density at trailing edge is observed.

IOP Conf. Series: Materials Science and Engineering 1149 (2021) 012027 doi:10.1088/1757-899X/1149/1/012027

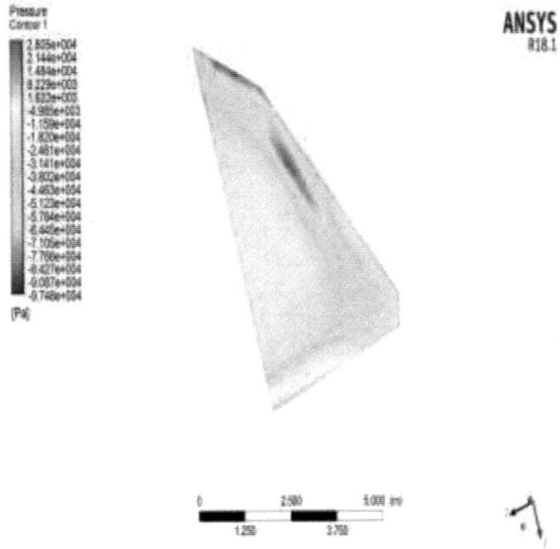

Figure 6. α = 15° Subsonic

Figure 7. α = 15° Supersonic

In Figure 6, at 15° attack angle, a powerful vortex is formed upon the leading edge and faints just in between the wing with high pressure at the trailing edge, although in Figure 7, the pressure density is observed to be high up on the sweeping edge, and very little pressure is observed at the leading edge.

3.4 *Variation of Values*

The graphs showing the variation of values between the subsonic and supersonic is shown below:

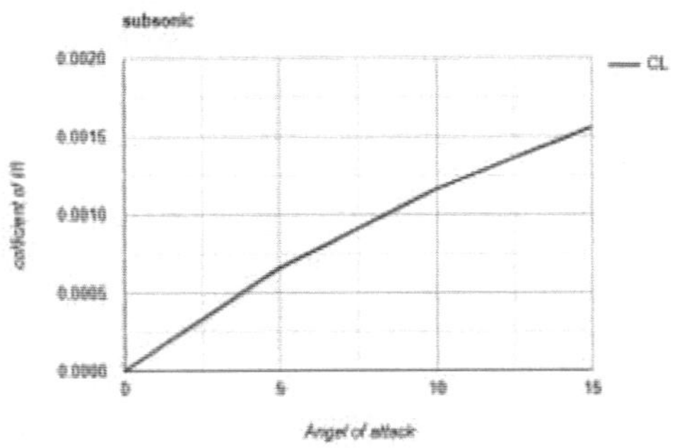

Figure 8. C_l vs Angle of Attack(Subsonic)

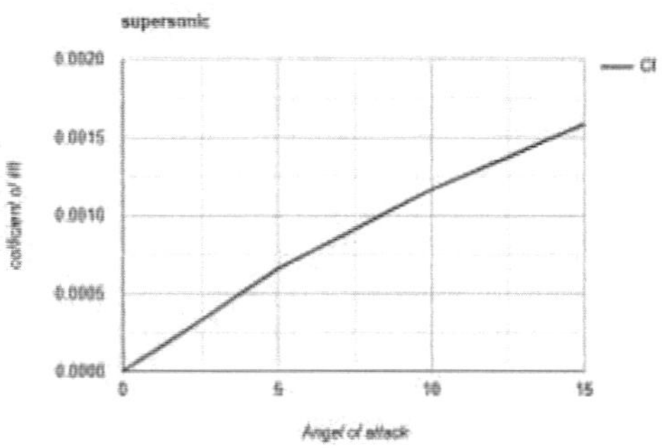

Figure 9. C_l vs Angle of Attack(Supersonic)

In Figure 8, the coefficient of lift is seen to increase gradually same in subsonic speed and supersonic speed in Figure 9, a little linear and smooth increase is observed in supersonic speed.

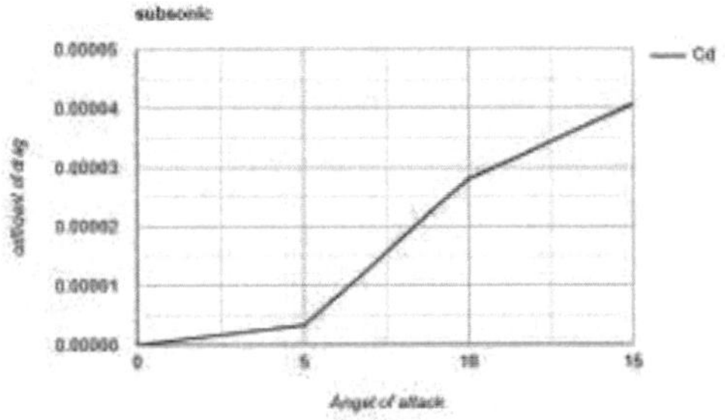

Figure 10. C_d vs Angle of Attack (Subsonic)

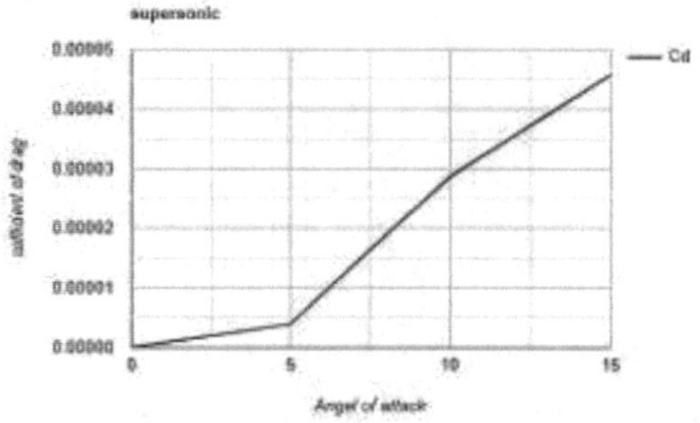

Figure 11. C_d vs Angle of Attack (Supersonic)

In Figure 10, the coefficient of drag in subsonic is seen to increase in steep variation whereas in Figure 11, for supersonic speed the variation is seen to be smoother and more stable. Thus, concluding that for higher speed the transition in the attack angle comes with a stable density increase of the vortex.

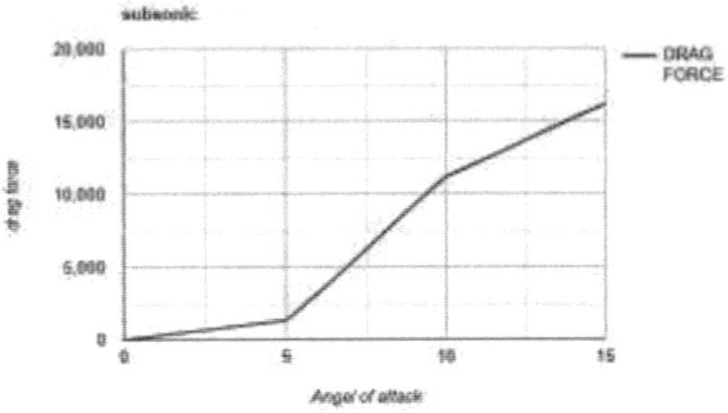

Figure 12. Drag Force vs α (Subsonic)

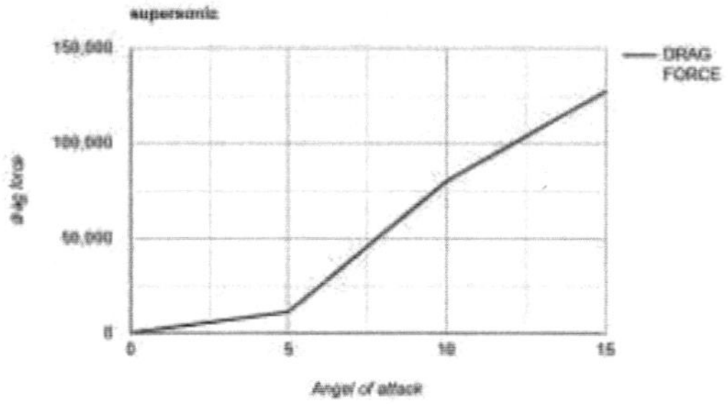

Figure 13. Drag Force vs α (Supersonic)

In Figure, 12 drag force is found to be increasing with the magnitude incrementation of attack angle and in Figure 13, at the supersonic level drag is seen to increase at a high rate with attack angle and following a linear slope in higher attack angle.

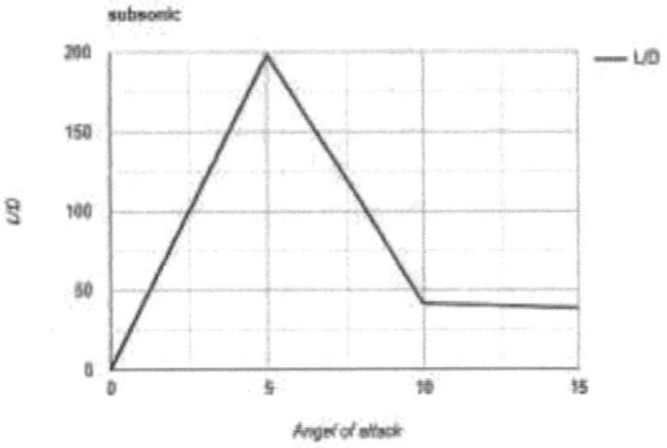

Figure 14. L/D Ratio vs α (Subsonic)

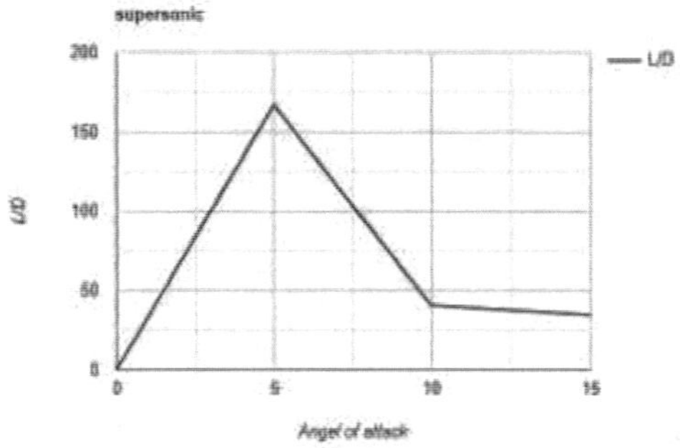

Figure 15. L/D Ratio vs α (Supersonic)

In Figure 14, the highest value of lift/drag ratio is high in subsonic speed, and in Figure 15, the peak value of lift/drag is low, this states that the drag is more when the speed is low.

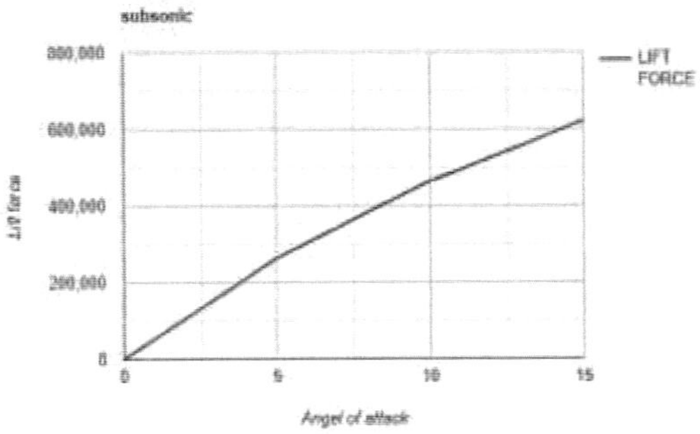

Figure 16. Lift Force vs α (Subsonic)

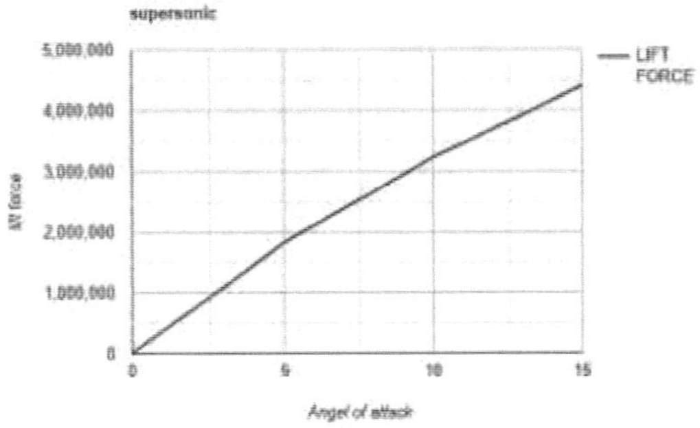

Figure 17. Lift Force vs α (Supersonic)

In Figure 16, the lift force is seen to be in constant variation in accordance to speed and angle of attack at subsonic speed the drag is seen to increase steeply whereas in Figure 17, for supersonic speed the lift increase is somewhat constant and linear and the value is quite high as compared to subsonic. Thus, the vortex is denser and more stable.

4.Conclusion

The study of flow visualization is done on subsonic and supersonic flow on compound delta wing at an attack angle of 5° to 15°. The results were discussed using the flow profile and pressure density upon the wing.

The result concludes that the formation of the primary vortex is highly dependent on the angle of attack. An increase in the angle of attack can boost the upstream movement hence, the formation of the primary vortex towards the aft portion of the wing. The lift force was seen to be very sensitive concerning attack angle and not consistent with the range of attack angle. A wide increase in lift force is observed with the increase of attack angle and drag force is also found to increase violently with the lift force. The result of lift by drag is found to be proportionally inverse to the attack angle. This could be concluded as the density of the vortex remains consistent with the higher angle of attack from 5° to 12° and the L/D ratio is superior.

For future scope of improvement regarding the stability of flight the plane should be operated in between the attack angle of 5° to 12°. A forward sweep should be added at the tail of the wing with strake to control the flow separation and to decrease the shock caused by variation in the angle of attack and speed.

The present physics study of compound delta wing for subsonic and supersonic flow can be helpful to study more about vortex formation.

References

[1] D. Hummel, "The Second International Vortex Flow Experiment (VFE-2): Objectives and Present Status. Collection of Technical Papers," *AIAA Applied Aerodynamics Conference,* 2007.

[2] H. H. K. T. M. Lee, "Vortex-induced vibration of a neutrally buoyant tethered sphere," *Journal of Fluid Mechanics,* vol. 719, pp. 97-128, 2013.

[3] R. G. M. V. I. Gursul, "Unsteady aerodynamics of nonslender delta wings," *Progress in Aerospace Sciences,* vol. 41, no. 7, pp. 515-557, 2005.

[4] X. &. Z. J. Zhang, "Off-Surface Aerodynamic Measurements of a Wing in Ground Effect," *Journal of Aircraft - J AIRCRAFT,* vol. 40, no. 4, pp. 716-725, 2001.

[5] F. S. A. M. A.-G. Ahmad Z. Al-Garni, "Experimental and Numerical Investigation of 65-deg Delta," *Journal of Aircraft.*

[6] K. F. A. O. Genta IMAI, "COMPUTATIONAL ANALYSES OF SUPERSONIC FLOWS OVER A DELTA WING AT HIGH ANGLES OF ATTACK," *25TH INTERNATIONAL CONGRESS OF THE AERONAUTICAL SCIENCES ,* 2006.

[7] C. &. R. A. &. C. J. &. B. J. &. M. G. Venier, "Comparing ANSYS Fluent ® and OpenFOAM ® simulations of Geldart A, B and D bubbling fluidized bed hydrodynamics.," *International Journal of Numerical Methods for Heat and Fluid Flow ahead-of-print(ahead-of-print),* 2019.

[8] K. R. R. M. C. A. J. L. Mehdi Ghoreyshi, "Vortical flow prediction of a diamond wing with rounded leading edges," *Aerospace Science and Technology,* vol. 57, pp. 103-117, 2016.

[9] Z. Z. Y. C. H. Tong, "Multi-objective aerodynamic optimization of supercritical wing with substantial pressure constraints," *54th AIAA Aerospace Sciences Meeting,* 2015.

[10] D. E. Alexander, "The biomechanics of solids and fluids: the physics of life," *European Journal of Physics,* vol. 37, no. 5, 2016.

IOP Conf. Series: Materials Science and Engineering 1149 (2021) 012028 doi:10.1088/1757-899X/1149/1/012028

Optimization of turning process parameters of SS-321 using taguchi based grey relational analysis

Gaurav Kumar[1*], Mohd Atif Wahid[2], Ankit Tomer[3], Mukesh Kumar[3] and Om Prakash Singh[3]
[1]M.M. Deemed to be University, Department of Mechanical Engineering, Ambala-133207, India.
[2]Delhi Technical Campus, Department of Mechanical Engineering, Greater Noida-201306, India.
[3]Vidya College of Engineering, Department of Mechanical Engineering, Meerut-251001, India.
[*]**gaurav.me86@gmail.com**

Abstract. Tool wear, high surface roughness & MRR (Material Removal Rate) are the major problems associated with turning of SS-321. The turning parameters such as feed rate, depth of cut and cutting speed have the major effects on surface roughness, MRR, and cutting forces. . An increase in magnitude of cutting forces leads to increase in the power requirements and simultaneously it decreases efficiency, which is the great concern in current scenario. In this turning work on SS-321, Taguchi based L_{27} orthogonal array was designed and experiments were performed accordingly. The data analysis was done by Taguchi, ANOVA and Grey relational analysis. The optimized results and improvements by using this method were also validated by a verification test. Results revealed that turning process with parameters such that feed rate, depth of cut and cutting speed at 0.3 mm/rev, 0.3 mm and 120 m/min respectively yields the optimum multi-performance characteristics. However, it was determined by ANOVA (Analysis of Variances) that depth of cut is a process parameter that affect most the multi-function features with percentage contribution of 64.86% followed by feed (16.84%) and speed (10.20%). Under these circumstances, the optimum parameters set seems to favorably affect the average values of all the three response parameters (surface roughness, cutting forces, MRR) so that the average values of output characteristics arrive at optimum level.

1. Introduction

Turning is a method of a method of machining used for extracting excess material. The metal removal is the removal of a piece of metal from the workpiece to produce a finished product with the desired dimensions, shapes and characteristics of the surface roughness. Engineers are always involved in determining the best suitable parameters with optimized output and performance with the minimum use of resources [1].

In general, the selection of appropriate process parameter is not so easy because it depends mostly on the expertise of operator and the machining process parameters given for the material by the machine tool manufacturers. Therefore, optimization of process parameters is very important because it plays a key role in economy and efficiency of machining process [2].

Surface roughness is very crucial to evaluate the accuracy of machining processes [3]. The system vibration directly affects the work material surface finish. In turning process vibrations are obtained due to different forces acting on cutting tool [4]. Our main concern now is to reduce costs and time, improve quality and economic productivity. Better surface finish and less tool wear are challenges when using materials of high strength and resistance to wear while turning. Optimized cutting parameters should be used for overcoming the above problems [5]. Multi-response optimization in industrial applications is therefore important. It is better than optimizing single-response technology as all factors are influenced by all input factors at once. Taguchi technique with GRA (Grey Relational Analysis) is being used to optimize the process parameters of turning on CNC lathe with surface roughness, cutting forces and MRR as multi-performance characteristics. It has been used effectively in developing high quality in the field of automotive, aerospace etc. at low cost [6]. Goel et al. [7] provided a successful methodology for optimizing the HSLA steel slab milling process based on Taguchi with GRA for multi-function features. Siddiquee et al [8] performed the experiments on AISI 321 steel using Taguchi method for optimizing deep drilling process parameters.

Literature surveys have shown a plethora of work on the effects of process parameters on surface roughness, material removal rate and cutting forces has been reported. There is, however, little literature on the effect on SS-321 of the turning process parameters. Many researchers have noted that the Taguchi approach and grey relationship analysis have become powerful tools for increasing productivity in the development process and have found a wide range of applications through different machining processes. High quality goods can therefore be manufactured easily and at low cost.

In this study 27 experiments were conducted by using taguchi and GRA for optimization of three turning parameters (depth of cut, speed and feed) for machining of SS-321 on CNC lathe machine and the multi-function features were surface roughness, material removal rate and cutting forces as radial force, cutting force and feed force.

2. Experimental Details

As per designed L_{27} orthogonal array 27 turning experiments were completed using CNC lathe machine and to obtain the optimum parameters for multi-function features ANOVA was applied. The all three turning parameters (speed, depth of cut, feed) and their different value selected for experiments is shown in table 1.

Table 1. Machining parameter and their levels

Notation	Parameter	Level 1	Level 2	Level 3
A	Depth of cut	.5	1	1.5
B	Speed (m/min.)	60	90	120
C	Feed (mm/rev)	.2	.25	.3

The various performance parameters for experimental investigations were measured by using Mitotoyo surface roughness tester. It is a stylus and skid-type instrument in which the profile of surface irregularities is traced through a sharply pointed stylus. The stylus movement cut-off length was .8 mm. In this experiment Kistler made 3-component 9129AA dynamometer with charge amplifier are used for measuring the cutting forces i.e., radial force (F_r), cutting force (F_c) and feed force (F_f).

The metal removal rate is an important factor for productivity. The metal removal rate was calculated in turning process by formula as shown below:

$$Material\ Removal\ Rate = \frac{1000 \times f \times Vc \times d}{60}\ mm^3/sec \tag{1}$$

Where:

V_C = Cutting speed in m/min, f= Feed in mm/rev, d= Depth of cut in mm

The measured values of surface roughness (Ra), MRR, radial force, cutting force and feed force see in table 2.

Table 2. Values of response variables

Experiments	A	B	C	Ra(μm)	MRR (mm^3/s)	Radial Force (F_r)	Cutting Force (F_c)	Feed Force (F_f)
1.	1	1	1	1.57	100.0	187.40	220.40	95.35
2.	1	1	2	0.98	125.0	194.4	274.8	121.6
3.	1	1	3	1.05	150.0	231.6	435.9	134.2
4.	1	2	1	0.96	150.0	205	282.8	99.58
5.	1	2	2	1.32	187.5	226.8	261.1	121.8
6.	1	2	3	1.28	225.0	256	496.5	134
7.	1	3	1	1.02	200.0	214.5	301.5	114.9
8.	1	3	2	1.12	250.0	233.3	279.6	117.3
9.	1	3	3	0.95	300.0	269	597.7	147.1
10.	2	1	1	2.00	200.0	189.1	355	175.7
11.	2	1	2	2.06	250.0	213.5	511.7	191.4
12.	2	1	3	2.66	300.0	237.4	786.6	198.7
13.	2	2	1	1.54	300.0	201.5	450.2	174.8
14.	2	2	2	2.07	375.0	198.2	345.3	178.5
15.	2	2	3	3.09	450.0	239	435.2	188.9
16.	2	3	1	1.82	400.0	214.1	454.5	176.7
17.	2	3	2	2.14	500.0	269.6	820.5	219.3
18.	2	3	3	2.45	600.0	243.8	501.9	193
19.	3	1	1	2.41	300.0	208.2	526.9	455.9
20.	3	1	2	2.44	375.0	241.6	772.3	499
21.	3	1	3	3.39	450.0	236.4	988	798
22.	3	2	1	2.10	450.0	222.8	610.6	484.1
23.	3	2	2	2.60	562.5	231.1	899.9	450.7
24.	3	2	3	3.16	675.0	265.4	1268	768.7
25.	3	3	1	1.69	600.0	249.5	709.8	558.5
26.	3	3	2	2.94	750.0	268.6	904.4	752.2
27.	3	3	3	3.15	900.0	271.9	1397	852.5

3. Calculations (Analysis Method)

3.1. Signal to noise(S/N) ratio and Data pre-processing

Taguchi offers a systematic, simple and efficient optimization technique. Depending on the features, the three S/N ratios used in taguchi method are of three types, first one is smaller is good, second greater is good, and the third is nominal is good [9,10].

Smaller is good:

$$\eta_{ij} = -10\log\left[\frac{1}{n}\sum_{j=1}^{n} {y^2}_{ij}\right] \quad (2)$$

Greater is good:

$$\eta_{ij} = -10\log\left[\frac{1}{n}\sum_{j=1}^{n}\frac{1}{{y^2}_{ij}}\right] \quad (3)$$

Nominal is good:

$$\eta_{ij} = -10\log\left[\frac{1}{ns}\sum_{j=1}^{n} {y^2}_{ij}\right] \quad (4)$$

Where,

y_{ij} means the "ith"- experiment at 'jth" - test, "n' is Total number of test and "s" refers to standard deviation. Some other formulations of the S/N ratio are also available, which can be obtained in references.

As per above discussion criterion smaller is good was selected for the S/N ratios of roughness and cutting forces, whereas the greater is good for MRR [11].

Data pre-processing is used to convert the original series in number which can be compare. So, the results of experiments were processed and normalized in between 0 to 1. There are three different approaches to achieve normalization. Normalization information found at [12,13,14,15].

3.2. Grey relational coefficient and grey relational grade
Grey relational coefficient is determined to demonstrate the relationship between ideal and actual effects of the normalized experiments. It can reflect as follows:

$$\xi_i(k) = \frac{\Delta_{min} + \xi . \Delta_{max}}{\Delta_{oi}(k) + \xi . \Delta_{max}} \quad (5)$$

Where $\Delta_{0i}(k)$ is the deviation sequence of the reference sequence $x_0^*(k)$ and the comparability sequence $x_i^*(k)$, called as,

$$\Delta_{0i}(k) \, ||x_0^*(k) - x_i^*(k)|| \quad (6)$$

$$\Delta_{max} = \max_{\forall j \in i} \min_{\forall k} ||x_0^*(k) - x_i^*(k)|| \quad (7)$$

$$\Delta_{min} = \max_{\forall j \in i} \min_{\forall k} ||x_0^*(k) - x_i^*(k)|| \quad (8)$$

Generally, identification faction ζ is taken as 0.5 and it ranges from 0 to 1.

3.3. Analysis of mean (ANOM)
The average of multi-function features i.e., grey relational grade is calculated using ANOM to identify the effect of every parameter individually. Grey relational grades are discussed below:

$$\gamma_i = \frac{1}{n}\sum_{k=1}^{n} \xi_i (k) \quad (9)$$

However, since in realistic implementations the effect of each element on the device isn't exactly the same, equation (8) may be changed to:

$$\gamma_i = \frac{1}{n}\sum_{k=1}^{n} w_k . \xi_i (k) \quad \sum_{k=1}^{n} w_k \quad (10)$$

where "w_k" represents the normalized weight of factor k. Equations (9) and (10) given the same weight, are identical.

4. Results and discussion

4.1. Optimal parameter combination
Grey relational coefficient and grades calculated for all three multi performance features (i.e., roughness, cutting forces, MRR) and the combination of parameters which give the optimum results for turning is obtained by grey relational analysis. Grey relational grades and their responses for different turning parameters at their different values shown in figure 1. As we can see from figure 1, A_3, B_3 and C_3 display the maximum value of grey relational grades for each turning process parameter. Thus, the optimized multi-function features are obtained at combination A_3B_3 C_3 for each turning parameter.

Table 3. Response Table for Grey Relational Grade

Symbol	Turning Parameter	Machining	Level 1	Level 2	Level 3	Max – Min
A	Depth of Cut		0.41444	0.518433	0.738	0.323101

B	Cutting speed	0.50268	0.538237	0.629497	0.126817
C	Feed rate	0.482629	0.541943	0.645841	0.163212

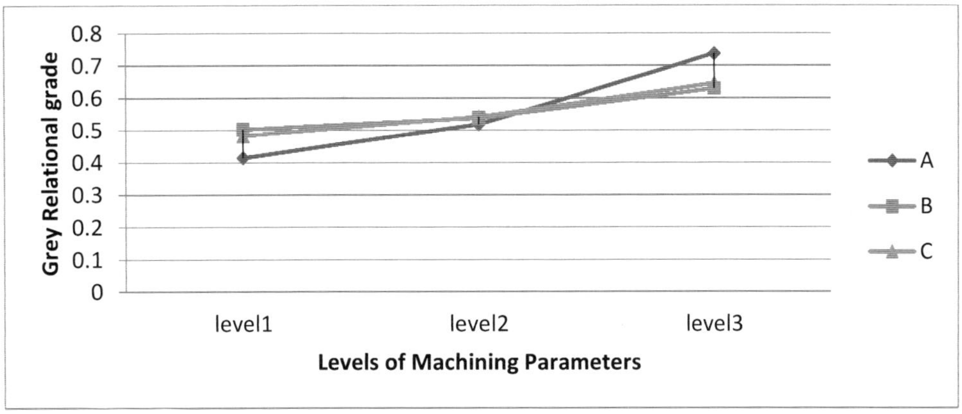

Figure 1. Grey relational grade graph

The differences in maximum and minimum value of grey relational grades are shown in Table 3 and which depict that difference for factor A (depth of cut) is highest followed by C (Feed rate) and B (cutting speed) as 0.323101,0.163212 and 0.126817 respectively. This shows that factor A has major impact on multi-function features. Figure 1 demonstrates the effect of multi-function features on turning parameter levels.

4.2. Analysis of variance

ANOVA is an approach to statically determine the influencing effects of different factors individually. As it is a complex design and performance of process is to be determine for roughness, cutting forces and material removal rate simultaneously and it is not possible by taguchi method only so ANOVA is also applied to find the contribution of each parameter. ANOVA also include F test to indicate the impact of a parameter on multi-function features. If value of F test less than four than the effects of particular parameter is insignificant. Here as shown in table 4 all the process parameters are significant because their F test value is more than four and out of which parameter A (depth of cut) is most effective with F test value 115.60.

Table 4. Result of Analysis of Variance

Symbol	Parameters	DF	SS	Mean square	F ratio	Contribution (%)
A	Depth of Cut	2.00	0.489	0.377	115.60	64.86
B	Speed	2.00	0.077	0.038	11.79	10.20
C	Feed rate	2.00	0.122	0.061	18.81	16.27
Error		20.00	0.065	0.003		8.65
Total		26.00	0.754			100

IOP Conf. Series: Materials Science and Engineering 1149 (2021) 012028 doi:10.1088/1757-899X/1149/1/012028

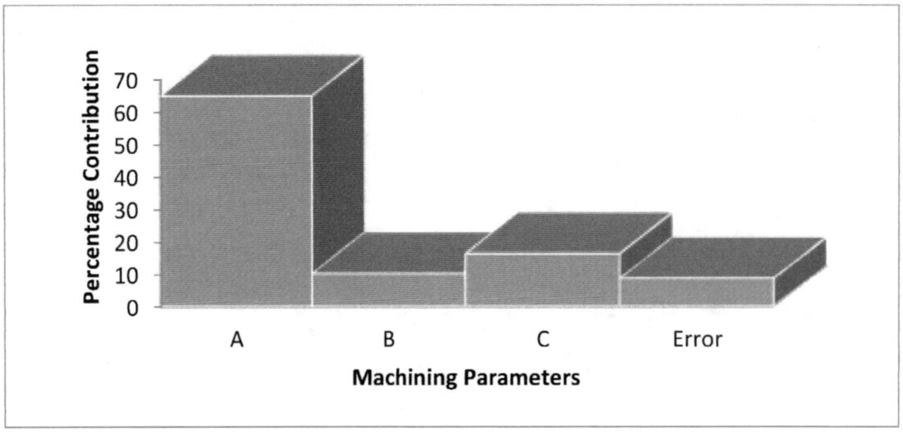

Figure 2. Percentage Contribution of machining parameters

So, it is very clear from ANOVA results as shown in table 4 that turning parameter A (depth of cut) is a most important parameter with 64.86% contribution, where as in decreasing order C (feed rate) and B (speed) have their contributions 16.27% and 10.20% respectively. Contribution of all the turning parameters is also shown in figure 2.

4.3. Verification test

A verification test was completed to compare and validate the optimized turning parameters results. As the increase in grey relational grades represent the improvement in multi-function features for turning parameters and it was observed that the grey relational grades increase from 0.899 (predicted value) to 0.979 (experimental value) which indicate an improvement of 8.899% (see table 5).

Table 5. Results of verification test

Level	Optimal Machining Parameter		% Change
	Prediction $A_3B_3C_3$	Experiment $A_3B_3C_3$	
Surface roughness		0.897	
Material Removal Rate		1.000	
Radial Force		1.000	
Cutting Force		1.000	
Feed Force		1.000	
Grey relational grade	0.899	0.979	8.899

5. Conclusion

The optimized machining parameters are obtained as $A_3B_3C_3$ where depth of cut1.5mm, speed-120m/min. and feed rate-0.30mm/rev. As the % contribution of depth of cut is higher (64.86%) followed by speed (10.20%) and feed rate (16.27%) respectively under these circumstances, the parameter set $A_3B_3C_3$ seems to be favorably affecting the average values of all response parameters (Surface roughness, Material removal rate, radial force, cutting force and feed force) so that the average values of output characteristics arrive at optimum level. This method achieves a multi-performance improvement of 8.89.

Acknowledgement

The authors are grateful for the support, inspiration, and facilities given by Vidya Knowledge Park, Meerut to do this research.

References

[1] Black K M 1994 An industry view of engineering education *Journal of Engineering Education* **83(1)** pp 26-28

[2] Saravanan R S R S, Sankar R S, Asokan P, Vijayakumar K and Prabhaharan G 2005 Optimization of cutting conditions during continuous finished profile machining using non-traditional techniques *The International Journal of Advanced Manufacturing Technology* **26(1)** pp 30-40

[3] Palanikumar K, Karunamoorthy L and Karthikeyan R 2006 Assessment of factors influencing surface roughness on the machining of glass fiber-reinforced polymer composites *Materials & Design* **27(10)** pp 862-871

[4] Choudhury S K and Sharath M S 1995 On-line control of machine tool vibration during turning operation *Journal of Materials Processing Technology* **47(3-4)** pp 251-259

[5] Ranganathan S and Senthilvelan T 2011 Multi-response optimization of machining parameters in hot turning using grey analysis *The International Journal of Advanced Manufacturing Technology* **56(5-8)** pp 455-462

[6] Singh O P, Kumar G and Kumar M 2019 Role of Taguchi and grey relational method in optimization of machining parameters of different materials: a review *Acta Electronica Malaysia (AEM)* **3(1)** pp19-22

[7] Goel P, Khan Z A, Siddiquee A N, Kamaruddin S and Gupta, R.K 2012 Influence of slab milling process parameters on surface integrity of HSLA: a multi-performance characteristics optimization *The International Journal of Advanced Manufacturing Technology* **61(9-12)** pp 859-871

[8] Siddiquee A N, Khan Z A, Goel P, Kumar M, Agarwal G and Khan N Z 2014 Optimization of deep drilling process parameters of AISI 321 steel using Taguchi method *Procedia Materials Science* **6** pp 1217-1225

[9] Vikram K A, Kanth T K, Shabana A M V and Lakshmi V V K 2019 Experimental study of surface roughness and cutting forces on stainless steel grade materials *International Journal of Mechanical Engineering and Technology* **10(2)** pp 87-96

[10] Tomadi S H, Ghani J A, Haron C C, Ayu H M and Daud R 2017 Effect of cutting parameters on surface roughness in end milling of AlSi/AlN metal matrix composite *Procedia Engineering* **184** pp 58-69

[11] Singh O P, Kumar G and Kumar M 2019 Multi Performance optimization of shoulder milling process parameters of AA6063 T6 aluminium alloy by Taguchi Based GRA *International Journal of Innovative Technology and Exploring Engineering* **8** pp 420-425

[12] Siddiquee A N, Mallick Z and Khan Z A 2010 Grey relational analysis coupled with principal component analysis for optimization design of the process parameters in in-feed center less cylindrical grinding *The International Journal of Advanced Manufacturing Technology* **46(9-12)** pp 983-992

[13] Khan Z A, Siddiquee A N and Kamaruddin S 2010 Feasibility study of use of recycled high-density polyethylene and multi response optimization of injection moulding parameters using combined grey relational and principal component analyse *Materials & Design* **31(6)** pp 2925-2931

[14] Kumar M, Kumar G, Singh O P and Tomer A 2021 Multiperformance Optimization of Parameters in Deep Drilling of SS-321 by Taguchi-Based GRA *In Recent Advances in Mechanical Engineering Springer Singapore* pp 675-681

[15] Kumar G, Kumar M and Tomer A 2021 Optimization of End Milling Machining Parameters of SS 304 by Taguchi Technique *In Recent Advances in Mechanical Engineering Springer Singapore* pp 683-689

Optimization of turning process variables for surface integrity using taguchi technique

Gaurav Kumar[1], Ankit Tomer[2], Mukesh Kumar[2], Pankul Goel[3]* and Om Prakash Singh[2]

[1]M.M. Deemed to be University, Department of Mechanical Engineering, Ambala-133207, India.
[2]Vidya College of Engineering, Department of Mechanical Engineering, Meerut-251001, India.
[3]IMS Engineering College, Department of Mechanical Engineering, Ghaziabad-201009, India.
* **pankul.goel@imsec.ac.in**

Abstract. The selection of proper machining parameters is crucial for the production of a product of desired quality. The Taguchi method will accomplish this. For this work, using tin coated tungsten carbide method, SS-321 work parts are placed onto CNC lathe machine. Stainless steel machining (SS-321) has issues such as worn-out tool and high roughness to the surface. The key reasons for choosing SS-321 grade as workpiece material is the issue of the machining process related with super finishing of Stainless Steel. For this work twenty-seven experiments were carried out on stainless steel (SS-321), as per Taguchi's L_{27} orthogonal array. Data analysis was carried out using Taguchi response table method and ANOVA. Following this method, a confirmation experiment was performed to assess the surface integrity change. Turning machining parameter levels to reduce surface roughness were $A_1B_3C_1$ i.e., depth of cut 0.5 mm, cutting speed of 120 m / min and feed of 0.2 mm / rev. ANOVA table represented the parameters which mostly affected the surface roughness with 76.27% contribution of depth of cut which is maximum followed by feed (7.72%), and effect of speed (3.48%) is insignificant. The research also shows that this approach can efficiently boost the surface finish of the turning process.

1. Introduction

Surface roughness (SR) plays a vital role in various fields and plays an important role in evaluating processing accuracy [1]. Our main concern now is to reduce costs and time, improve quality and economic productivity. Better surface finish and less tool wear are a challenge when using materials with high strength, resistance to corrosion and resistance to wear while turning. Optimized cutting parameters should be employed to solve the above problems [2].

Turning is one of the excess metal removal methods used for machining. The metal removal is the removal of a piece of metal from the workpiece to produce a finished product with the desired lengths, shapes and characteristics of surface roughness. The task for engineers is to search the best parameters for the desired output, and optimize performance with the resources available [3]. Most of the components like rail road cars, mining, various kinds of vehicles etc. are constructed from machined SS-321. Siddiquee et al [4] conducted the experiments on SS-321 steel by taguchi method for optimizing deep drilling process parameters. Dhar et al. [5] submitted that the roughness of the SS-4340 surface decreases by using minimal lubrication. Sijo M.T et al. [6] concluded that optimization of the turning process parameter is time consuming and highly complex. During turning the cutting parameters were optimized by taguchi techniques. Om Prakash Singh et al [7,8] proposed the taguchi and grey relational analysis for optimizing the shoulder milling parameters of AA 6063T6 Aluminium alloys.

A plethora of research on the impact of process parameters on surface roughness has been published in literature surveys. There is, however, little literature on the effect on SS-321 of the turning process parameters. Many researchers have noted that the taguchi approach and grey relationship analysis have become powerful tools for increasing productivity in the development process and have found a wide range of applications through different machining processes. High quality goods can therefore be manufactured easily and at low cost.

In this study 27 experiments were conducted by using taguchi for optimization of three turning parameters (depth of cut, speed and feed) for machining of SS-321 on CNC lathe machine for improving surface roughness.

2. Experimental Details

2.1 Work piece Material

Machined SS-321 material was used to perform experimental work. General purpose SS-321 has been chosen as the standard because it is most flexible and commonly used. SS-321 fragment samples used during testing are as shown in figure1 and the chemical constituents present in SS-321 are shown in table 1.

Figure 1. Stainless steel (AISI-321) pieces

Table 1. Chemical constituents of Stainless Steel (SS-321) Material

Elements	Concentration (by weight%) Minimum	Concentration (by weight %) Maximum	Concentration (by weight %) Actual
Carbon		00.080	00.070
Manganese		2.000	1.780
Silicon		1.000	1.000
Phosphorus		00.045	00.380
Sulphur		00.030	0.260
Nickel	9.000	12.000	10.380
Chromium	17.000	19.000	17.200
Copper		0.750	00.380

2.2 CNC Lathe Machine

The turning operations were performed on the SS-321 pieces by the CNC Lathe machine as shown in figure 2. Mitotoyo tester was used for testing the surface roughness as shown in figure 3.

Figure 2. CNC lathe Machine

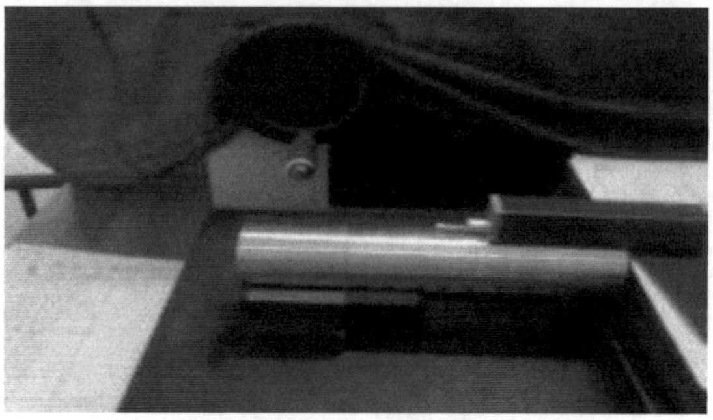

Figure3. Surface Roughness Tester

Different operating conditions for turning with their specifications are shown in table 2.

Table 2. Operating conditions

Condition	Specification
Work piece material	SS-321
Geometry of work piece	50mm in dia. and 200mmin length
Turning machine	CNC Turing machine (Make: Taiwan in 2011 Model: Leadwell, T-6; Tool movement: X-150 mm, Z-350mm, and 5.5-KW/4500 rpm)
Cutter used	Tin coated solid carbide tool (nose radius 0.8mm)
Measuring tool	Mitotoyo surface roughness tester

2.3 Experimental Design

The experimental work was carried out according to the experimental configuration of the L_{27} orthogonal array on a CNC turning machine. Optimal order of turning parameters for turning was obtained by using ANOVA. As shown in table 3, three key turning parameters i.e., doc, speed and feed and their levels were determined during machining.

Table 3. Machining parameters and their levels

Factor identifier	Machining parameter	Level 1	Level 2	Level 3
A	Depth of Cut (mm)	0.5	1	1.5
B	Speed (m/min.)	60	90	120
C	Feed (mm/rev.)	0.2	0.25	0.3

3. Results and discussion

3.1 Analysis of Signal to Noise Ratio

In this study, equation one was used for minimization of surface roughness which is based on lower is the good criterion.

Lower-is-Good:

$$S/N_{\mathrm{dB}} = -10\log\left[\frac{1}{\mathrm{m}}\sum_{\mathrm{y}=1}^{\mathrm{m}} \mathrm{R}^2{}_{\mathrm{xy}}\right] \quad (1)$$

IOP Conf. Series: Materials Science and Engineering 1149 (2021) 012029 doi:10.1088/1757-899X/1149/1/012029

Where, R_{xy} is the value of surface roughness for x^{th} observation at the yth test, m is the number of total observations taken. Other formulae of Signal to Noise ratio calculations can be found in references [9, 10, 11]. Table 4 represents the S/N ratios for 27 experiments. Calculations of mean value (θ_i) of S/N ratios was find out by equation 2 as given below [12,13,14].

$$\theta_i = \frac{1}{M}\sum_{j=1}^{M} \gamma(j) \qquad (2)$$

Where j=1,2………, M (here M=27) and $\gamma(j)$ is S/N ratio for j^{th} experiment condition. From table 5 and figure 4 based on the maximum value of S/N ratio, an optimum level for each turning parameter was $A_1B_3C_1$ which means depth of cut 0.5 mm, speed 120 m/min and feed of 0.2 mm/rev. was noticed.

Table 4. Values of S/N ratio

Experiment No.	doc	speed	feed	Ra(μm)	S/N Ratio $\gamma(j)$
1	1	1	1	1.57	-3.918
2	1	1	2	0.92	0.724
3	1	1	3	1.26	-2.007
4	1	2	1	0.97	0.265
5	1	2	2	1.285	-2.178
6	1	2	3	1.11	-0.906
7	1	3	1	0.645	3.809
8	1	3	2	0.71	2.975
9	1	3	3	0.8	1.938
10	2	1	1	2.005	-6.042
11	2	1	2	2.06	-6.277
12	2	1	3	2.86	-9.127
13	2	2	1	1.545	-3.779
14	2	2	2	2.07	-6.319
15	2	2	3	3.09	-9.799
16	2	3	1	1.82	-5.201
17	2	3	2	2.145	-6.629
18	2	3	3	2.655	-8.481
19	3	1	1	2.41	-7.640
20	3	1	2	2.445	-7.766
21	3	1	3	3.09	-9.799
22	3	2	1	2.105	-6.465
23	3	2	2	2.605	-8.316
24	3	2	3	3.165	-10.007
25	3	3	1	1.695	-4.583
26	3	3	2	2.74	-8.755
27	3	3	3	3.335	-10.462
			Avg.	1.967	-134.74

Table 5. Average S/N ratio response

level	doc	speed	feed
1	0.0778	-5.761	-3.728
2	-6.850	-5.278	-4.726

3	-8.199	-3.932	-6.516
Delta	8.277	1.829	2.788
Rank	1	3	2

The total mean S/N ratio = -4.99

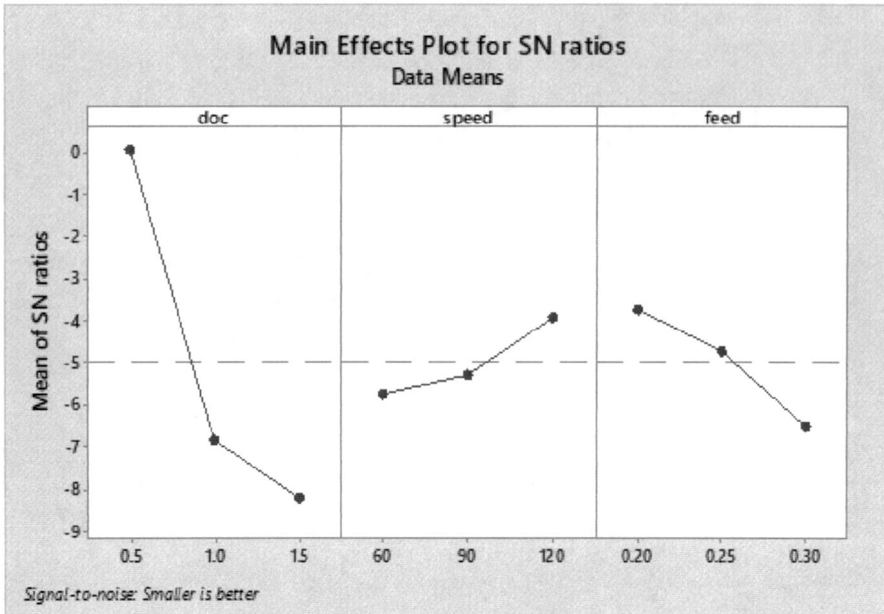

Figure 4. Response Graph

It was observed from table 5 and figure 4, that the S/N ratios for depth of cut and feed decreases with increase in their levels where as for speed value of S/N ratios increases with increase in its levels. Values of delta for all parameters is also shown in table 5 which depict that doc has maximum value of delta i.e., 8.277 and hence ranked first, feed and speed are ranked second and third with reduced values as 2.788 and 1.829 respectively and it was also observed that S/N are maximum at level1 for doc, level 3 for speed and level1 for feed so optimum arrangement of parameters is $A_1B_3C_1$.

3.2 Analysis of variance

Evaluation of S / N ratios was completed by ANOVA at a confidence level of 95 percent see table 6. Contribution in percentage of each cutting parameter to the surface roughness is also evaluated by ANOVA. The depth of cut indicates a greater contribution of 76.27 % from the table of ANOVA. The error was noticed here at 12.54%.

Table 6. Results of ANOVA

Source	Dof	Sum of square	Mean square	F-Value	Contribution
doc	2	355.00	177.502	60.84	76.27%
speed	2	16.18	8.088	2.77	3.48%
feed	2	35.93	17.965	6.16	7.72%
Error	20	58.35	2.918		12.54%
Total	26	465.46			100.00%

Table 6 shows that the speed given has little impact on the surface roughness for the given sample, since the speed F value is less than 4. ANOVA's results also show that the depth of cut (76.27%) is the most important turning parameters likewise feed rate (7.72%) and speed (3.48%) (see figure 5).

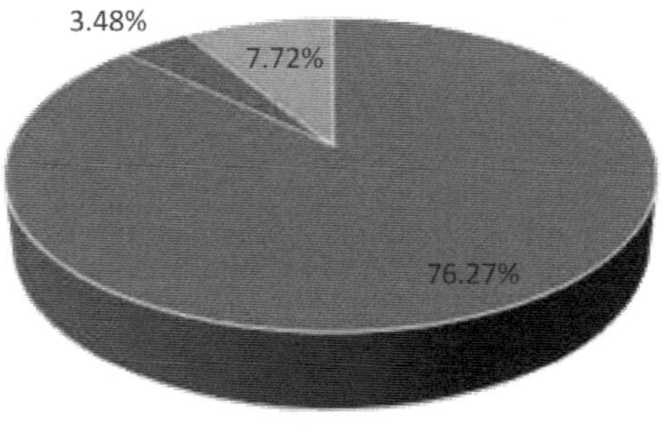

Figure 5. Percentage contribution

3.3 Confirmation Experiments
When the optimal level of process parameters has been established, the next step is to use this optimal level to check the performance characteristic improvement. The outcomes of the confirmation experiment are shown in Table 7 using optimal turning process parameters.

Table 7. Optimum parametric conditions

	Prediction	Experiment	% change
	$A_1B_3C_1$	$A_1B_3C_1$	
Surface roughness	0.428	0.645	50.70 %
S/N Ratio (db)	7.38	3.80	48.50%

From table 7 it was observed that signal to noise ratio changed from 7.38 to 3.80 (a decrement of 48.50%) on account of raise in value of roughness from 0.428 to 0.645 μm (an increment of 50.70%) which represents that the optimal combination ($A_1B_3C_1$) are good enough for reducing the machined surface roughness.

4. Conclusions
The optimization of turning operations is done by taguchi method in this paper. As described earlier, taguchi technology design gives an efficient, well arranged and easy technology for the optimization. Optimized level of parameters ($A_1B_3C_1$) in turning shows minimum surface roughness (i.e., depth of cut1.5 mm, speed-120m / min. and feed rate-0.20mm / rev). The determination of the optimum cutting parameters is verified by the confirmation experiment. The percentage contribution of depth of cut is higher (76.27%), followed by feed (7.72%) and speed (3.48%). Therefore, the use of this method can significantly improve surface roughness. So, taguchi is most helpful to optimize process parameters in the industry in order to reduce cost and time.

Acknowledgement

The authors are thankful for the motivation, support and facilities given by Vidya Knowledge Park, Meerut, India for carrying out this study.

References

[1] PalaniKumar K, Karunamoorthy L and Karthikeyan R 2006 Assessment of factors influencing surface roughness on the machining of glass fiber-reinforced polymer composites *Materials & Design* **27(10)** pp 862-871

[2] Ranganathan S and Senthilvelan T 2011 Multi-response optimization of machining parameters in hot turning using grey analysis *The International Journal of Advanced Manufacturing Technology* **56(5-8)** pp 455-462

[3] Black K M 1994 An industry view of engineering education *Journal of Engineering Education* **83(1)** pp 26-28

[4] Siddiquee A N, Khan Z A, Goel P, Kumar M, Agarwal G and Khan N Z 2014 Optimization of deep drilling process parameters of AISI 321 steel using Taguchi method *Procedia Materials Science* **6** pp 1217-1225

[5] Dhar N R, Kamruzzaman M and Ahmed M 2006 Effect of minimum quantity lubrication (MQL) on tool wear and surface roughness in turning AISI-4340 steel *Journal of Materials Processing Technology* **172(2)** pp 299-304

[6] Sijo M T and Biju N 2011 Taguchi method for optimization of cutting parameters in Turning Operations *International Journal on Manufacturing and Material Science* **1(1)** pp 44

[7] Singh O P, Kumar G and Kumar M 2019 Multi Performance optimization of shoulder milling process parameters of AA6063 T6 aluminium alloy by Taguchi Based GRA *International Journal of Innovative Technology and Exploring Engineering* **8** pp 420-425

[8] Singh O P, Kumar G and Kumar M 2019 Role of Taguchi and grey relational method in optimization of machining parameters of different materials: a review *Acta Electronica Malaysia (AEM)* **3(1)** pp19-22

[9] Kuram E and Ozcelik B 2017 Optimization of machining parameters during micro-milling of Ti6Al4V titanium alloy and Inconel 718 materials using Taguchi method *Proceedings of the Institution of Mechanical Engineers. Part B: Journal of Engineering Manufacture* **231(2)** pp 228-242

[10] Sosa PB, Makwana RD and Acharya GD 2018 Optimization of Machining Parameters on End Milling of EN 8 Back Shaft for Power Press *Trends in Mechanical Engineering and Technology* **8(3)** pp49-58

[11] Roy R K 2010 A primer on the Taguchi method *Society of Manufacturing Engineers*

[12] Chang CL, Tsai CH and Chen L 2003 Applying grey relational analysis to the decathlon evaluation model *International Journal of Compute Internet Manage* **11(3)** pp 54–62

[13] Kumar M, Kumar G, Singh O P and Tomer A 2021 Multiperformance Optimization of Parameters in Deep Drilling of SS-321 by Taguchi-Based GRA In *Recent Advances in Mechanical Engineering Springer Singapore* pp 675-681

[14] Kumar G, Kumar M and Tomer A 2021 Optimization of End Milling Machining Parameters of SS 304 by Taguchi Technique in *Recent Advances in Mechanical Engineering Springer Singapore* pp 683-689

IOP Conf. Series: Materials Science and Engineering 1149 (2021) 012030 doi:10.1088/1757-899X/1149/1/012030

Assessment of appropriate areas for the establishment of Solar Thermal Power Plant in Rajasthan

Sajal Agarwal[1*], Harshit Kumar Gautam[2] and Altaf Hasan Tarique[3]

[1,2,3] School of Mechanical Engineering, Galgotias University Greater Noida, 226001, Uttar Pradesh, India.

[*]**sajalagrawal275@gmail.com**

Abstract. In past years, cost of photovoltaic has decreased and investments of solar power plant have increased. In spite of this, practical studies remain paramount as cost of installation is comparatively high. One of pivotal parameters in studies is the proper location selection. So, in this paper we mix Analytical hierarchy process (AHP) and multi-criteria decision making method (MCDM) to get the results of the ideal position of sunlight based energy station in Jodhpur, Rajasthan. Using MCDM the models or factors referenced will be weighted to assess possible destinations to find a sunlight based plant. Examination and count of the loads of these variables will be directed utilizing AHP. For this study Sardarpura, Kabir Nagar and Mandore are selected as examples. These places are located in same plane. The decision of such measures is set apart by the impact introduced to the general objective; for this situation in climatic models and geomorphological. In study, it is found that Kabir Nagar is ideal position for establishment of solar power plant. This result is based on criteria, feeder capacities, solar energy potential and earth slopes. The result are ratified by considering the total amount of area and sunshine duration.

1. Introduction

1.1. Sustainable energy in India

The momentum monetary circumstance, the increase in worldwide energy interest and judicious utilization of the accessible assets drives the quest for energy choices to address the issues of the present without trading; the objective of feasible improvement is empowering progresses in the investigation into new methodologies to upgrade the assets and innovations. Keeping that in mind, a few investigations have been created here. In India, the important development regulation of ozone depleting substance discharges set by the consistence of the protocols in European Union white book [1]. Among the various kinds of environment friendly power, sun oriented thermal power is demonstrating dependable and, despite the fact that it has not arrived at adequate development, critical endeavors are being fabricated in innovation research highlighting lower fabricating prices and higher productivity. The significant degree of sun oriented force created in the Region of Rajasthan is on the grounds that this locale has probably the most elevated level of potential or sun based radiation in the country; explicitly nearby Jodhpur. The normal yearly worldwide radiation in the vast majority of its domain surpasses 2400 KW/m^2 [2].

1.2. Multi-criteria decision making methods applications

The reason or extreme objective of a MCDM technique is to explore various choices in the light of numerous standards and clashing goals [3]. Few renowned MCDM includes the Analytic Hierarchy Process—AHP, its primary component is issuing the choice, whose pinnacle is frontal intent of the issue and the potential choices to be assessed are situated at the base [4]. The AHP procedure will be utilized to decide the heaviness of the models. Another strategy generally utilized is Order Proclivity Approach by Similarity to Order Preference, this technique is at present used to recognize arrangements that are just conceivable to an exemplary arrangement applying for it some proportion of space, and hence showed arrangements are called settles. Order Proclivity Approach will be utilized in estimating the conveying limit that will fit the various areas to introduce sun based thermal energy stations. The utilization of MCDM has been led in many applications and controls [5], made an audit of the uses of the AHP model incorporated with different method.

2. Multi-Criteria Decision Making techniques (MCDM)

We analyse the choices that we can pick, just as the standards on which said options are to be assessed. This, which from the outset sight is by all accounts basic, frames segment of the entire order that is called multi-criteria decision making (MCDM) [6-8]. MCDM is a system that comprises finest option within a bunch of attainable other options. A MCDM issue with m other options and n measures can be communicated in grid design where A_1; A_2; ···; A_m are achievable other options; C_1; C_2; ···; C_n are assessment models; z_{ij} is the presentation estimation of elective A_i under rule C_j; and w_j is the heaviness of rule C_j. To appropriately decide the heaviness of every rule or calculate included the ultimate result of the subsequent layers, the AHP technique has been utilized inside the MCDM [9]. It is a couple shrewd correlation method of the measures that depends on a square framework where the quantity of lines and sections characteristic the quantity of rules to gauge. The AHP has broadly applied in taking care of an assortment of issues, among which are the applications identified with energy arranging and the conveying limit of environmentally friendly power offices. Then again, to assess different choices as indicated by their appropriateness, the TOPSIS strategy will be utilized. This technique involves activity in getting to know the distances to the ideal as well as the counter ideal points. The philosophy was picked in light of the fact that it doesn't need an appraisal by the master for every one of the other options; they can be assessed straightforwardly from the data set given by the GIS evaluation of every rule for every other option.

2.1. Analytical Hierarchy Process (AHP)

AHP [9] has four core ideas: organizing the intricate choice issue as a progression of objective, standards and choices, set-wise examination of components on every stage of the pecking order concerning every measure on the former level, lastly vertically orchestrating the decisions over the various levels of the chain of importance. AHP depends on a firm hypothetical establishment. The essential hypothesis of AHP might be improved as follows: we accept that we have n unique and autonomous choices ðA_1; A_2; ···; A_{np} and that they have the loads ðw_1; w_2; ···; w_{np}, separately. The decision maker doesn't have a clue about the estimations of w_i; I ¼ 1; 2; ···; n ahead of time, however he/she is fit for making a couple shrewd examination between the various other options. Likewise, we accept that the evaluated decisions given by the leader on sets of choices (A_i, A_j) are addressed in a (n x n) lattice as given in Figure 1.

$$A = \begin{matrix} A_1 \\ A_2 \\ \vdots \\ \vdots \\ A_n \end{matrix} \begin{bmatrix} a_{11} & a_{12} & \dots & a_{1n} \\ a_{21} & a_{22} & \dots & a_{2n} \\ \vdots & \vdots & \dots & \vdots \\ \vdots & \vdots & \dots & \vdots \\ a_{n1} & a_{n2} & \dots & a_{nn} \end{bmatrix}$$

Figure 1. Comparison matrix of AHP

The qualities allotted to a_{ij} as indicated by the scale are ordinarily in the span 1-9 [9]. Albeit the AHP procedure considers the assessment and examination of choices and accordingly the end-product of the choice issues represented. This article will be centered around the underlying advances that will set up the loads of the measures on the grounds that the investigation and assessment of options will be created with the TOPSIS strategy.

2.2. Order Proclivity Approach by Similarity to Ideal Solution method

In MCDM, various choices must be assessed and looked at utilizing a few rules. The point of MCDM is to offer help to the leader during the time spent settling on the decision between options. Along these lines, functional issues are regularly described by a few clashing models, and there might be no arrangement which fulfills all standards all the while. Consequently, the arrangement is a trade-off arrangement as per the chief's inclinations. In this sense, TOPSIS depends on the idea that the picked option ought to have the most limited separation from the PIS (Positive Ideal Solution) and afar from the (NIS) Negative Ideal Solution. Order Proclivity Approach by Similarity to Ideal Solution method comprises of accompanying advances. The algorithm (Figure 2.) of the TOPSIS method is given below.

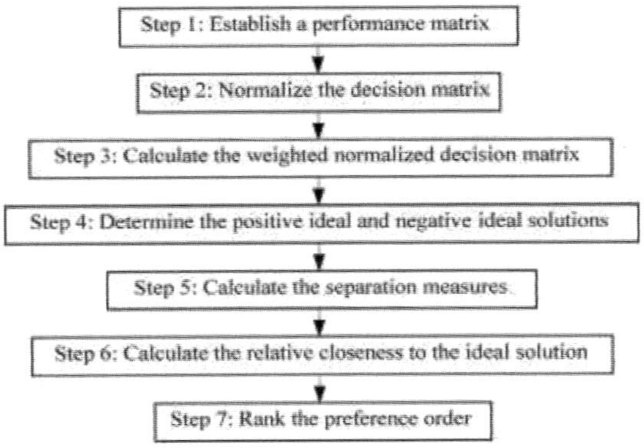

Figure 2. Algorithm of TOPSIS method

3. Problem Identification

3.1. Solar Energy Potential

Solar energy potential is one of most pivotal norm for finding of SPP location. PV-type area energy generation (kWh/year) and sunshine duration (hours) are sub-norm of it. It is necessary to acquire precise data whether SPP is to be established [10]. In this study, amount of sunshine available and energy of PV/area generation of each area is get from Solar Energy Potential Atlas (Figure 3-8). Data of these areas of Sardarpura, Kabir nagar and Mandore. According to it, the priority order among areas for solar energy potential is as follows: Kabir Nagar, Mandore and Sardarpura.

3.2. Feeder Capacity of Distribution Center

Feeder lines actually transmits power from any power station to its corresponding sub station. Data is obtained from the notification of Directorate General of Electricity Transmission Corporation [11]. The feeder capacity examined while installing energy generation facility in a region. Sardarpura has a number of 4, Kabir Nagar has a number of 7 and Mandore has a number of 12 allocated capacity of feeders.

3.3. Surface Slope

By the use of pythagoras theorem, the tangent of the angle will give rise in the water surface. Usually, land with slope higher than 4% have a lower priority because panels shadow next row and affect the efficiency of system adversely [12-13]. The precedence order between areas for slope is as follows: Mandore, Sardarpura and Kabir Nagar.

4. Results and Discussions

By the use of Saaty's Fundamental Scale a matrix is formed for pair wise comparison from 1 to 9 [14-15]. A matrix (m x n) for alternatives using each criterion and matrix (n x n) for criteria where m represents the alternatives available and n represents the criteria available. The sunshine duration (Figure 3,5,7) and the total amount of energy/PV area (Figure 4,6,8) of Sardarpura, Kabir Nagar, Mandore respectively are gives below.

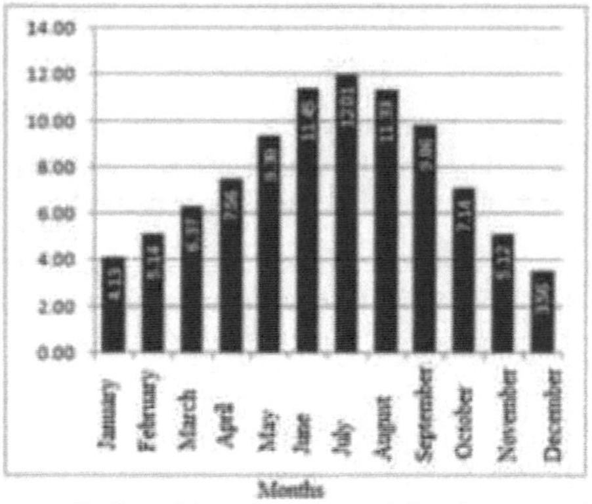

Figure 3. Sunshine duration of Sardarpura [17]

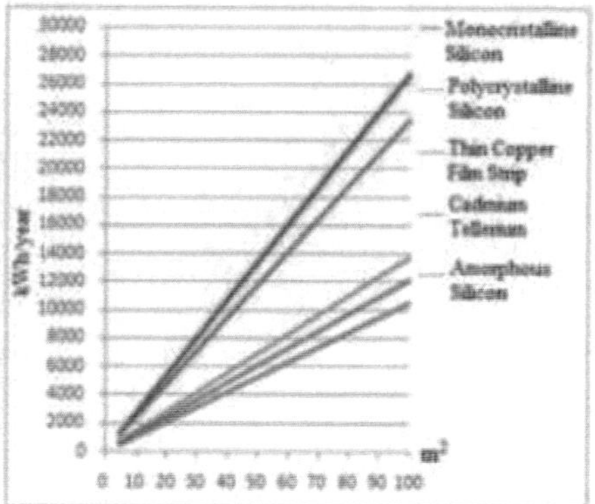

Figure 4. Total amount of Energy/PV area of Sardarpura [17]

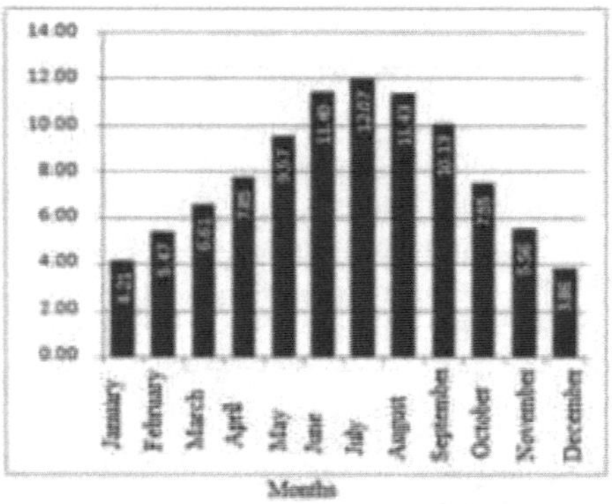

Figure 5. Sunshine duration of Kabir Nagar [17]

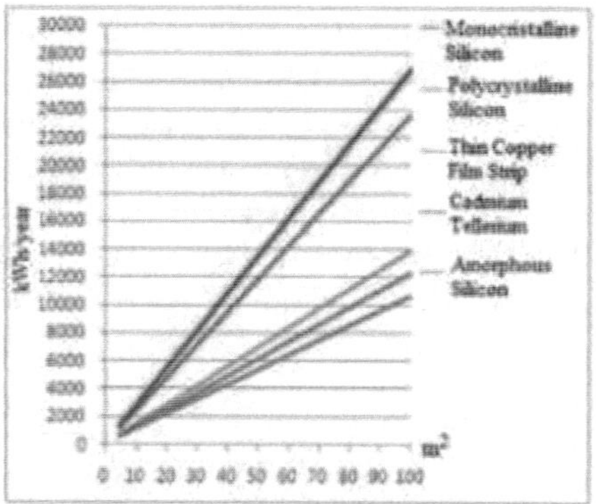

Figure 6. Total amount of Energy/PV area of Kabir Nagar [17]

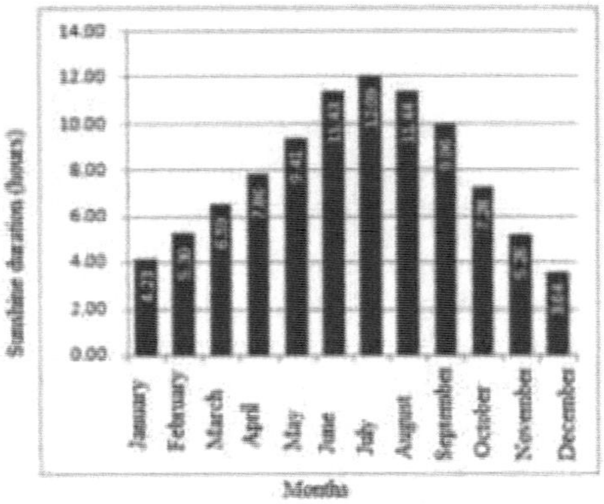

Figure 7. Sunshine duration of Mandore [17]

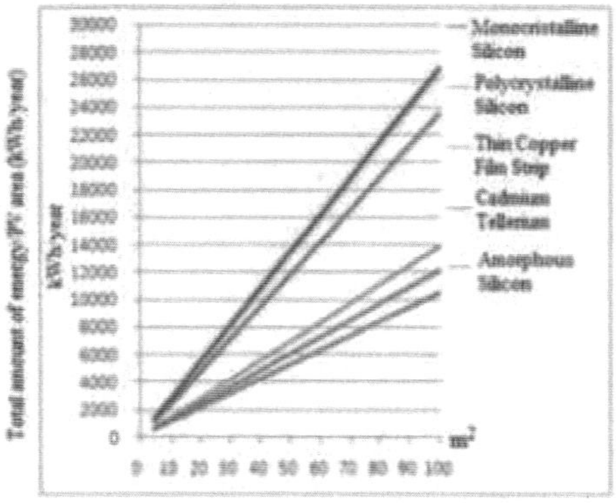

Figure 8. Total amount of Energy/ PV area of Mandore [17]

Sunshine duration of Mandore is highest in the month of July i.e. 12.09 hours (Figure 7.) while in December sunshine duration of Sardarpura is lowest i.e. 3.56 hours (Figure 3.). Apart from this total energy is almost same in all areas. After analyzing all the data by using AHP we found out that the areas situated in Jodhpur region of Rajasthan are most suitable areas and has gigantic capability for establishments of sun based thermal energy station since Jodhpur has one of the greatest level sun powered radiation in the country (Figure 4,6,8). Now, Kabir Nagar has better other criteria like location (distance to road, distance to city, distance to power lines and distance to substation), planning regulations, protected areas, road networks, railways, waterways, mountains, etc. After Kabir Nagar, Mandore and then Saradarpura is best for sun based thermal energy station (Figure 3,5,7). When a developer wants to plant a solar based thermal power plant with specific criteria (sunshine hours, total energy, feeder capacity, surface slope, solar energy potential etc.), the starting point is to select the best area based on such criteria. However, the analysis conducted has weaknesses that could be strengthened by including linguistic labels in the methodology which could be applied in the definition of certain factors whose nature is qualitative (agrological capacity, orientation etc.).

Table 1. Saaty's Fundamental Scale

Intensity of Importance	Definition
1	Equal Importance
2	Weak
3	Moderate importance
4	Moderate plus
5	Strong
6	Strong importance
7	Very strong
8	Very strong plus
9	Extreme strong

Relative weights are calculated by taking average of the rows. Percentage wise distribution of decision points (Table 1) are found by multiplying decision matrix with weighted vector of criteria. Consistency Index (CI) is computed using eequation:

$$CI= \lambda-n/n-1 \quad (1)$$

where λ is Eigen value of paired comparison matrix.

IOP Conf. Series: Materials Science and Engineering 1149 (2021) 012030 doi:10.1088/1757-899X/1149/1/012030

Table 2. RI Values

n	RI	n	RI
1	0	7	1.35
2	0	8	1.40
3	0.52	9	1.45
4	0.89	10	1.49
5	1.11	11	1.51
6	1.25	12	1.54

Now Consistency Ration (CR) is calculated and multiplication of decision matrix and weighted vector of criteria will give percentage distribution of decision points.

CR= CI/RI (2)

Random Index (RI) (Table 2.), if CR is less than 0.1 the comparison will be accepted or else will be up for review [16].

Table 3. Solar Energy Potential comparison matrix

Solar energy potential	Sadarpura	Kabir nagar	Mandore
Sadarpura	1	1/7	1/5
Kabir nagar	7	1	3
Mandore	5	1/3	1

Table 4. Surface slope comparison matrix

Surface slope	Sadarpura	Kabir nagar	Mandore
Sadarpura	1	5	1/3
Kabir nagar	1/5	1	1/7
Mandore	3	7	1

Table 5. Feeder capacity comparison matrix

Feeder capacity	Sadarpura	Kabir nagar	Mandore
Sadarpura	1	1/5	1/7
Kabir nagar	5	1	1/3
Mandore	7	3	1

Table 6. Criteria comparison matrix

Criteria	Sadarpura	Kabir nagar	Mandore
Sadarpura	1	5	3
Kabir nagar	1/5	1	1/2
Mandore	1/3	2	1

Table 3. shows comparison matrix for solar energy potential. Kabir Nagar has the largest capacity for establishing sun based power plant followed by Mandore and Sardarpura respectively. Table 4. shows comparison matrix for surface slope The province with the smallest slope is Mandore has smallest slope followed by Sardarpura and Kabir Nagar respectively. Table 5. shows comparison matrix for allocated maximum feeder capacity Sardarpura has a number of 4, Kabir Nagar has a number of 7 and Mandore has a number of 12 allocated capacity of feeders. Therefore, Mandore is selected as the primary place for feeder capacity criteria followed by Kabir Nagar and Sardarpura respectively. Table 6. shows the comparison matrix for criteria. The order of priority is as solar energy potential, feeder capacity and surface slope, respectively. The weights of each criteria are 0.65 for solar energy potential, 0.12 for surface slope and 0.23 for feeder capacity. The values of CR are calculated as less than 0.10 for all comparison matrices. After all the steps of AHP carried out,

percentage distribution of alternatives is obtained. Kabir Nagar has 49%, Mandore has 41% and Sardarpura has 10% share.

5. Conclusions

By using AHP we found out that the area of Kabir Nagar has the largest capacity for establishing sun based power plant followed by Mandore and Sardarpura respectively. Kabir Nagar, situated in Jodhpur region of Rajasthan is the most suitable city and has gigantic capability for establishments of sun based thermal energy station since Jodhpur has one of the greatest level sun powered radiation in the country. Kabir Nagar, Jodhpur has a territory of 22.1 square kilometers (8.5 sq mi). It is found at 26.29°N, 72.98°E. It has a normal height of 7 meters (22 feet). Mandore is selected as the primary place for feeder capacity criteria followed by Kabir Nagar and Sardarpura respectively.

References

[1] European Commission. Energy for the future: renewable power plants White Book for a Strategy and a Communitarian Plan of Action. COM 1997, 599. Brussels; 1997.
[2] Gómez-López MD, García-Cascales MS, Ruiz Delgado E. Situations and pro blems of renewable energy in the Region of Murcia, Spain. Renewable and Sustainable Energy Reviews 2010;14:1253–62.
[3] Voogd H. Multicriteria evaluation for urban and regional planning. London: Pion; 1983.
[4] Saaty TL. The analytic hierarchy process: planning, priority setting, resource allocation. New York, NY: McGraw-Hill; 437.
[5] Ho W. Integrated analytic hierarchy process and its applications—a literature review. European Journal of Operational Reserarch 2008;186: 211–28.
[6] Hwang CL, Yoon K. Multiple attribute decision methods and applications. Berlin Hiedelberg: Springer; 1981.
[7] Chen SJ, Hwang CL. Fuzzy multiple attribute decision making: methods and applications. Berlin: Springer-Verlang; 1992.
[8] Zeleny M. Multiple criteria decision making. New York: McGraw-Hill; 1982.
[9] Saaty TL. Group decision making and the AHP. New York: Springer Verlag; 1989.
[10] Sindhu, S., Nehra V., Luthra, S., 2017, "Investigation of feasibility study of solar farms deployment using hybrid AHP-TOPSIS analysis: Case study of India", Renewable and Sustainable Energy Reviews, vol. 73, 496-511.
[11]EnergyInstitutewebsite,2017,[Online].Available:http://enerjienstitusu.com/2016/03/21/teias-ges-restrafo-merkezi-kapasiteleri-mart-2016/
[12] Noorollahi, E., Fadai, D., Shirazi, M. A., Ghodsipour, S. H., 2016, "Land Suitability Analysis for Solar Farms Exploitation Using GIS and Fuzzy Analytic Hierarchy Process (FAHP)-A Case Study of Iran," Energies, 9, 643; doi:10.3390/en9080643.
[13] Wheatbelt Development Commission, 2010, "Site Options for Concentrated Solar Power Generation in the Wheatbelt," Final Report, Northam, Australia.
[14] Aragones-Beltran, P., Chaparro-Gonzalez, F., Pastor-Ferrando, J. P., Pla-Rubio, A., 2014, "An AHP (Analytic Hierarchy Process)/ANP (Analytic Network Process)-based multi-criteria decision approach for the selection of solar-thermal power plant investment projects," Energy, vol. 66, 222-238.
[15] Saaty, T. L., 1994, "Fundamentals of decision making and priority theory with the Analytic Hierarchy Process", 1st ed. RWS Publications.
[16] Chen, C., 2006, "Applying the Analytical Hierarchy Process (AHP) Approach to Convention Site Selection," Journal of Travel Research, vol. 45, 167-174.
[17] International Journal of Energy Applications and Technologies, Year 2017, Vol. 4, No.3, pp. 122-128.

Optimization of sand- defects process parameters for reducing its defect using taguchi technique

Vikas Yadav[1], Gaurav Kumar[1*], Mukesh Kumar[1] and Peeyush Vats[2]

[1]Vidya College of Engineering, Department of Mechanical Engineering, Meerut-251001, India.
[2]IIMT College of Engineering, Department of Mechanical Engineering, Greater Noida-201308, India.

gaurav.me86@gmail.com

Abstract. Foundry or casting manufacturing process is largely used in manufacturing of automobile and general-purpose machine components. Still this method is popular due to low cost, easiness in manufacturing and easy availability of input materials and resources. It involves a number of processes like sand preparation, pattern making, melting, pouring, cooling etc. Due to influence of so many factors defect arises by one factor or many factors. Still trial and error method are used by casting units to remove defects. Therefore, a more disciplined approach is required to find the root cause of problem. In this paper taguchi's method and design of experiment is used to analyze sand and mould related defects. Standardization (optimization) attempt is done to obtain optimum values of process parameters by taguchi's method whose experimental approach of design is robust. The parameters affecting the process are chosen as sand grain size, clay content, moisture percentage, pouring temperature, pouring time, green strength, permeability, mould hardness, number of vent holes, number of ramming etc. An orthogonal array L_{27} is selected which is based on taguchi. In this study a cast iron alloy (FG260) retainer used in tractor is focused.

1. Introduction

Sand casting is a centuries old manufacturing process. It successfully serves the requirements of industries. In the last few decades so many different types of casting are fulfilling different demands of transportation and agricultural sector. This process can manufacture a very complex shape with great accuracy. Various components manufactured by green sand casting are affected by different parameters. These parameters influence the quality of product. Parameters are of two types one is controllable parameter while another is uncontrollable also called noise factor.

Kassie A A and Assfaw S B [1] have focused gas and shrinkage defect to optimize. In process parameters binder ratio, mold permeability, pouring temperature and deoxidant are studied. Here factorial experiment is used. Various inspection done are process inspection, visual inspection, chemical composition inspection etc. Muzammil M et al [2] have optimized gear blank casting by taguchi's robust design technique. The parameters selected are moisture content, sand grain size, fluidity of metal, gating design ramming & clay content. Kumaravadivel A and Natarajan U [3] used process window approach for optimization of control factors. They have taken help of taguchi's method for analyzing critical parameters influencing the process. The results are confirmed in foundry which were obtained by taguchi's method & RSM. Process window approach helps in approving final process parameters. Dabade U A and Bhedasgaonkar R C [4] optimized the process parameters of a casted part is done which is casted through green sand molding process. Different critical parameters selected are mould hardness (horizontal), green compression strength, permeability & moisture content. For experimentation Taguchi method is adopted having L_{18} orthogonal array. ANOVA is used for analysis. Minitab software helps in analysis of means plot also. Different types of defects i.e., sand related defect, method related defect, solidification and filling related defects are optimized. Change in design of gating & feeding reduces shrinkage porosity by 15%. Kumar S et al [5] have used six sigma approach to reduce green sand-casting defects. The methodology used is in sequence of defining, measuring, analyzing, improving & controlling Six sigma approach helps in defect minimization at minimum cost. It also gives optimal levels of factors for which responses are affected by noise factors. Manikanda Prasath K et al [6] have used simulation and taguchi experimental design for better yield of casting. In the second section, the gating system was optimized to reduce the shrinkage defect and weight of the gating system. Tiwari S K et al [7] studied the castings process of mild steel used in big foundry. Quality concerns are improved by optimization of process parameters of green sand casting. Signal to noise ratio in ANOVA analysis helps in finding optimum value of parameter. Also, each factor's contribution towards defects is determined. Choudhari C M et al [8] have studied about casting of cover plate (square shaped) which is suffering from shrinkage defect. Simulation is done for improvement in feeding and other important areas using a software known as AUTOCAST-X, the cover plate material was LM6 and silica sand was taken as mould material. Designing of pattern, gating and feeder is done. Solidification related defects such as hot spots etc are identified. Bentonite is used which helps in retaining moisture and also holds sand firmly. Reilly C et al [9] have studied about oxide formation on the free surface of castings and their modeling. Optimization is done by software related to computational fluid dynamics. Defects are modeled and quantified. This study is helpful in complex shaped products. This relates to porosity defect causes analysis. Nimbulkar S L and Dalu R S [10] studied defects like porosity and incomplete filling gating and riser system design is done. Then it is numerically simulated using AUTO-CAST X1 software. Then validated with experimental results. After that ultrasonic testing is done and results are compared. Simulation showing the process of solidification gives stages of freezing from inner region of casting to external environment. It also helps to identify regions solidified in last. It also identifies metal flow rates, hot spots etc., and vertical gating and feeding systems are not up to the mark for thick castings because of many defects of solidification present in the casting. In their proposed work the horizontal gating and feeding systems and the riser are placed symmetrically, and the flow is uniform and the gases escape comfortably into the environment. Khandelwal H and Ravi B [11] have investigated and optimized chemically bonded sand than green sand. Through the lab experimentation they have decided the effect on mechanical properties such as core hardness, compression strength and shear strength by the size of sand grain, percentage of binder and the curing time according to the three different levels of given parameters. Raghupathy R and Amirthagadeswaran K S [12] completed the optimization of process parameters of a grey cast pump adapter is done by DOE and CAE. optimized parameters are mold hardness, moisture content, clay content etc. each parameter is having is having three levels. Feeding and gating parameters are redesigned by using a simulation software PROCAST.

A plethora of research on the impact of process parameters on sand casting defects have been published in literature surveys. A research gap is an unfilled area where the previous research has not paid attention. It may be the method of research, method of collecting the data, selection of factors, method of sampling, selection of software etc. Sand casting is affected by so many parameters such as pouring temperature, pouring time etc. In this study some more significant parameters and selection of levels are considered for better output. Here signal to noise ratio of taguchi's method helps in determining effect of uncontrollable factors or noise factors. In this paper design of experiment is used to identify causes of defects in a product called retainer used in tractor's axle. Also, optimum values of parameters are found out.

2. Experimental Details

Among the different types of defects analysis methodologies taguchi method which comes from design of experiments is selected for analysis purpose. Here different defects of different nature and types i.e.

gas related defects, sand related defects and mould related defects are analyzed. Taguchi method of design of experiments is suitable when a number of process parameters is involved. Taguchi method is considered as robust as in this method variations are minimum and it is not affected by variation in component or outside conditions [13]. The success of taguchi's method lies in proper selection of control factors and their levels. A set of experiments is done with the help of orthogonal arrays. In this research work defects of a product named "Retainer" used in escort tractors is studied. This involves different parameters at different levels which affects the quality. In sand casting different types of defects such as gas defect or sand related defects are observed. The percentage rejection is reduced by setting different process parameters at different levels and optimum values are found.

2.1 Material selection and problem identification

The company Pragati Techno cast is a private limited company, which is the manufacturer and supplier of tractor parts. There is one of the tractor parts called as retainer made of cast iron (FG 260). The nominal composition of FG 260 is: Carbon-3.5%, Silicon- 1.8%, Sulphur- 0.15%, Phosphorous-0.15, Manganese- 0.9%. The total number of retainers produced in a single day is approximately 40 and an average of 4 retainers got rejected due to defects as shown in Figure 1. Due to this the productivity and the quality of this retainer was low. So, there is a need to reduce the number of rejections of retainers.

Figure 1. Retainer

2.2 Setting the control factors with their levels

Finding the important factors affecting the process with suitable range of levels is the next step in defect analysis, based on the brainstorming with the group of people like lab-in-charge, quality manager, furnace supervisor, workers working at furnace and mold checking supervisor. The parameters affecting the process and their significance in the green sand moulding and casting of "Retainer" is illustrated as follows in Table 1.

Table 1. Parameters affecting the sand-casting process with their levels

Control Factors	Range of Factors	First Level	Second Level	Third Level
Water Content (A)	4-8	4	6	8
Sand grain size (B)	60-70	60	65	70
Clay Content (C)	12-20	12	16	20
Pouring Temp. (D)	1385-1430	1385	1410	1430
Pouring Time (E)	20-25	20	23	25
Green Strength (F)	1000-1500	1000	1250	1500
Permeability (G)	230-270	230	250	270
Mold Hardness (H)	65-80	65	75	80
Vent Holes (I)	8-10	8	9	10
Ramming (J)	2-6	2	4	6

2.3 Experimental Design

Orthogonal arrays give the no. of experimentations. In factorial design no. experiments becomes very large. Hence fractional factorial suggested by taguchi is followed. In full factorial design if x factors are considered at y levels, then total number of experiments are y^x. Orthogonal arrays conduct a limited number of experimentations to know the most important factors affecting the process. It is calculated by equation 1:

$$X = 1 + Y\,(L\text{-}1) \tag{1}$$

where X is no. of experiments to be conducted. Y = no. of parameters. L = no. of levels.

Based on calculation orthogonal array is selected. After selecting orthogonal array experiments are conducted. Here L_{27} orthogonal array is selected as per calculations. Table 2 shows the orthogonal array L_{27}[14].

Table 2. Design of Experiment with L_{27} Orthogonal array

Trial no.	(A)	(B)	(C)	(D)	(E)	(F)	(G)	(H)	(I)	(J)	% Defects
1	4	60	12	1385	20	1000	230	68	8	2	7.77
2	4	60	12	1385	23	1250	250	75	9	4	7.51
3	4	60	12	1385	25	1350	270	80	10	6	7.34
4	4	64	16	1410	20	1000	230	75	9	4	7.40
5	4	64	16	1410	23	1250	250	80	10	6	5.64
6	4	64	16	1410	25	1350	270	68	8	2	8.64
7	4	70	19	1430	20	1000	230	80	10	6	6.20
8	4	70	19	1430	23	1250	250	68	8	2	7.55
9	4	70	19	1430	25	1350	270	75	9	4	6.90
10	6	60	16	1430	20	1250	270	68	9	6	6.80
11	6	60	16	1430	23	1350	230	75	10	2	6.80
12	6	60	16	1430	25	1000	250	80	8	4	7.40
13	6	64	19	1385	20	1250	270	75	10	2	6.30
14	6	64	19	1385	23	1350	230	80	8	4	6.90
15	6	64	19	1385	25	1000	250	68	9	6	8.10
16	6	70	12	1410	20	1250	270	80	8	4	6.30
17	6	70	12	1410	23	1350	230	68	9	6	8.07
18	6	70	12	1410	25	1000	250	75	10	2	8.40
19	8	60	19	1410	20	1350	250	68	10	4	7.97
20	8	60	19	1410	23	1000	270	75	8	6	7.11
21	8	60	19	1410	25	1250	230	80	9	2	7.30
22	8	64	12	1430	20	1350	250	75	8	6	7.01
23	8	64	12	1430	23	1000	270	80	9	2	8.07
24	8	64	12	1430	25	1250	230	68	10	4	7.90
25	8	70	16	1385	20	1350	250	80	9	2	6.90
26	8	70	16	1385	23	1000	270	68	10	4	7.54
27	8	70	16	1385	25	1250	230	75	8	6	7.42

3. Results and discussion

Experimental works has been carried out in a foundry unit which is engaged in manufacturing cast iron products for different automobile and other sectors. From discussion with factory personnel the average rejection per month in the foundry is 8-10 %. The rejection in the product retainer is not visible after casting but holes are observed after machining on lathe machine. Here small cavities are observed during machining.

3.1 Analysis of Signal –to- Noise ratio

The robustness of design is indicated by 'Signal to noise ratio'. Target value is given by signal while unwanted value is noise. System's robustness is evaluated by higher 'S/N ratio'. It is also a measure of good quality product. The table below shows signal to noise ratio values for different factors. Taking "lower the better" the 'signal to noise ratio' is calculated by equation 2 [15]:

$$S/N = -10 \log x \left(\sum (1/Y_i^2)\right)/n \qquad (2)$$

Where, n = Number of repetitions and Y_i = value of quality characteristics at ith experimental run. The S/N ratio and mean for 27 trials are shown in Table 3.

Table 3. S/N Ratio and Mean

Trial no.	A	B	C	D	E	F	G	H	I	J	K	S/N ratio	Mean
1	4	60	12	1380	20	1000	230	68	8	2	7.77	-17.8084	7.77
2	4	60	12	1380	23	1200	250	75	9	4	7.51	-17.5128	7.51
3	4	60	12	1380	26	1400	270	82	10	6	7.34	-17.3139	7.34
4	4	64	16	1410	20	1000	230	75	9	4	7.40	-17.3846	7.40
5	4	64	16	1410	23	1200	250	82	10	6	5.64	-15.0256	5.64
6	4	64	16	1410	26	1400	270	68	8	2	8.64	-18.7303	8.64
7	4	68	20	1440	20	1000	230	82	10	6	6.20	-15.8478	6.20
8	4	68	20	1440	23	1200	250	68	8	2	7.55	-17.5589	7.55
9	4	68	20	1440	26	1400	270	75	9	4	6.90	-16.7770	6.90
10	6	60	16	1440	20	1200	270	68	9	6	6.80	-16.6502	6.80
11	6	60	16	1440	23	1400	230	75	10	2	6.80	-16.6502	6.80
12	6	60	16	1440	26	1000	250	82	8	4	7.40	-17.3846	7.40
13	6	64	20	1380	20	1200	270	75	10	2	6.30	-15.9868	6.30
14	6	64	20	1380	23	1400	230	82	8	4	6.90	-16.7770	6.90
15	6	64	20	1380	26	1000	250	68	9	6	8.10	-18.1697	8.10
16	6	68	12	1410	20	1200	270	82	8	4	6.30	-15.9868	6.30
17	6	68	12	1410	23	1400	230	68	9	6	8.07	-18.1375	8.07
18	6	68	12	1410	26	1000	250	75	10	2	8.40	-18.4856	8.40
19	8	60	20	1410	20	1400	250	68	10	4	7.97	-18.0292	7.97
20	8	60	20	1410	23	1000	270	75	8	6	7.11	-17.0374	7.11
21	8	60	20	1410	26	1200	230	82	9	2	7.30	-17.2665	7.30
22	8	64	12	1440	20	1400	250	75	8	6	7.01	-16.9144	7.01
23	8	64	12	1440	23	1000	270	82	9	2	8.07	-18.1375	8.07
24	8	64	12	1440	26	1200	230	68	10	4	7.90	-17.9526	7.90
25	8	68	16	1380	20	1400	250	82	9	2	6.90	-16.7770	6.90
26	8	68	16	1380	23	1000	270	68	10	4	7.54	-17.5474	7.54
27	8	68	16	1380	26	1200	230	75	8	6	7.42	-17.4081	7.42

The response table for means is shown in Table 4

IOP Conf. Series: Materials Science and Engineering 1149 (2021) 012031 doi:10.1088/1757-899X/1149/1/012031

Table 4. Response Table for Means

Level	A	B	C	D.	E	F	G	H	I	J
1	7.217	7.333	7.597	7.309	6.961	7.554	7.307	7.816	7.344	7.526
2	7.230	7.329	7.171	7.426	7.243	6.969	7.387	7.206	7.450	7.313
3	7.469	7.253	7.148	7.181	7.711	7.392	7.222	6.894	7.121	7.077
Delta	0.252	0.080	0.449	0.244	0.750	0.586	0.164	0.921	0.329	0.449
Rank	7	10	5	8	2	3	9	1	6	4

3.2 ANOVA analysis

Analysis of variance (ANOVA) gives relative effect of different parameters. Contribution of each factor on the response collectively is calculated by ANOVA. F-ratio gives contribution of each factor & its importance compared to other factors. F – Test value at 95% confidence level helps in deciding important parameters controlling or influencing the process. The F-Statistics is defined as the ratio of variation between the samples to the variations within the sample. The F statistic is dependent on the proportion of the average square. The term "average square" can be more confounding, but it is only an estimate of population differences that accounts for the degrees of freedom (DOF) used to calculate that estimate.

Table 5. Rejection analysis at 95% confidence limit using ANOVA

Factors	DOF	SS	MS	F ratio	P value	Contribution	Best level
A	2	0.3626	0.18129	0.68	0.540	Significant	4%
B	2	0.0364	0.01819	0.07	0.934	Non-significant	
C	2	1.1494	0.57471	2.17	0.196	Significant	20%
D	2	0.2691	0.13454	0.51	0.626	Non-significant	
E	2	2.5829	1.29145	4.87	0.055	Significant	20 sec
F	2	1.6452	0.82260	3.10	0.119	Significant	1200 kg/cm^2
G	2	0.1217	0.06086	0.23	0.802	Non-significant	
H	2	3.9520	1.97600	7.45	0.024	Significant	82
I	2	0.5076	0.25378	0.96	0.436	Significant	10
J	2	0.9077	0.45383	1.71	0.258	Non-significant	
Error	6	1.5920	0.26533				
Total	26	13.1265					

Higher value of F-value informs that change in the process parameters have a good influence on performance characteristic and more significant the parameter will be. In the ANOVA at a given confidence level if a particular factor's value becomes below tabulated F-value for that confidence level, then that factor or parameter is not important and that parameter is pooled. The rejection analysis at 95% confidence level is shown in Table 5. The significant and non-significant process parameters may be separated out on the basis of Table 5.

The plots of S-N ratio have been represented in the Figure 2. The greater values of S/N ratio identify the controlling factors which mitigates the effects of the noise factors. A two-stage optimization process offered by Taguchi experiments. The signal-to-noise ratio are identified the control factors which decreases the variability stage 1. The various controlling factors are identified which move the mean to target and have a negligible effect on the S/N ratio in stage 2. Under different noise conditions, the S/N ratio decides how the response varies relative to the nominal or target value.

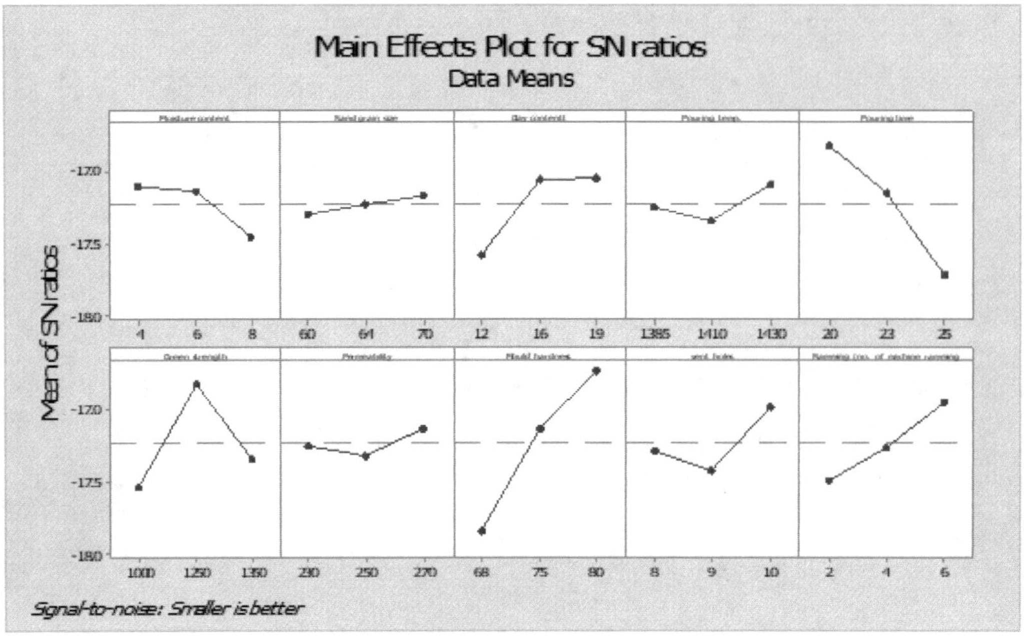

Figure 2. S-N ratio plots

There are 10 process parameters and each process parameter has 3 different levels. Figure 2 depicts the variation in process parameters along with the level. According to the Figure 2, the moisture content reduces as the level increases. It is clear that the moisture content is a significant factor in sand-casting process as the value of S/N ratio decreases the percentage composition of moisture content is also decreasing. From level 1 to 2, the moisture content decreases slowly but from level 2 to level 3, the moisture content reduces drastically. In this study, smaller is the better is taken for S/N ratio. From the above figure it can be concluded that level 3 is best for moisture content. The sand grain size increases as its level increases. The clay content increases from level 1 to level 2 and from level 2 to level 3, it is almost regular as it can be observed that the S/N ratio increases from level 1 to level 2. It is also observed that initially the pouring temperature decreases and then increases. The similar effect has been observed on the green strength also, initially from level 1 to level 2, the green strength is increasing then after it is decreasing from level 2 to level 3. In the case of clay content level 1 is found to the best. In the case of pouring time the level 3 is most accepted. Similarly, the variation in each process parameter can be explained with the change in its level.

4. Conclusions

There are two types of finding in this study. First one is to sort out the contributing and non-contributing process parameters and the second one is the optimized process parameters to minimize the sand casting defects. The results of this study, show that water content, clay content, pouring time, Green strength, mould hardness and ventholes are the contributing parameters while sand grain size, pouring temp, permeability and ramming are the non-contributing factors and the optimized values of process parameters of the sand found by taguchi method are: Moisture content-4%, sand particle size – 70, clay content-18%, pouring temperature-1430°C, pouring time- 20 seconds, green strength- 1350 g/cm^2, number of vent holes- 10, number of ramming- 6. By adopting taguchi optimization method, the percentage rejection of castings is reduced from 8 % to 5.095%.

References

[1] Kassie A A and Assfaw S B 2013 Minimization of casting defects *IOSR journal of engineering (IOSRJEN)* **3(5)** pp 31-38

[2] Muzammil M, Singh P P and Talib F 2003 Optimization of gear blank casting process by using Taguchi's robust design technique *Quality Engineering* **15(3)** pp 351-359
[3] Kumaravadivel A and Natarajan U 2013 Optimization of sand-casting process variables—a process window approach *The International Journal of Advanced Manufacturing Technology* **66(5-8)** pp 695-709
[4] Dabade U A and Bhedasgaonkar R C 2013 Casting defect analysis using design of experiments (DoE) and computer aided casting simulation technique *Procedia Cirp* **7** pp 616-621
[5] Kumar S, Satsangi P S and Prajapati D R 2013 Improvement of Sigma level of a foundry: a case study *The TQM Journal*
[6] Manikanda Prasath K, Madesh R, Birundha P and Pranav R 2014 Effective Yield Improvement of Casting using Taguchi Technique and Computer Aided Casting Simulation Technique pp 75-81
[7] Tiwari S K, Singh R K and Srivastava S C 2016 Optimisation of green sand-casting process parameters for enhancing quality of mild steel castings. *International Journal of Productivity and Quality Management* **17(2)** pp127-141
[8] Choudhari C M, Narkhede B E and Mahajan S K 2014 Methoding and simulation of LM 6 sand casting for defect minimization with its experimental validation *Procedia Engineering* **97** pp 1145-1154
[9] Reilly C, Green N R and Jolly M R 2013 The present state of modeling entrainment defects in the shape casting process *Applied Mathematical Modelling* **37(3)** pp 611-628
[10] Nimbulkar S L and Dalu R S 2016 Design optimization of gating and feeding system through simulation technique for sand casting of wear plate *Perspectives in Science* **8** pp 39-42
[11] Khandelwal H and Ravi B 2016 Effect of molding parameters on chemically bonded sand mold properties *Journal of Manufacturing Processes* **22** pp127-133
[12] Raghupathy R and Amirthagadeswaran K S 2017 Parametric optimization and solidification analysis of grey cast pump adapter castings using DOE and CAE–a case study *Australian Journal of Mechanical Engineering* **15(1)** pp 27-35
[13] Singh O P, Kumar G and Kumar M 2019 Role of Taguchi and grey relational method in optimization of machining parameters of different materials: a review *Acta Electronica Malaysia (AEM)* **3(1)** pp19-22
[14] Satishkumar P, Dharmalingam S, Raja K, Lingadurai K and Padmanaban G 2015 Investigation on Electrochemical Micro Machining of Al 6061-6% wt Gr based on Taguchi design of experiments *Int J Chem Tech Res* **7(1)** pp 203-211
[15] Kumar G, Kumar M and Tomer A 2021 Optimization of End Milling Machining Parameters of SS 304 by Taguchi Technique in *Recent Advances in Mechanical Engineering Springer Singapore* pp 683-689

Application of taguchi method coupled with GRA for optimization of drilling process parameters

KhushbooSharma[1], Gaurav Kumar[1] and Mukesh Kumar[1*]

[1]Vidya College of Engineering, Department of Mechanical Engineering, Meerut-251001, India.

mukesh.ankwan@gmail.com

Abstract. In present article, an experimental study is performed for optimizing the machining parameters of AA6082. The control factors like cutting fluid (present - absent), speed (520 rpm – 420 rpm), feed rate (120 mm/min – 80 mm/min) and hole-depth (16 mm – 24 mm) are taken into account to optimize the material removal rate and surface roughness. Taguchi's L_{18} orthogonal array was chosen for experimental trials. Grey relational analysis is employed to find, which process parameters affect the surface finish most with their percentage contribution individually. The results of ANOVA show that among all the control factors, feed rate (74.82%) is the highest contributing factor followed by cutting fluid (11.42%). This study also shows that by using the taguchi approach, the surface finish of the machining process can be upgraded.

1. Introduction

For the sake of cost-effectiveness, increasing the quality grade of the machined component is the biggest difficulty in the global manufacturing industry. Over the past few years, the demand for the good surface finish in various machining operations has increased. Surface quality became a significant factor to judge the productivity of machined components.

Various machining operations like drilling, turning, milling etc. are widely used in manufacturing industries. Among these various machining processes Drilling is a versatile machining process used in machining shops and industries to get the desired product. Material is removed using a rotating cutting tool called drill bits in the form of burr in this process. Bit tool rotating at high speed is used to make a hole in solid material. A good surface finish is desired in every production process. It is one of the most important criteria for the evaluation of quality of the drilled hole. The small roughness can lead to affect the process and decrease the reliability of the product. This roughness factor gives a continuous need to study the optimization techniques of various parameters in the machining process to minimize the manufacturing cost and maximize the quality in terms of roughness.

Andrey Belyakov [1] discussed the mechanical behavior of various structural metals and their alloys. Various investigations regarding the micro structural changes in metallic materials when they are subjected to various machining processes are shown. Om Prakash Singh et al [2] performed an experiment by applying the GRA method to optimize the shoulder milling operation parameters. The study represents that in the optimization of cutting parameters of Al 6063, feed rate contributes the most. Arshad Noor Siddiquiee et al [3] emphasized on optimizing the drilling operation with the help of the taguchi approach. The experimental results revealed speed is the most important factor for surface roughness. The taguchi technique has been widely used in various analyses and by applying it, the time required for investigations can be saved. Avinash A Thakre and Julie Z Zhang et al [4-5] used the taguchi technique to optimize the surface quality in different machining processes. Adem Cicek et al [6] investigated the drilling parameters and observed the effects on circularity lapse in drilling of SS 316 by taguchi technique under cryogenic treatment and observed that in reduction of roughness and hole size errors speed is a most leading factor. G Campatelli et al [7] studied the cutting forces of the milling process for aluminium 6082 and tried to develop a cutting force model for tangential cutting coefficient. Akhil K.T et al [8] tried to optimize the drilling characteristics using GRA. Glass reinforced polymer is chosen as material after noticing the increment in its use in automotive industry. M Subramanian et al [9] made a statistical model for investigation of cutting forces using RSM. Aluminium 7075 T6 was shoulder milled and a dynamometer was used to measure cutting forces. M S Sukumar et al [10] applied the taguchi approach for identification of the optimum conditions in milling of aluminium 6061. Lirong Zhou et al [11] suggested a multi-objective model to minimize the operation time and energy consumed for the end milling process of AISI 1045. Mohammad Aamir et al [12] reviews the details on drilling forces, drill tool geometry, drilling parameters, chip formation, hole metrics, analysis of tool wear, surface roughness, drill tool material and burr formation during machining of aluminium alloys. Gaurav Kumar et al [13] tried to optimize the end milling machining parameters while milling SS304. It is observed that coolant is the highest contributing factor. Mukesh Kumar et al [14] tried to optimize the machining parameters of deep drilling of SS321 by using the taguchi technique with GRA. The results revealed an improvement of 5.56%.

The literature survey shows that the Aluminium is the first choice of each and every industry especially for electrical and aerospace industry. Al 6082 is replacing 6061 in major industries. Rapidly increasing usage of 6082 is the major reason for selection of this material as a work piece. But there are certain problems related to machining of aluminium as it is a soft metal and selection of right cutting tool, cutting speed, evacuation of chip formed etc is important. While drilling of aluminium, heat generation is very large which lead to rubbing of tool with the material. This led to sticking of tool with the work piece and gives very poor surface finish and increases the rejection rate. In this article, the effects of various drilling parameters like cutting fluid, feed rate, hole-depth and speed on surface roughness were studied. The optimization was done using the taguchi technique.

2. Experimental Details

2.1 Work piece Material

Machined AA 6082 was used for the experiment. The sheet of 200×200×25 mm is taken to carry out the experiment which is shown in the Figure 1 a piece of 24 mm×24 mm is cut out from the sheet for composition test. The chemical constituents of the AA6082 is also shown below in Table 1.

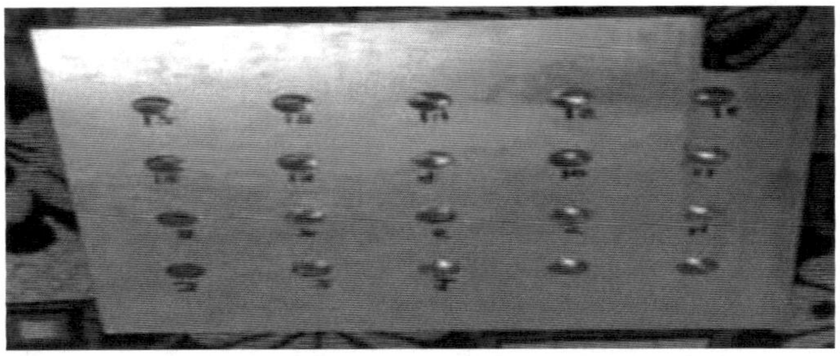

Figure 1. AA6082 Sheet

Table 1. Chemical constituents of AA6082 Material

Element	% Constituents
Silicon	0.7 65
Iron	0.312
Copper	0.024

Manganese	0.492
Magnesium	0.607
Zinc	0.029
Titanium	0.032
Chromium	0.030
Aluminium	Balance
Lead	0.001
Tin	0.005
Nickel	0.012

2.2 VMC Machine

All the experiments were performed on the AA6082 sheet by the VMC machineavailable at Anushree Electricals Pvt. Ltd. Meerutas shown in Figure 2 and Mitotoyo tester was used for testing the surface roughness as shown in Figure 3.

Figure 2. VMC Machine

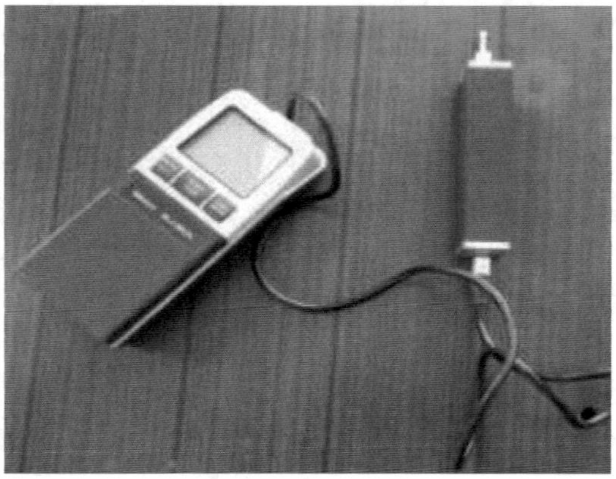

Figure3. Mitotoyo Tester

The working conditions of the experiment are shown below in Table 2.

Table 2. Operating conditions

Conditions	Specifications
Specimen	AA 6082
Size of specimen	200mm×200mm×25mm
Diameter of drilled holes	11 mm
Machine	VMC milling machine
Tool	End mill 11mm
Measuring instrument	Mitotoyo tester
Cutting fluid	Sherol BN

Work piece was cut into 5 pieces after machining with the aid of a power hacksaw to obtain the surface roughness measurements of eighteen drilled holes. The cross section of drilled holes after machining is shown in the Figure 4 and 5.

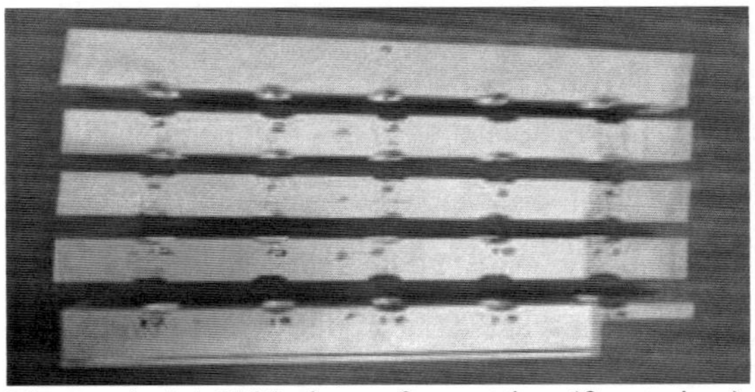

Figure 4. AA6082 sheet after cutting (front view)

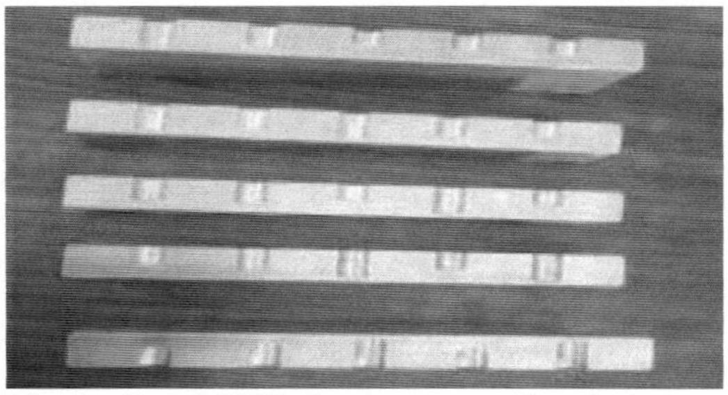

Figure 5. AA6062 sheet cross-sectional view

2.3 Experimental Design

L_{18} orthogonal array is chosen for the design of experiment. On a VMC milling machine, all of the experiments were carried out. To find the optimized combination of multi performance factors, ANOVA was used. The four main machining parameters like cutting fluid, speed, feed and hole depth and feed were observed during drilling process as shown in Table 3.

Table 3. Process parameters and their levels

	Process Parameters	Levels		
		1	2	3
A	Cutting fluid	Present	Absent	-
B	Speed (rpm)	520	470	420
C	Feed (mm/min)	120	100	80
D	Hole – Depth (mm)	16	20	24

All 18 experiments were done according to taguchi's experimental design. Mitotoyo's tester was used to determine the roughness of drilled holes. The sampling length was taken 5.5 mm. MRR is calculated by equation 1 as given below.

$$\text{MRR} = \frac{\pi \times D \times D \times F}{4 \times 60}\ \text{mm}^3/\text{sec} \quad (1)$$

F = feed rate (mm/min); D = diameter of drilled hole (mm), diameter of drilled holes is kept constant i.e., 11mm.

The calculated material removal rate and measured roughness evaluation is given in the Table 4.

Table 4. Values of MRR and roughness

EXPERIMENT NO.	A	B	C	D	Ra(μm)	MRR (mm³/sec)
1	1	1	1	1	1.7	189.97
2	1	1	2	2	0.875	158.3
3	1	1	3	3	1.5	126.64
4	1	2	1	1	0.865	189.97
5	1	2	2	2	1.76	158.3
6	1	2	3	3	1.205	126.64
7	1	3	1	2	1.21	189.97
8	1	3	2	3	0.715	158.3
9	1	3	3	1	1.35	126.64
10	2	1	1	3	1.51	189.97
11	2	1	2	1	1.475	158.3
12	2	1	3	2	1.96	126.64
13	2	2	1	2	1.685	189.97
14	2	2	2	3	2.2	158.3

15	2	2	3	1	1.69	126.64
16	2	3	1	3	2.65	189.97
17	2	3	2	1	1.92	158.3
18	2	3	3	2	2.325	126.64

3. Analysis method

3.1 Analysis of Signal to Noise Ratio

The influential factors like noise factors and design parameters influence the product quality in taguchi's design method. The Signal-to-Noise ratio takes the account of both; variability of results and mean, and generally depends on the characteristics to be optimized. The signal to noise ratios used in analysis are "nominal-the-better, lower-the-better and higher-the better". The detailed equations are available in [15].

3.2 Grey relational generation

Grey relational generation is also called normalization. While using grey relational analysis one has to be cautious about preprocessing of data, as this analysis might give false conclusion if there exists a difference in factors, directions and goals. The transferring of real sequence to a parallel sequence is called data preprocessing. So, the standardization or normalization of the results are done in the range (0 - 1). The detailed equations are given in [16]

3.3 Normalization

After standardization, the grey coefficient is evaluated with the help of processed data to show the association of desired and real results which can be depicted as

$$\xi_i(y) = \frac{\Delta_{minimum} + \zeta . \Delta_{maximum}}{\Delta_{oi}(y) + \zeta . \Delta_{maximum}} \quad (2)$$

Where $\Delta_{0i}(y)$ is the Deviation order of reference series $x_0^*(y)$ and comparability order$x_i^*(y)$, named as:

$$\Delta_{0i}(y) = \|x_0^*(y) - x_i^*(y)\| \quad (3)$$

$$\Delta_{max} = \max_{\forall j \in i} \min_{\forall y} \|x_0^*(y) - x_i^*(y)\| \quad (4)$$

$$\Delta_{minimum} = \max_{\forall j \in i} \min_{\forall y} \|x_0^*(y) - x_i^*(y)\| \quad (5)$$

ζ is distinctive coefficient: $\zeta \varepsilon [0,1]$. Normally $\zeta = 0.5$ is taken.

After calculating the grey coefficient, ANOM is done to attain the grey relational grade. ANOM is the average of the normalized data. To find the grey relational grade, equation is given below

$$\gamma_i = \frac{1}{n}\sum_{y=1}^{n} \xi_i(y) \quad (6)$$

In actual scenario, the influence of each factor is different. So, equation (6) can be modified as:

$$\gamma_i = \frac{1}{n}\sum_{y=1}^{n} w_y \xi_i(y) \sum_{y=1}^{n} w_y \quad (7)$$

where, w_y = the standardized weight of factor y.

4. Results and discussions

The grey relational grade and coefficient obtained by analysis are investigated and presented in Table 5 and the optimization of drilling parameters is done.

Table 5. Values of grey relational grade

Experiment No.	Grey relational coefficient		Grade	Order
	Surface roughness	MRR		
1	0.5956	1.0000	0.7978	2
2	0.3715	0.5265	0.4490	14
3	0.5348	0.3333	0.4341	15
4	0.3691	1.0000	0.6845	6
5	0.6150	0.5265	0.5707	10
6	0.4537	0.3333	0.3935	18
7	0.4550	1.0000	0.7275	5
8	0.3333	0.5265	0.4299	16
9	0.4925	0.3333	0.4129	17
10	0.5377	1.0000	0.7688	4
11	0.5276	0.5265	0.5270	11
12	0.6840	0.3333	0.5087	12
13	0.5908	1.0000	0.7954	3
14	0.7777	0.5265	0.6521	7
15	0.5924	0.3333	0.4629	13
16	1.0000	1.0000	1.0000	1
17	0.6696	0.5265	0.5980	8
18	0.8323	0.3333	0.5828	9

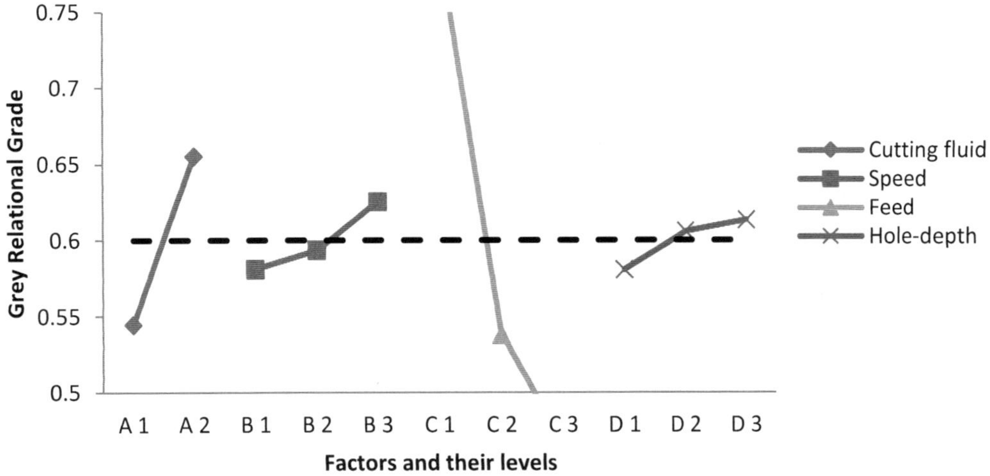

Figure 6. Response graph

It was observed from Figure 6 that the grey relational grade for feed decreases with increase in their levels, the value of grey relational grade increases with increase in levels of cutting fluid, hole-depth and speed. The dotted line shows the mean of grey relational level. Values of max-min for all parameters is also shown in Table 6 which depict that feed has maximum value of max-min i.e., 0.3299 and hence ranked first, cutting fluid, speed and hole-depth are ranked second, third and fourth with reduced value as 01107, 0.0443 and 0.0326respectively and it is also observed that grey relational grade are maximum at second level of cutting fluid, third level of speed, first level of feed and third level of hole-depth. So, $A_2B_3C_1D_3$ is the best combination of machining factors for optimization of drilling process parameters.

Table 6. Response table

Symbol	Level1	Level2	Level3	Max-Min
A	0.5444	0.6551		0.1107
B	0.5809	0.5932	0.6252	0.0443
C	0.7957	0.5378	0.4658	0.3299
D	0.5805	0.6057	0.6131	0.0326

4.1 Analysis of Variance

Analysis of variance (ANOVA) is incorporated to find the significant effects of cutting parameters on the performance characteristics. To check these impacts, there is a tool F – test. When F>4, The variation in machining parameters has a high impact on quality characteristics.

Table 7. Results of ANOVA

Symbol	Machining parameters	SS	DOF	Mean square	F - ratio	% contribution
A	Cutting fluid	0.0551	1.0000	0.0551	9.7346	11.42
B	Speed	0.0063	2.0000	0.0031	0.5545	1.30
C	Feed rate	0.3610	2.0000	0.1805	31.8923	74.82
D	Hole Depth	0.0035	2.0000	0.0017	0.3087	0.72
Error		0.0566	10.0000	0.0057		11.73
Total		0.4825	17.0000			100.0000

Table 7 gives the data of the contribution of taken cutting parameters on the characteristics of drilling process. Speed and hole-depth have very little effect on the performance characteristics.

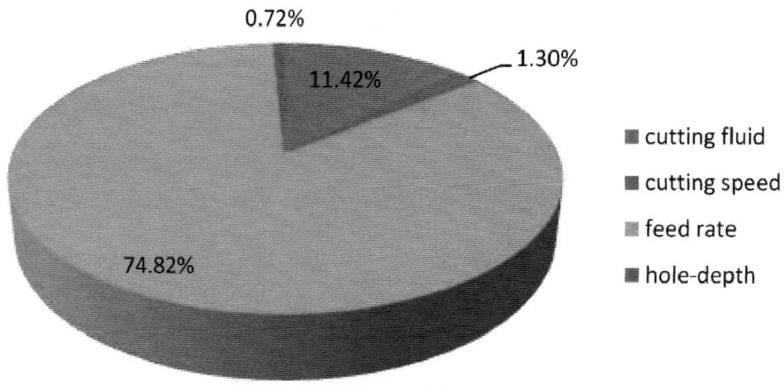

Figure 7. Contribution of process parameters

The results of ANOVA proves that the feed rate has the maximum contribution (74.82 %) followed by cutting fluid (11.42%), speed (1.30%) and hole depth (0.72%) which is shown in Figure 7.

4.2 Confirmation test

The optimization is done using the taguchi's method. Table 8 shows that the actual results and predicted results are in good agreement.

Table 8. Confirmation test results

	Prediction	Experiment	% change
Level	$A_2B_3C_1D_3$	$A_2B_3C_1D_3$	
Surface roughness		1	
MRR		1	
Grey relational grade	0.8898	1	12.39

5. Conclusions

The article is focused on taguchi's technique to optimize the drilling process parameters since it is an easy and systematic way to do so. Findings for finest optimum combination of multi-performance parameters of drilling of AA 6082 is $A_2B_3C_1D_3$ that is with cutting fluid absent, speed 420 rpm, feed of 120 mm/min and hole-depth 24 mm. The results obtained from ANOVA prove that the feed is the highest impactful factor chased by cutting fluid, speed and hole-depth. By applying this method, an improvement of 12.39% is observed.

The future scope is that more optimization methods like ANN model, genetic algorithms, cuckoo search, RSM, etc can be applied for analysis. There is always a chance of betterment.

References

[1] Belyakov A 2018 Microstructure and mechanical properties of structural metals and alloys

[2] Singh O P, Kumar G and Kumar M 2019 Multi Performance optimization of shoulder milling process parameters of AA6063 T6 aluminium alloy by Taguchi Based GRA *International Journal of Innovative Technology and Exploring Engineering* **8** pp 420-425

[3] Siddiquee A N, Khan Z A, Goel P, Kumar M, Agarwal G and Khan N Z 2014 Optimization of deep drilling process parameters of AISI 321 steel using Taguchi method *Procedia Materials Science* **6** pp 1217-1225

[4] Thakre A A 2013 Optimization of milling parameters for minimizing surface roughness using Taguchi's approach *International Journal of Emerging Technology and Advanced Engineering* **3(6)** pp 226-230

[5] Zhang J Z and Chen J C 2009 Surface roughness optimization in a drilling operation using the Taguchi design method *Materials and manufacturing processes* **24(4)** pp 459-467

[6] Cicek A, Kıvak T and Samtaş G 2012 Application of Taguchi method for surface roughness and roundness error in drilling of AISI 316 stainless steel *Strojniški vestnik-Journal of Mechanical Engineering* **58(3)** pp165-174

[7] Campatelli G and Scippa A 2012 Prediction of milling cutting force coefficients for Aluminum 6082-T4 *Procedia Cirp* **1** pp 563-568

[8] Akhil K T, Shunmugesh K, Aravind S and Pramodkumar M 2017 Optimization of drilling characteristics using grey relational analysis (GRA) in glass fiber reinforced polymer (GFRP) *Materials Today: Proceedings* **4(2)** pp 1812-1819

[9] Subramanian M, Sakthivel M, Sooryaprakash K and Sudhakaran R 2013 Optimization of cutting parameters for cutting force in shoulder milling of Al7075-T6 using response surface methodology and genetic algorithm *Procedia Engineering* **64** pp 690-700

[10] Sukumar M S, Ramaiah P V and Nagarjuna A 2014 Optimization and prediction of parameters in face milling of Al-6061 using Taguchi and ANN approach *Procedia Engineering* **97** pp 365-371

[11] Zhou L, Li J, Li F, Mendis G and Sutherland J W 2018 Optimization parameters for energy efficiency in end milling *Procedia CIRP* **69** pp 312-317

[12] Aamir M, Giasin K, Tolouei-Rad M and Vafadar A 2020 A review: Drilling performance and hole quality of aluminium alloys for aerospace applications *Journal of Materials Research and Technology* **9(6)** pp 12484-12500

[13] Kumar G, Kumar M and Tomer A 2021 Optimization of End Milling Machining Parameters of SS 304 by Taguchi Technique In *Recent Advances in Mechanical Engineering Springer Singapore* pp 683-689

[14] Kumar M, Kumar G, Singh O P and Tomer A 2021 Multiperformance Optimization of Parameters in Deep Drilling of SS-321 by Taguchi-Based GRA In *Recent Advances in Mechanical Engineering Springer Singapore* pp 675-681

[15] Siddiquee A N, Khan Z A and Mallick Z 2010 Grey relational analysis coupled with principal component analysis for optimisation design of the process parameters in in-feed centreless cylindrical grinding *The International Journal of Advanced Manufacturing Technology* **46(9)** pp 983-992

[16] Khan Z A, Kamaruddin S and Siddiquee A N 2010 Feasibility study of use of recycled High Density Polyethylene and multi response optimization of injection moulding parameters using combined grey relational and principal component analyses *Materials & Design* **31(6)** pp 2925-2931

AUTHOR INDEX